Managing Boreal Forests in the Context of Climate Change

Impacts, Adaptation and Climate Change Mitigation

Managing Boreal Forests in the Context of Climate Change
Impacts, Adaptation and Climate Change Mitigation

Seppo Kellomäki

University of Eastern Finland, School of Forest Sciences
FI-80101 Joensuu
Finland

CRC Press is an imprint of the
Taylor & Francis Group, an **informa** business

A SCIENCE PUBLISHERS BOOK

Cover illustration reproduced by kind courtesy of Alpo Hassinen

CRC Press
Taylor & Francis Group
6000 Broken Sound Parkway NW, Suite 300
Boca Raton, FL 33487-2742

© 2017 by Taylor & Francis Group, LLC
CRC Press is an imprint of Taylor & Francis Group, an Informa business

No claim to original U.S. Government works

Printed on acid-free paper
Version Date: 20161121

International Standard Book Number-13: 978-1-4987-7126-9 (Hardback)

This book contains information obtained from authentic and highly regarded sources. Reasonable efforts have been made to publish reliable data and information, but the author and publisher cannot assume responsibility for the validity of all materials or the consequences of their use. The authors and publishers have attempted to trace the copyright holders of all material reproduced in this publication and apologize to copyright holders if permission to publish in this form has not been obtained. If any copyright material has not been acknowledged please write and let us know so we may rectify in any future reprint.

Except as permitted under U.S. Copyright Law, no part of this book may be reprinted, reproduced, transmitted, or utilized in any form by any electronic, mechanical, or other means, now known or hereafter invented, including photocopying, microfilming, and recording, or in any information storage or retrieval system, without written permission from the publishers.

For permission to photocopy or use material electronically from this work, please access www.copyright.com (http://www.copyright.com/) or contact the Copyright Clearance Center, Inc. (CCC), 222 Rosewood Drive, Danvers, MA 01923, 978-750-8400. CCC is a not-for-profit organization that provides licenses and registration for a variety of users. For organizations that have been granted a photocopy license by the CCC, a separate system of payment has been arranged.

Trademark Notice: Product or corporate names may be trademarks or registered trademarks, and are used only for identification and explanation without intent to infringe.

Library of Congress Cataloging-in-Publication Data

Names: Kellomäki, Seppo, author.
Title: Managing boreal forests in the context of climate change : impacts, adaptation and climate change mitigation / Seppo Kellomäki.
Description: Boca Raton, FL : CRC Press, 2017. | "A Science Publishers Book." | Includes bibliographical references and index.
Identifiers: LCCN 2016044381| ISBN 9781498771269 (hardback : alk. paper) | ISBN 9781498771276 (e-book)
Subjects: LCSH: Taiga ecology. | Taigas--Climatic factors. | Climate change mitigation.
Classification: LCC QH541.5.T3 K45 2017 | DDC 577.3/7--dc23
LC record available at https://lccn.loc.gov/2016044381

Visit the Taylor & Francis Web site at
http://www.taylorandfrancis.com

and the CRC Press Web site at
http://www.crcpress.com

Preface

Responses of forest ecosystems to climate change, mainly to increasing CO_2 and temperature and changes in precipitation, have been studied intensively since the early 1990s. Many findings about the impact of climate change on the physiological performance of trees are based on experiments in growth chambers and greenhouses. Such studies provide valuable information for modeling, which is used to assess climate change impacts on forests on larger spatial and temporal scales. However, many questions are still open about how to manage forests in a sustainable way under climate change. In the boreal zone, in particular, there is a clear need to modify current management practices in order to meet the rapid warming likely to occur in this century.

Climate change is global but the impacts on forests are local, depending on the environmental conditions and management of forests for different purposes. In northern Europe, including Finland, the forests are mainly boreal, and large parts are managed for timber and biomass. In this context, the goal of this book is to analyze the way these forests are growing and developing under climate change, and consider how to meet climate change in the management, and how to mitigate climatic warming in forestry (Chapter 1). Experimental findings and model simulations provide information about how to reduce the abiotic and biotic risks in forest production, and how to utilize the opportunities provided by climatic warming in forest-based production.

The book is divided in seven parts and 20 chapters. In Part I, the main features of boreal forests are addressed in both global (Chapter 2) and local (Chapter 3) contexts. The focus is on the boreal coniferous forests under regular management for timber and biomass and carbon sequestration. The likely impacts and changes in growing conditions due to climate change are described in Part II. Climate change is further addressed in Chapters 4 and 5, with the focus on northern Europe at high latitudes, where climate warming is pronounced and has large impacts on the structure and functions of forest ecosystems and forests.

Part III considers the responses of selected boreal tree species to climate change, including the main eco-physiological (Chapters 6–9) and ecological processes (Chapter 10), controlling the growth and development of trees and the properties of timber and wood (Chapter 11). Findings from experimental studies have been used to compile a process-based forest ecosystem model that is sensitive to climate change and the management described under Part IV in Chapters 12 and 13. The model has been used in Part V, along with other models, to identify how climate change affects the productivity of managed boreal forests (Chapter 14) and the growth and development of forests under climate change (Chapter 15). In Chapter 16, rule-based and optimized management are applied over the whole forest area in Finland (26 million hectare) to estimate how climate change affects the potential growth and production of timber and biomass. Furthermore, climate change is likely to affect risks of abiotic and biotic damage, as discussed in Chapter 17.

There is an evident need to avoid climate change-induced damage, as discussed in Part VI, in separate chapters for adaptive management (Chapter 18) and the mitigation of climate change (Chapter 19). Chapter 20 summarizes the main findings about how climate change is likely to affect managed boreal forests, and how to manage these forests to avoid different risks and utilize the opportunities that climate change is likely to offer in forestry.

I acknowledge the support of my colleagues when compiling this book. Especially, the role of Prof. Heli Peltola, Mr. Harri Strandman and Mr. Hannu Väisänen has been decisive in different phases in preparing the book.

Joensuu, Finland **Seppo Kellomäki**
May 2016 Email: seppo.kellomaki@uef.fi

Contents

Preface   v

1. Introduction   1

PART I: Boreal Forests in Global and Local Context

2. Global Boreal Forests   13
3. Managed Boreal Forests in Local Context   22

PART II: Climate Change and Impact on Growing Conditions

4. Global Climate Change   37
5. Climate Change in the Local Context, with Changes in Growing Conditions   44

PART III: Impact of Climate Change on the Eco-physiological Performance of Selected Boreal Tree Species

6. Carbon Uptake and Climate Change   63
7. Response of Respiration to Climate Change   89
8. Response of Transpiration to Climate Change   106
9. Response of Whole Tree Physiology to Climate Change   115
10. Growth and Structure of Trees under Climate Change   126
11. Properties of Plant Material under Climate Change   141

PART IV: Eco-physiological Approaches to Modeling Responses of the Boreal Forest Ecosystem to Climate Change

12. Impact Mechanisms Linking the Dynamics of Forest Ecosystem to Climate Change   157
13. Integrating Climate Change Impacts on Ecosystem Dynamics for Management Studies   165

PART V: Responses of Boreal Forest Ecosystem to Climate Change and Management

14. Impact of Climate Change on the Productivity of Boreal Forests   203
15. Impact of Climate Change on the Growth and Development of Boreal Forests   218
16. Management of Boreal Forests for Timber and Biomass under Climate Change   233
17. Disturbances and Damage Affecting Boreal Forests under Climate Change   245

PART VI: Management of Boreal Forests under Climate Change for the Adaptation and Mitigation of Climate Change

18. Management of Forests for Adaptation to Climate Change 273
19. Management of Forests for the Mitigation of Climate Change 298

PART VII: Managed Boreal Forests under Climate Change—Summary and Perspectives

20. Climate Change and Managed Boreal Forests 345

Appendix: Units and Conversions 353

Index 355

Introduction

ABSTRACT

This book addresses the impact of climate change on management of boreal forests. It uses the findings based on experiments and model simulations mainly in Finland and Sweden, where Scots pine (*Pinus sylvestris* L.), Norway spruce [*Picea abies* (L.) Karst.] and birch (*Betula pendula* Roth and *B. pubescens* Ehrh.) are dominant species. Process-based modeling has been used in studying the effects of management on the productivity of the forest ecosystem and the production of timber and biomass and the sequestration of carbon. The adaption of boreal forest and forestry to climate change and the potentials in mitigating climate change are discussed.

Keywords: climate change, boreal forests, ecosystem goods and services, managed boreal forests, forest management, adaption to climate change, mitigation of climate change

1.1 Forests for Human Well-being

Global forests

The global forest area is about 4033 million hectares, which is about 31% of the global land area (FAO 2010). Forests range from high-density tropical rain forests to low-density northern or high-altitude boreal forests. Thermal condition is the primary factor separating tropical, temperate and boreal forests from the Equator toward the poles. Humidity provides the further classification of forest into subgroups or sections, with various combinations of thermal and humid conditions as shown in Fig. 1.1. The same zonal pattern could theoretically exist in both hemispheres, but the uneven distribution of the continents means that certain patterns can only be found in the northern hemisphere.

Goods and services provided by forests

Forest ecosystems and forests (Box 1.1) provide many goods and services (or ecosystem services) for humans. According to the Millennium Ecosystem Assessment (2005), ecosystem services fall into four categories: supporting, provisioning, regulating and cultural services affecting human

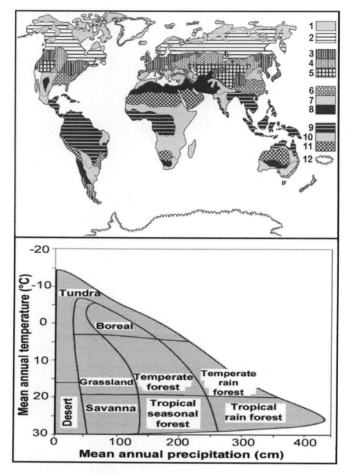

Fig. 1.1 Above: Bioclimatic regions based on thermal and humid conditions (Walters 1979). Permission of Springer. Legend: 1 = arctic zone, 2 = boreal zone, 3–5 temperate zone (3 = humid region, 4 = semiarid region, 5 = arid region), 6–8 = subtropical zone (6 = humid region, 7 = semiarid region, 8 = arid region), 9–11 = tropical zone (9 = humid region, 10 = semiarid region, 11 = arid region), 12 = glaciers and permafrost and snow. Below: Climate (temperature, precipitation) boundaries of the main forest domains of the world, based on Whittaker (1975).

Box 1.1 Forest goods and services in ecosystem context

Following Odum (1971), the ecosystem refers to the interaction between living organisms and their non-living environments: "any unit that includes all the organisms (i.e., the community) in a given area interacting with the physical environment so that a flow of energy leads to clearly defined trophic structure, biotic diversity and material cycles (i.e., exchange of matter between living and non-living parts) within the system is an ecological system or ecosystem". In this context, a forest including the sites, trees and other organisms (structure) is an ecosystem, where trees and other green plants intercept solar energy flowing through food webs (functioning). The ecosystem structure includes the living mass distributed among the genotypes occupying the site and the resources available in the site for the regeneration and growth of trees and other organisms. Populations/communities of trees and other organisms control the cycles of nutrients, carbon and water globally in biosphere and locally in sites. Sites with trees and other organisms are spatial units, whose functioning produces provisioning services, such as wood and non-wood products, and their structure provides regulating and cultural services, such as climatic regulation and aesthetic values (Millennium Ecosystem Assessment 2005).

well-being. Supporting services represent the basic structure and functioning of forest ecosystems; i.e., the interaction between genotypes and environment produces provisioning, regulating and cultural services (Fig. 1.2). Provision services include concrete services such as timber and ground water, whereas cultural services involve the amenity values and recreation opportunities.

In many developed countries, timber and biomass are the main provision services, but even berries and mushrooms and the forest environment play an important role (Krebs et al. 2009). Throughout the world, the availability of different ecosystem services varies depending on the properties of ecosystems and their dynamics. These factors also determine how sensitive and adaptive the forest ecosystems are to possible climate change. The importance of forests and forest ecosystems is increasing, because they provide biomass that substitutes for fossil fuels and fossil-intensive materials, thus retaining carbon in ecosystems and wood-based products outside ecosystems for mitigating climate change.

Ecosystem goods involve the direct use of materials and resources forming the ecosystem structure, or their use for manufacturing different goods. Ecosystem goods include: (i) inorganic matter (C, N, CO_2, H_2O, etc.); and (ii) organic matter (proteins, carbohydrates, lipids, etc.) produced by plants in both primary production (forming organic matter from inorganic matter), and in secondary production (micro- and macro-organisms consuming primary production). Ecosystem services further include the indirect use of ecosystem structure in modifying the properties of the environment for: (i) reducing the impact of energy in radiation, heat, mechanical forces (wind, snow load and gravity (e.g., landslide)), noise, etc.; (ii) enhancing environmental health by absorbing chemicals and particles from the atmosphere (e.g., air impurities); (iii) enhancing the amenity of the environment and creating a functional environment for different human activities; and (iv) maintaining cultural heritage and maintaining and conserving ecosystem functions and biodiversity.

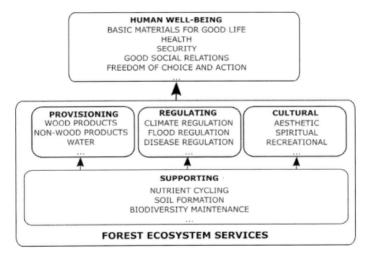

Fig. 1.2 Ecosystem goods and services contributing to human well-being, modified by Seppälä et al. (2009) from the Millennium Ecosystem Assessment (2005). Reproduced by permission of IUFRO.

Management of forests for goods and services

The sustainable production of goods and services is based on management that controls the ecosystem to function in such a way that it creates structures optimal for producing the desired goods and services. Production [P (i, j)] is based on the interaction between environment [E (j)], and genotype [G (i)]:

$$P(i,j) = G(i) + E(j) + G(i) \times E(j) \tag{1.1}$$

In general, trees are long lived, meaning that the production cycle in forestry extends over decades, and that forests are thus highly vulnerable to climate change. In this context, management aims to optimize genotype/environment interaction to avoid the harmful impacts of climate change but still satisfy production needs. This is possible by controlling the long-term functional and structural development of forest ecosystems (succession) to help them produce the goods and services focused in the management goals. Through their management, the physiological and ecological performance of tree populations (and populations of other species) is directed to produce the ecosystem structures that are needed to produce the specified goods and services (Kellomäki et al. 2009).

Sustainability is widely used to indicate the success of management in providing varying ecosystem goods and services. Sustainability (Sustainable Forest Management (SFM)) involves the management applying the concept of sustainable development in forestry implying: "the use of forests and forest lands in a way, and at a rate, that maintains their biodiversity, productivity, regeneration capacity, vitality and their potential to fulfill, now and in the future, relevant ecological, economic and social functions, at local, national, and global levels, and that does not cause damage to other ecosystems." (State of Europe's Forests 2003). The ecological limits for sustainable management are described as: (i) absorbing energy; (ii) converting energy to chemical energy and matter; and (iii) the distribution of matter between different organisms (biodiversity) (Fig. 1.3). In general, management is ecologically sustainable if the energy absorption, and the production and the distribution of matter for different organisms follow the processes typical for unmanaged ecosystems. The management may alter the habitat distribution and biogeochemical cycles but still within the limits set by the long-term dynamics (succession) of the forest ecosystem.

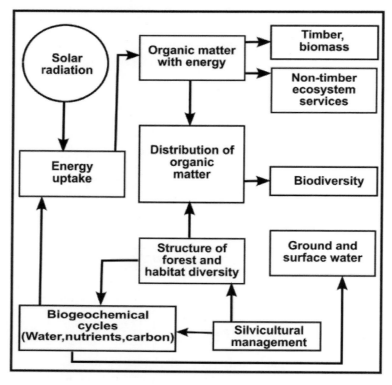

Fig. 1.3 Factors affecting the sustainability of forest management and the relationships between various ecosystem goods and services.

1.2 Global Climate Change and Forest Growth

Impact mechanisms

According to the International Panel for Climate Change (IPCC), climate change refers to any change in climate over time, whether due to natural variability or as a result of human activity (IPCC 2007a). Climate change is most related to the global increase of atmospheric concentrations of carbon dioxide (CO_2). The increasing concentrations of other greenhouse gases (GHG), such as methane (NH_4) and nitrous dioxide (N_2O), trap extra heat in waters, lands and atmosphere, with a consequent increase in temperature and changes in precipitation and other climatic factors. By 2100, the annual mean temperature is expected to increase globally by 2–6°C (IPCC 2007a). The increase will be especially pronounced in the high latitudes of the northern hemisphere. The temperature in northern Europe, for example, may increase by 1–2°C in summer (June, July, August) and by 2–3°C in winter (December, January, February) over the next 50 years. An increase in precipitation of 5–40% is further expected (IPCC 2007b).

Modeling is widely used to identify how climate warming is likely to affect the functioning and structure of trees, e.g., such as photosynthesis, respiration or transpiration in leaves/needles responding to the elevating CO_2 and/or temperature. The basic mechanisms behind the eco-physiological responses and their changes work at the level of cells/tissues, but the organ-level findings can be scaled up to the level of whole trees, and further to the level of the populations and communities of trees occupying sites and landscapes. The tree-level responses within the ecosystem level are further affected by within-population competition and/or the competition between the populations of different tree species occupying sites. The ecosystem responses to climate change integrate genotypic responses and the availability of resources related to changes in environmental conditions. In the ecosystem context, scenario analyses may be used to relate the future dynamics of the birth, growth and mortality of trees to the changing climate. Scenario analysis refers, for example, to model calculations in order to analyze how trees are growing and developing under higher atmospheric CO_2 and warming climate.

Forest ecosystems are temporally and spatially hierarchical systems, where the metabolism of trees controlled by the climatic and edaphic factors link the dynamics of ecosystem to climate change. The magnitude of responses, and even their direction, could vary in different time scales depending on the biogeochemical feedbacks, including surface energy fluxes, hydrology, and carbon and nutrient cycles. For example, experiments based on the free-air CO_2 enrichment (FACE) show that a 50% increase in atmospheric CO_2 may increase forest growth by 23% over several years (Norby et al. 2005), thus slowing the rate of increase in atmospheric CO_2. This is likely to take place in boreal forests, where climatic warming may increase forest growth if not limited by a short supply of nitrogen (Jarvis and Linder 2000). On the other hand, the increase of net primary production in boreal forests is probably reducing surface albedo, thus increasing warming, and subsequently tree growth and the turnover of soil carbon (Bonan 2008).

Climate change impacts on boreal forests

Under the changing climate, the area of boreal forests in the south is likely to be replaced by temperate forests (Soja et al. 2007), and boreal forests are likely to replace tundra in the north. Simulations in the early 1990s suggested that the area of boreal forest may be reduced by 40% (Monserud et al. 1993), but there was a great deal of variability between the models. This may, however, imply a 500–1000 km shift of boreal forests northwards (Kauppi and Posch 1985), but the limits between the boreal and temperate zones are of transition ones from the dominance of coniferous species to the dominance of deciduous species as presented in Fig. 1.4 for Europe. In northern Europe, the dominance of deciduous species is likely to increase in southern areas, whereas in the northern areas the dominance of coniferous species will remain (Falk and Hempelmann 2013). The changes in species composition indicate changes in the vegetation functional types rather than the shift of the boreal vegetation zone northwards (Bonan 2008). Such changes are likely to substantially affect the management of European boreal forests, which are very important for commercial use and environmental conservation.

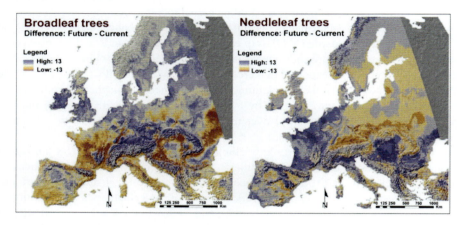

Fig. 1.4 Expected changes in plant functional types due to climate change as indicated by the difference between the current and future dominance of deciduous (left) and coniferous (right) species throughout Europe (Fitzgerald and Lindner 2013). Courtesy of Pensoft Publishers. Combined results are based on several models and several climate scenarios, using the A1B emission scenario. Plant functional types are used in climatology to classify plants according to their physical, phylogenetic and phenological characteristics for use in land use studies and climate models.

1.3 Management of Forests for Adaptation to Climate Change

In general, forests may adapt autonomously to climate change, but their adaptive capacity may be dependent on proper management or adaptive management. In adaptive management, the selection of management strategy is related to the responses of ecosystem processes involving varying process hierarchies and time resolutions (Fig. 1.5). At the level of cells/organs, the physiology of trees reacts quickly, from seconds to hours, to changes or variability in environmental factors (e.g., air temperature, CO_2) (Eamus and Jarvis 1989). Physiological responses themselves may further adapt to changes in environmental conditions. This implies that time resolution may remain the same but the response level may change under elevated CO_2 and temperature (Wang et al. 1996). At the level of a whole tree, reactions are substantially slower, and at the level of populations/communities of trees, changes are detectable in years or decades (Kellomäki et al. 2008). Slow adaptation at the tree and stand levels integrates the changes in tree populations/communities with the long-term changes in the biogeochemical cycles controlling the succession of forest landscape over decades.

In general, adaptive management aims at adjusting management to meet the requirements "to moderate or offset the potential damage or to take advantages of opportunities created by a given climate change" (IPCC 2001). Adaptive management is thus a strategy to help the structure and the consequent functioning of forest ecosystems to resist the harmful impact of climate change and to utilize the opportunities created by climate change. A variety of measures (e.g., planting, thinning, fertilizing, etc.) may be applied to control the successional process for adaptation based on monitoring and learning from the outcomes of current management (Innes et al. 2009) (Fig. 1.6). In management, the choice and combination of necessary measures are subjected to single populations/communities of trees and sites, but management needs cross-landscape integration to maintain the sustainability of timber and biomass and non-timber resources. A forest policy responsive to a wide variety of economic, social, political and environmental circumstances is necessary in order to avoid the risk of unacceptable losses due to future changes in climate and socio-economic context (Spittlehouse and Steward 2003).

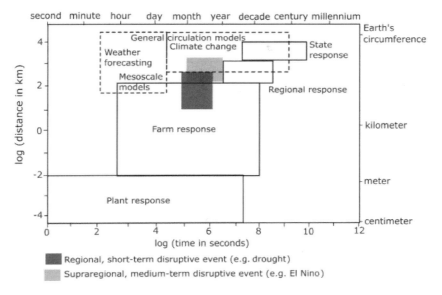

Fig. 1.5 Temporal and spatial scales of responses to climate change, based on Parry and Carter (1984).

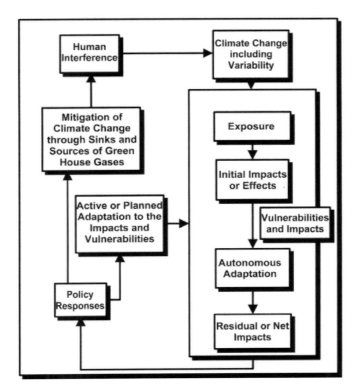

Fig. 1.6 Outline for identifying requirements for adaptation, with defining adaptive strategy and measures, based on IPCC (2001). Courtesy of Cambridge University Press.

1.4 Management of Forests for the Mitigation of Climate Change

Climate change mitigation is closely linked to adaptation (Fig. 1.6). In this respect, forests and forest ecosystems have a substantial effect on the global carbon cycle in atmosphere/biosphere interaction, which has a strong effect on radiative forcing and subsequent climatic warming. For example, the gross photosynthesis in global forests cycles 8% of the total amount of CO_2 in the atmosphere, comprising 50% of terrestrial photosynthesis (Mahli et al. 2002). According to Canadell and Raupach (2008), terrestrial ecosystems, mainly forests, bind three billion Mg of anthropogenic carbon (3 Pg C yr^{-1}) annually, which is 30% of CO_2 emissions due to the use of fossil fuels and net deforestation. The potential of forest ecosystems to mitigate CO_2 emissions emphasizes the fact that four billion hectares of forest ecosystems (30% of the global land area) hold double the amount of carbon as that in the atmosphere.

Radiative forcing is the difference between the energy in solar radiation absorbed by the earth and energy radiated back to space. Mitigation includes reducing sources and increasing sinks of greenhouse gases, especially those for CO_2, for controlling radiative forcing. Canadell and Raupach (2008) identify four major strategies to mitigate carbon emissions and radiative forcing in forests and forestry: "(i) to increase forested land area through reforestation; (ii) to increase the carbon density of existing forests at both stand and landscape scales; (iii) to expand the use of forest products that sustainably replace fossil-fuel CO_2 emissions; and (iv) to reduce emissions from deforestation and degradation." Biophysical factors, including reflectivity, evaporation and surface roughness, change temperatures even more than carbon sequestration, with an impact on radiative forcing. Jackson et al. (2008) claim that the carbon sequestration and accumulation in forest ecosystems may be "counteracted in boreal and other snow-covered regions, where darker trees trap more heat than snow". Such findings suggest that the biophysical properties of forests may provide additional opportunities for forcing down global warming in forestry (Betts 2000).

In general, the maximum sustainable drain of timber and biomass reduces carbon in managed forest ecosystems to a half of that in unmanaged forests, depending on site fertility, tree species and management. On the other hand, 50–70% of timber and the above-ground biomass is used for wood-based materials, products and energy to reduce carbon emissions, while the remaining harvest residues (e.g., foliage, branches, top part of stem, stump, roots) are left to decay in forest or harvest for energy use (Eriksson et al. 2007, Kilpeläinen et al. 2013). Management and harvest change the structure of forest cover, with changes in canopy and soil albedo (reflecting power of a surface). In the boreal conditions, the impact of management and harvest on carbon stocks and albedo are generally opposite: increasing carbon stocks tend to reduce albedo, while reducing carbon stocks increase albedo (Bonan 2008).

1.5 Scope of the Book

This book addresses the impacts of climate change on boreal forest ecosystems, forestry and management, with the focus on the production of timber and biomass and the sequestration of carbon. It is based on experimental and model-based studies carried out mainly in Finland and Sweden, where forests are dominated by Scots pine (*Pinus sylvestris* L.), Norway spruce [*Picea abies* (L.) Karst.] and birch (*Betula pendula* Roth and *B. pubescens* Ehrh.), with high commercial and environmental values for societies. Responses of boreal forests to climate change, mainly to increasing CO_2 and temperature, have been studied intensively since the early 1990s, but proper management of boreal forests to meet the rapid warming likely to occur in this century is still poorly known.

References

Betts, R. A. 2000. Offset of the potential carbon sink from boreal forestation by decrease in surface albedo. Nature 408: 187–190.
Bonan, G. B. 2008. Forests and climate change: forcings, feedbacks, and the climate benefits of forests. Science 320: 1444–1449.
Canadell, J. G. and M. R. Raupach. 2008. Managing forests for climate change mitigation. Science 320: 1456–1457.
Eamus, D. and P. G. Jarvis. 1989. The direct effects of increase in the global atmospheric CO_2 concentration in natural and commercial temperate trees and forests. Advances in Ecological Research 19: 1–55.
Eriksson, E., A. R. Gillespie, L. Gustavsson, O. Langvall, M. Olsson, R. Sathre et al. 2007. Integrated carbon analysis of forest management practices and wood substitution. Canadian Journal of Forest Research 37: 671–681.
Falk, W. and N. Hempelmann. 2013. Species favourability shift in Europe due to climate change: a case study for *Fagus sylvatica* L. and *Picea abies* (L.) Karst. based on an ensemble of climate models. Journal of Climatology 2013: 1–18.
FAO. 2010. Global Forest Resource Assessment 2010. FAO Forestry Paper 163: 1–340.
Fitzgerald, J. and M. Lindner. 2013. Adapting to Climate Change in European Forests—Results of the MOTIVE Project. Pensoft Publishers, Sofia, Bulgaria.
Innes, J., L. A. Joyce, S. Kellomäki, B. Louman, A. Ogden, J. Parrotta and I. Thompson. 2009. Management for adaptation. *In*: R. Seppälä, A. Buck and P. Katila (eds.). Adaptation of Forests and People to Climate Change. A Global Assessment Report. IUFRO World Series 22: 135–186.
IPCC. 2001. Climate Change 2001: Impacts, Adaptation, and Vulnerability. Contribution of Working Group II to the Third Assessment Report of the Intergovernmental Panel on Climate Change. Cambridge University Press, Cambridge, UK.
IPCC. 2007a. Climate Change 2007: The Physical Science Basis. Contribution of Working Group I to the Fourth Assessment. Report of the Intergovernmental Panel on Climate Change. Cambridge University Press, Cambridge, UK.
IPCC. 1997b. Climate Change 2007: Contribution of Working Group II to the Fourth Assessment Report of the Intergovernmental Panel on Climate Change, Cambridge University Press, Cambridge, UK.
Jackson, R. B., J. T. Randerson, J. G. Canadell, R. G. Anderson, R. Avissar, D. D. Baldocchi et al. 2008. Protecting climate with forests. Environmental Research Letters 3: 1–5.
Jarvis, P. J. and S. Linder. 2000. Constraints to growth of boreal forests. Nature 405: 904–905.
Kauppi, P. and M. Posch. 1985. Sensitivity of boreal forest to possible climatic warming. Climatic Change 7(1): 45–54.
Kellomäki, S., H. Peltola, T. Nuutinen, K. T. Korhonen and H. Strandman. 2008. Sensitivity of managed boreal forests in Finland to climate change, with implications for adaptive management. Philosophical Transactions of the Royal Society B363: 2341–2351.
Kellomäki, S., V. Koski, P. Niemelä, H. Peltola and P. Pulkkinen. 2009. Management of forest ecosystem. pp. 252–373. *In*: S. Kellomäki (ed.). Forest Resources and Sustainable Management. Gummerus Oy, Jyväskylä, Finland.
Kilpeläinen, A., H. Strandman, S. Kellomäki and J. Seppälä. 2014. Assessing the net atmospheric impacts of wood production and utilization. Mitigation and Adaptation Strategies for Global Change 19(7): 955–968.
Krebs, C. J., R. Boonstra, K. Cowcill and A. J. Kenney. 2009. Climatic determinants of berry crops in the boreal forests of the southwestern Yukon. Botany 87: 401–408.
Louman, B., A. Fischlin, P. Gluck, J. Innes, A. Lucier, J. Parrotta, H. Santos, I. Thompson and A. Wreford. 2009. *In*: R. Seppälä, A. Buck and P. Katila (eds.). Adaptation of Forests and People to Climate Change. A Global Assessment Report. IUFRO World Series 22: 15–27.
Mahli, Y. P., P. Meir and S. Brown. 2002. Forests, carbon and global change. Philosophical Transactions of the Royal Society A360: 1567–1591.
Millennium Ecosystem Assessment. 2005. Ecosystems and Human Well-being: Synthesis. Island Press, Washington, DC, USA.
Monserud, R. A., N. M. Tchebakova and R. Leemans. 1993. Global vegetation change predicted by modified Budyko model. Climatic Change 25: 59–83.
Odum, E. P. 1971. Fundamentals of Ecology. W.B. Saunders Company, Philadelphia, USA.
Norby, R. J., E. H. DeLucia, B. Gielen, C. Calfapietra, C. P. Giardina, J. S. King et al. 2005. Forest response to elevated CO_2 in conserved areas across a broad range of productivity. PNAS 102(50): 18052–18056.
Parry, L. M. and T. R. Carter. 1984. Assessing the impact of climatic change in cold regions. Report of a workshop held in Villach, Austria, 19–23 September, IIASA, Austria.
Soja, A. J., N. M. Tchebakova, N. H. F. French, M. D. Flannigan, H. H. Shugart, B. Stocks et al. 2007. Climate-induced boreal forest changes: predictions versus current observations. Global and Planetary Change 56: 274–296.
Spittlehouse, D. L. and R. B. Stewart. 2003. Adaptation to climate change in forest management. BC Journal of Ecosystems and Management 4(1): 1–11.
State of Europe's Forests. 2003. The MCPFE Report on Sustainable Forest Management in Europe. Jointly prepared by the MCPFE Liaison Unit Vienna and UNECE/FAO. Edited and published by Ministerial Conference on the Protection of Forests in Europe Liaison Unit Vienna, Vienna, Austria.
Walters, H. 1979. Vegetation of the Earth and Ecological Systems of the Geo-Biosphere. Springer-Verlag, New York, USA.
Wang, K. -Y., S. Kellomäki and K. Laitinen. 1996. Acclimation of photosynthetic parameters in Scots pine after three years of exposure to elevated temperature and CO_2. Agricultural and Forest Meteorology 82: 195–217.
Whitaker, R. H. 1975. Communities and Ecosystems. Macmillan, New York, USA.

PART I
Boreal Forests
in Global and Local Context

Global Boreal Forests

ABSTRACT

Boreal forests are circumpolar at high latitudes (45°–70°) in the northern hemisphere. They cover 1640 million ha (27% of the global forest area), mainly in North America (Canada, Alaska), the Nordic countries (Finland, Sweden, Norway, Iceland) and Russia. Boreal forests are characterized by cool coniferous and deciduous species growing in mosaics of successional and subclimax stands and communities susceptible to varying environmental conditions and disturbances (fire, wind, snow, pests). They are largely outside harvest and management except in the Nordic countries, where boreal forests have been used systematically for the production of timber and biomass for decades.

Keywords: Global boreal forests, boreal tree species, boreal forest resources, carbon sequestration, biodiversity

2.1 Bioclimatic Limits

Boreal forests in a global context

Boreal forests (forests and woodlands) are circumpolar at high latitudes (45°–70°). They are characterized by the dominance of coniferous species, but mixed with deciduous species on fertile sites (Bonan and Shugart 1989). The northern boreal forest is limited by tundra, whereas the southern boreal forest gradually changes to temperate mixed and deciduous forest and even grassland/steppe in the most continental parts of North America and Eurasia (Sirois 1992). However, the limits of the treeless arctic tundra are undefined, especially in Alaska, Canada and Siberia. In these regions, large areas between the treeless tundra and the boreal forest are forest tundra, where the soil temperature regularly remains below 0°C in summer. This is due to the cold winters and thin snow cover that results in deep soil frost (Walter 1979). Forest tundra is especially vulnerable to climate change, which probably causes the boreal forest to move northwards in the warming climate. Similarly, the limits of the southern boreal forest are undefined, with a gradual transition from the main dominance of coniferous trees to the main dominance of deciduous trees. The southern boreal forests are also vulnerable to warming climate, with their likely replacement by temperate forest or even grasslands/steppes (IPCC 2007a,b).

Boreal forest covers about 1,640 million ha (40% of the global forest area of 4,033 million ha) (FAO 2006, 2010, Wulder et al. 2007), mainly in North America (Canada, Alaska), the Nordic countries (Finland, Sweden, Norway, Iceland) and Russia (Fig. 2.1) in such a way, that two thirds of all boreal forests are in Eurasia, mostly in Russia (69% of the total global boreal forest area). In many places in Kazakhstan, Mongolia, China and Japan, the forests at higher elevations have a boreal character, but the total area of boreal forest in these countries is small (Bergeron et al. 2010). Boreal forest accounts for 33% of the total circumpolar region, and most of the carbon in this area accumulates in peatlands (bogs, mires) and permafrost soils, of which nearly half is covered by forest (DeLuca and Boisvenue 2012). Boreal landscape is a mixture of closed forests and open woodlands, which bogs, mires, lakes and rivers split into separate patches of varying size. The boreal landscape is thus characterized by a great variability of habitats, facilitating large species and genetic richness. The forests themselves are a mosaic of successional and subclimax plant communities, which are susceptible to varying disturbances (e.g., fire, wind, snow, pests). Boreal forests are largely unmanaged except in Nordic countries (Finland, Sweden, and Norway), where they have been used systematically for the production of timber and biomass for decades.

In the northern hemisphere, peatlands cover 250–350 million hectare of land in the boreal zone (Apps et al. 1993, Strack 2008). Peatland is waterlogged land, where the net primary production exceeds the decay of organic matter, leading to the accumulation of incompletely decayed organic matter in the soil. The peatlands cover 18–28% of the boreal land area, the percentage being higher in the maritime than in the continental parts (Kauppi et al. 1997). Globally, boreal peatlands are a major sink for carbon dioxide (CO_2), and an important source of atmospheric methane (CH_4). The carbon deposits in boreal peatlands (419 Pg) substantially exceed the amount of carbon bound in upland forest soil (199 Pg), plant detritus (32 Pg) and plant biomass (64 Pg) (Apps et al. 1993). In general, nitrous oxide (N_2O) emissions are low from natural peatlands, but they may substantially be increased by the use of peat soils in agriculture (Maljanen et al. 2001). In general, the uptake and losses of carbon in/from peatlands are related to the changes in the balance between the net

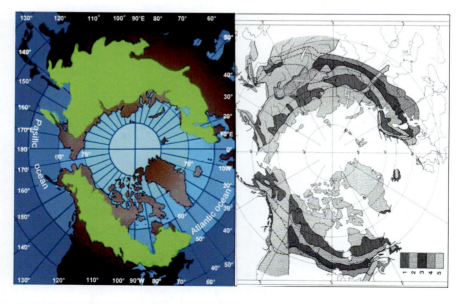

Fig. 2.1 Left: Distribution of the boreal forest around the North Pole as indicated in green. Map available at http://www.borealforest.org is based on Hare and Ritchie (1972). Permission of borealforest.org, Ulf Runesson. Right: Distribution of boreal forests into subgroups, with the legend: (1) northern, (2) central and (3) southern boreal forests; (4) hemi-boreal forests; and (5) artic and mountainous areas with patches of boreal forests (Hämet-Ahti et al. 1989). Permission of the Finnish Dendrological Society.

exchange of CO_2, emission of CH_4, and losses of dissolved carbon controlled by the ground water level and water flowing through mire ecosystems.

Climate and the boreal forest

In the area occupied by the boreal forest, the mean annual total radiation is less than half of that at the Equator (Fig. 2.2). The main part of radiation is received in the period from March 15 to September 15, thus including the main growing season. In Finland, for example, the mean total radiation in this period is about 4,300 MJ m^{-2} yr^{-1} at 60° N and about 3,900 MJ m^{-2} yr^{-1} at 70° N. The radiation is large enough to support forests, wherever low temperature does not limit the regeneration and growth of trees.

Boreal forests fall between the summer and winter limits of the arctic air mass (Gordon et al. 1989). In these conditions, the northern limit of forest cover is set by the 10°C July isotherm, whereas the southern limit is set by the 18°C July isotherm. In this range, the climate is characterized by short summers; i.e., the number of days where the daily mean temperature ≥ 10°C is 30–120 per year. The mean annual temperature varies from –5°C to +5°C, but in the continental parts (e.g., Siberian), the mean annual temperature may fall to –10°C. In the most extreme areas against tundra, the soil may be frozen throughout the year (permafrost). Winters are severe, with the mean daily temperature < 0°C for five to seven months (http://en.wikipedia.org/wiki/subarctic_climate, visited 20.1.2014) (Fig. 2.3). The within-year variability in temperature is also large, from –50°C to +40°C, and the mean daily summer temperature is 15–20°C. The mean annual precipitation is 200–750 mm, excluding most maritime areas (e.g., western parts of North American boreal forest), where the annual precipitation may exceed 1000 mm. In many parts of the boreal forest, the climate is humid and precipitation exceeds evaporation (e.g., in the Nordic countries). However, in most continental areas of the boreal forest (e.g., in Siberia), the climate is subhumid or semiarid due to limited precipitation (http://en.wikipedia.org/wiki/subarctic_climate, visited 20.1.2014). Under humid conditions, the most common soil formation is the podsolization due to the acid soil solution produced under a coniferous canopy. In many areas of the boreal forest, soils are still young (e.g., Nordic countries) due to glaciation, which disappeared only 10,000–15,000 years ago.

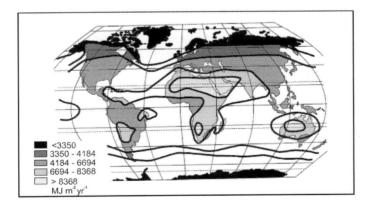

Fig. 2.2 Annual radiations over the globe (Larcher 1975). Permission of Springer.

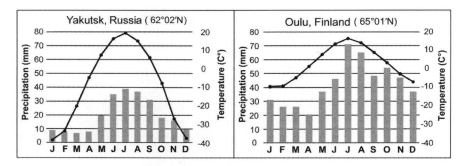

Fig. 2.3 Example of the boreal (subarctic) climate in terms of the within-year distribution of precipitation (columns) and temperature (line) in the continental boreal region in Yakutsk, Russia (left), and semi-maritime boreal region, Oulu, Finland (right), based on http://en.wikipedia.org/wiki/subarctic_climate, visited 20.1.2014.

2.2 Structure of Forest Vegetation and Forest Resources

Main tree species

The boreal forest is characterized by a relatively small number of tree species. The most important coniferous species include pines (*Pinus*), spruces (*Picea*), firs (*Abies*), larches (*Larix*), junipers (*Juniperus*), thujas (*Thuja*) and hemlocks (*Tsuga*), while the most common deciduous species include aspens (*Populus*), birches (*Betula*), willows (*Salix*) and alders (*Alnus*) (Table 2.1). Most boreal tree genera occur throughout the boreal zone representing transcontinental distributions across Eurasia or North America (Nikolov and Helmisaari 1992). The number of conifers is greatest in North America, but also high in the southern part of the Far East. The number of tree species is particularly small in the northwestern areas of Eurasia (Nordic countries), where Scots pine and Norway spruce dominate forested landscapes.

The boreal forest provides a large number of services locally, but globally it is timber and biomass, carbon sequestration and biodiversity that are the main services provided (Fischling et al. 2009). However, large parts (50%) of boreal forests are still outside commercial use and regular management (Olsson 2011). In Canada, for example, 80% of the boreal forest is still untouched by settlements or roads (Ruckstuhl et al. 2008). This is not true in the Nordic countries, where the main part of forest land (locally up to 90%) is used and managed for timber and biomass. Even the northern boreal forests are used in forestry in the conditions, where climatic warming will be pronounced by the end of this century. In this context, there are still many open questions how to adapt managed boreal forest to climate change (Innes et al. 2009), including the tree species choice and planting practices, the tending of seedling stands and proper tree species mixtures, the thinning practices and the source/sink dynamics in carbon sequestration for climate change mitigation (Kellomäki et al. 2005, 2008).

Forest resources

The boreal forest is in the control of several countries (states), which makes it difficult to obtain precise statistics of the boreal forest resources. For example, estimates of the area of boreal forest vary in the range 1,000–2,000 million ha. Wulder et al. (2007) estimates the total area of boreal forest and woodland to be 1,960 million ha, including 1,105 million ha of proper forests. Based on Kuusela (1990) and global forestry statistics (FAO 2001, 2006, 2010), the area of global boreal forest is approximately 1,470–1,640 million ha (Table 2.2). The total growing stock of stem wood is about 96,000 million m^3, of which 80,000 million m^3 are coniferous species (Fischlin et al. 2009).

Table 2.1 Main boreal tree species (Kuusela 1990).

North America		Eurasia	
Coniferous		**Coniferous**	
Pinus contorta Dougl.	Lodgepole pine	*Pinus sylvestris* L.	Scots pine
Pinus banksiana Lamb.	Jack pine	*Pinus cembra* L. var. sibirica	Siberian cembra
Pinus resinosa Ait.	Red pine	*Pinus pumila* Rig.	Japanese stone pine
Pinus strobus L.	Eastern white pine	*Larix sibirica* Ledeb.	Siberian larch
Pinus rigida Mill.	Pitch pine	*Larix gmelini* (Rup.) Litv.	Dahurian larch
Larix laricina (Du Roi) K.	Tamarack	*Picea abies* (L.) Karst.	Norway spruce
Picea mariana (Mill.) B.S.P.	Black spruce	*Picea abies* var. obovata Ledeb.	Siberian spruce
Picea glauca (Moench) Voss	White spruce	*Picea jezoensis* (Sieb. & Zucc.) Carr	Yeddo spruce
Picea rubens Sarg.	Red spruce	*Picea glehnii* (Fr. Schmidt) Mast.	Sachalin spruce
Tsuga canadensis (L.) Carr.	Eastern hemlock	*Abies sibirica* Ledeb.	Siberian fir
Thuja occidentalis (L.)	Eastern white cedar	*Abies nephrolepis* Maxim.	Khingan fir
Abies balsamea (L.) Mill.	Balsam fir	*Abies sachalinensis* Mast.	Sachalin fir
		Abies gracilis Komar.	Kamchatka fir
Broad-leaved		**Broad-leaved**	
Alnus rugosa (Du Roi) Spreng.	Hazel alder	*Alnus incana* L.	Grey alder
Alnus incana L. Moench	Speckled alder	*Alnus hirsute* Tures.	Manchurian alder
Betula neoalaskana Sarg.	Alaskan birch	*Alnus kamschaitic* Call. Kam.	Creeping alder
Betula papyrifera Marsh.	Paper birch	*Betula pendula* Roth.	White birch
Betula occidentalis Hook.	Water birch	*Betula pubescens* Ehrh.	Downy birch
Betula alleghaniensis Britton	Yellow birch	*Betula platyphylla* Sukacz.	Asian white birch
Populus balsamifera L.	Balsam poplar	*Betula ermani* Cham.	Erman's birch
Populus tremuloides Michx.	Trembling aspen	*Betula dahuric* Patt.	Dahurian birch
Populus deltoides Bartr.	Eastern cottonwood	*Betula japonica* Sieb.	Japanese white birch
Populus grandidento Michx.	Largetooth aspen	*Betula costata* Trautv.	Yellow birch
		Populus tremula L.	Aspen
		Populus suaveolensis	Mongolian poplar
		Populus sibolda	Siebold aspen

Table 2.2 Forest resources in the boreal regions and their share of global totals (Kuusela 1990, FAO 2001, 2006). Stock, growth and removal represent stem wood.

Region	Total values				Mean values		
	Forest area, million ha	Growing stock, million m^3	Annual growth, 1,000 m^3	Annual removal, 1,000 m^3	Stock, m^3 ha^{-1}	Annual growth, m^3 ha^{-1} yr^{-1}	Removal, m^3 ha^{-1} yr^{-1}
Alaska	48	1,300	3,600	3,000	160	0.8	0.06
Canada	436	23,000	356,000	152,000	110	1.7	0.3
Nordic countries	60	4,500	157,000	102,000	90	3.3	1.7
Russia	930	67,000	750,000	357,000	130	1.4	0.4
Total	1,474	96,000	1,270,000	614,000	120	1.6	0.4

On a global scale, the total boreal stem wood stock is about 45% of the total forests of the world, and it produces about 50% of all coniferous stem wood timber. Globally, mean stem wood stock in the boreal forest is about 120 m^3 ha^{-1}, but it varies substantially from region to region. In Nordic countries, the mean stock is about 90 m^3 ha^{-1}: these forests are intensively used and managed for timber production. In Alaska, the mean stock is about 150 m^3 ha^{-1}, including large areas of old-growth forest left outside management. Regardless of the main boreal regions, mean annual cutting removal in the late 1970s and early 1980s was substantially lower than mean annual growth (Kuusela 1990). Globally, the annual industrial round wood harvest from boreal forest is 500–600 million m^3 (Fischlin et al. 2009), which is 35–40% of the global round wood harvest.

2.3 Carbon Sequestration and Carbon Stocks

Keith et al. (2009) estimated that the mean carbon stock in boreal forest is 60–100 Mg C ha^{-1}; i.e., 60–70% in the below-ground biomass (roots and dead biomass), and the rest in the living biomass above ground. They estimated that the mean carbon residence time in boreal forest is 100 years, but the decay rate varies substantially depending on the temperature and moisture conditions (Raich and Schlesinger 1992). Fischlin et al. (2007) estimated that boreal forests over the world contain 703 Pg (10^{15} g) of carbon (trees, soils), which is 13% of the carbon in forests globally. Most carbon (70–90%) is in soils. Cold and wet soil allows only slow decay and the consequent development of deep organic soil. This is especially the case in peatlands, where waterlogged conditions further reduce decay. Globally, the boreal peatlands contain 400–500 Pg of carbon, which is one third of the world's organic soil carbon (Gorham 1991, Apps et al. 1993) or the equivalent of 40 ppm in atmospheric CO_2 concentration (Moore et al. 1998).

In boreal forests, the mean carbon budgets include the net sinks in the range 0.5–2.5 Mg C ha^{-1} yr^{-1} (Shvidenko and Nilsson 1993), and globally boreal forests are a carbon sink of 0.50 ± 0.08 Pg C yr^{-1} (Pan et al. 2011). These estimates are still fairly uncertain. The difference between annual growth and removal, for example, does not account for changes in the frequency and severity of disturbances. Kurz and Apps (1999) showed that the Canadian forest ecosystems changed from being a sink (0.075 Gt C yr^{-1}) from 1920–1970 to a source of 0.050 Gt C yr^{-1} in 1994, when changes due to disturbances (fire, harvest) were also included. The exclusion of management effects may also bias the estimates of the carbon sink/source relations. Intensive management may increase the carbon uptake more than it increases emissions, whereas less intensive management may create the opposite situation (Shepashenko et al. 1998). Factors such as nitrogen deposition, CO_2 fertilization and climate warming, with a consequent increase in growth, have not been included in the carbon sink/source analyses. These factors may increase carbon uptake and thus offset losses of carbon due to disturbances and cuttings (Schimel et al. 2000).

2.4 Biodiversity

The share of mature forests is still high in many parts of the boreal forest, providing habitats for species dependent on specific conditions of old-growth forests. Ruckstuhl et al. (2008) estimate that some 25% of the total boreal forest area is still fully natural, not involved in harvest and other commercial use of forest land. They note further that the boreal forest is "one of the few biomes in which the large predator-prey systems are still widely operational, invasive species are few by proportion, and large-scale natural disturbances such as wild fire and epidemic insects still occur extensively at their natural frequencies and patterns". However, the number of species per unit area is low at northern latitudes but the total species richness in the boreal forest is greater than in the tundra while less than in the temperate forests. The differences in the number of species between temperate and boreal forest and tundra are related to the productivity of land along the south-north temperature gradient across the vegetation zones (Waide et al. 1999).

Dynamics of boreal forest ecosystems are historically driven by disturbances such as fire, storms and the eruption of insect populations. Fire controls biodiversity by providing standing and lying dead wood in different phases of decay. This sustains resources and sites for a set of species typical of early post-fire communities, including birds, beetles, spiders, and vascular and non-vascular plants (Esseen et al. 1992). The success of many species is very closely linked to the regular fire cycle, with a consequent reduction in species richness if fires do not occur regularly. Similarly, harvest and management increase the fragmentation and reduce species richness. Bradshaw et al. (2009) estimate that 348 boreal species are threatened, and included in the IUCN (International Union for the Conservation of Nature) Red List (Table 2.3). Across all the taxa, the share of threatened species is highest in northern Europe. Bradshaw et al. (2009) emphasized the urgent need to preserve existing boreal forest and restore degraded areas in order to avoid a reduction in biodiversity and carbon sink, which the boreal forest provide in protection of the global environment.

Table 2.3 Taxonomic breakdown (%) of threatened species in the boreal forests (Bradshaw et al. 2009).

Taxon	% of species in IUCN Red List in 2008
Fungi and lichens	5
Plants	5
• Ferns	-
• Mosses	-
• Vascular plants	-
• Trees	-
Butterflies and moths	-
Birds	50
Mammals	35
Other vertebrates	5

2.5 Concluding Remarks

The boreal forest falls between the summer and winter limits of arctic air mass. This makes them especially vulnerable to climate warming, which is likely to modify tree species composition and enhance the growth and development of forests. On an annual basis, climatic warming in this century is likely to be 3–5°C, which may trigger heat stress and more frequent drought episodes, leading to widespread forest death due to outbreaks of insect pests and enhanced fire episodes (Olsson 2011). Proper management is therefore necessary to adapt the boreal forest to climate change and mitigate the harmful effects, that climatic warming is likely to have on their growth and development. The conversion of intact boreal forest into managed ones may reduce carbon stocks in boreal forests, while proper management may provide many opportunities to control the forest ecosystem/atmosphere interaction in mitigating climate change.

References

Apps, M. J., W. A. Kurz, R. J. Luxmore, L. O. Nilsson, R. A. Sedro, R. Schmidt et al. 1993. Boreal forests and tundra. Water, Air and Soil Pollution 70: 39–53.

Bergeron, Y., B. E. C. Bogdanski, G. P. Judah, T. Kuuluvainen, B. J. McAfee, A. Ogden et al. 2010. Sustainability of boreal forests and forestry in a changing environment. *In*: G. Mery, P. Katila, G. Galloway, R. I. Alforo, M. Kanninen, M. Lobovikov et al. (eds.). Forestry and Society—Responding to Global Drivers of Change. IUFRO World Series 25: 249–282.

Bonan, G. B. and H. H. Shugart. 1989. Environmental factors and ecological processes in boreal forests. Annual Review of Ecology and Systematics 20: 1–28.
Bradshaw, C. J., I. G. Warkentin and N. S. Sodhi. 2009. Urgent preservation of boreal carbon stocks and biodiversity. Trends in Ecology and Evolution 24(10): 541–548.
DeLuca, T. H. and C. Boisvenue. 2012. Boreal forest soil carbon: distribution, function and modelling. Forestry 85(2): 161–184.
Esseen, P., A., L. Ehnström, L. Eriksson and K. Sjöberg. 1992. Boreal forest—the focal habitats of Fennoscandia. pp. 252–325. *In*: L. Hansson (ed.). Ecological Principles of Nature Conservation. Applications in Temperate and Boreal Environments. Elsevier Applied Science, London, UK.
FAO. 2001. Forest resources assessment 2000. http://www.fao.org/forestry/11747/en/ Rome, Italy.
FAO. 2006. Global Forest Resources Assessment 2005. Progress towards Sustainable Forest Management. FAO Forestry Papers 147: 1–320.
Fischlin, A., G. F. Midgley, J. T. Prince, R. Leemans, B. Gopal, C. Turley, M. Rounsevell et al. 2007. Ecosystems, their properties, goods and services. *In*: M. L. Parry, O. F. Canziani, J. P. Putikof, J.P. van der Linden and C. E. Hanson (eds.). Climate Change 2007: Impacts, Adaptation and Vulnerability. Contribution of Working Group II to the Fourth Assessment Report of Intergovernmental Panel on Climate Change (IPCC). Cambridge University Press, Cambridge, UK.
Fischlin, A., M. Ayres, D. Karnosky, S. Kellomäki, S. Louman, C. Ong et al. 2009. Future environmental impacts and vulnerabilities. *In*: R. Seppälä, A. Buck and P. Katila (eds.). Adaptation of Forests and People to Climate Change—A Global Assessment Report. IUFRO World Series 22: 53–100.
Gordon, B., G. G. Bonan and H. H. Shugart. 1989. Environmental factors and ecological processes in boreal forests. Annual Review of Ecology and Systematics 20: 1–28.
Gorham, E. 1991. Northern peatlands: role in the carbon cycle and probable responses to climatic warming. Ecological Applications 1: 182–195.
Hämet-Ahti, L., A. Palmén, P. Alanko and P. M. A. Tigerstedt. 1989. Suomen puu- ja pensaskasvio. Yliopistopaino, Helsinki, Finland.
Hare, F. K. and J. C. Ritchie. 1972. The boreal microclimates. Geography Review 62: 333–365.
http://www.borealforest.org, visited 20.1.2014.
http://en.wikipedia.org/wiki/subarctic_climate, visited 20.1.2014.
Innes, J., L. A. Joyce, S. Kellomäki, B. Louman, A. Ogden, J. Parrotta et al. 2009. Management for adaptation. *In*: R. Seppälä, A. Buck and P. Katila (eds.). Adaptation of Forests and People to Climate Change. A Global Assessment Report. IUFRO World Series 22: 135–186.
IPCC. 2007a. Climate Change 2007: The Physical Science Basis. Contribution of Working Group I to the Fourth Assessment. Report of the Intergovernmental Panel on Climate Change. Cambridge University Press, Cambridge, UK.
IPCC. 1997b. Climate Change 2007: Contribution of Working Group II to the Fourth Assessment Report of the Intergovernmental Panel on Climate Change, 2007. Cambridge University Press, Cambridge, UK.
Kauppi, P., M. Posch, P. Hänninen, H. Henttonen, A. Ihalainen, E. Lappalainen et al. 1997. Carbon reservoirs in peatlands and forests in the boreal region of Finland. Silva Fennica 31(1): 13–25.
Keith, H., B. G. Mackey and D. B. Lindenmayer. 2009. Re-evaluation of forest biomass carbon stocks and lessons from the world's most carbon-dense forests. PNAS 106(28): 11635–11640.
Kellomäki, S., H. Strandman, T. Nuutinen, H. Peltola, K. T. Korhonen and H. Väisänen. 2005. Adaptation of forest ecosystems, forests and forestry to climate change. Finnish Environment Institute, FinAdapt Working Paper 4: 1–50.
Kellomäki, S., H. Peltola, T. Nuutinen, K. T. Korhonen and H. Strandman. 2008. Sensitivity of managed boreal forests in Finland to climate change, with implications for adaptive management. Philosophical Transactions of the Royal Society B363: 2341–2351.
Kurz, W. A. and M. J. Apps. 1999. A 70-year retrospective analysis of carbon fluxes in the Canadian forest sector. Ecological Applications 9(2): 526–547.
Kuusela, K. 1990. The Dynamics of Boreal Coniferous Forests. Gummerus Kirjapaino Oy, Jyväskylä, Finland.
Larcher, W. 1975. Physiological Plant Ecology. Springer-Verlag, Berlin, Germany.
Maljanen, M., J. Hytönen and P. J. Martikainen. 2001. Fluxes of N_2O, CH_4 and CO_2 on afforested boreal agricultural soils. Plant and Soil 231: 113–121.
Moore, T. R., N. T. Roulet and J. M. Waddington. 1998. Uncertainty in predicting the effect of climatic change on the carbon cycling of Canadian peatlands. Climatic Change 40: 229–245.
Nikolov, N. and H. Helmisaari. 1992. Silvics of the circumpolar forests tree species. *In*: H. H. R. Shugart, R. Leemans and G. B. Bonan (eds.). A System Analysis of the Global Boreal Forest. Cambridge University Press, New York, USA.
Olsson, R. 2011. To manage or protect. Air Pollution and Climate 26: 1–67.
Pan, Y., R. A. Birdsey, J. Fang, R. Houghton, P. E. Kauppi, W. A. Kurz et al. 2011. A large and persistent carbon sink in world's forests. Science 333: 988–993.
Raich, J. W. and W. H. Schlesinger. 1992. The global carbon dioxide flux in soil respiration and its relationship to vegetation and climate. Tellus 44B: 81–99.
Ruckstuhl, K. E., E. A. Johnson and K. Miyanishi. 2008. Introduction. The boreal forest and global change. Philosophical Transactions of the Royal Society B363: 2245–2249.

Schimel, D. S., I. J. House, K. A. Hibbard, P. Bousquet, P. Ciais, P. Peylin et al. 2001. Recent patterns and mechanisms of carbon exchange by terrestrial ecosystems. Nature 414: 169–173.

Shepashenko, A. D., A. Shvidenko and S. Nilsson. 1998. Phytomass (live biomass) and carbon of Siberian forests. Biomass and Bioenergy 14(1): 21–31.

Shvidenko, A. and S. Nilsson. 2003. A synthesis of the impact of Russian forests on the global carbon budget for 1961–1998. Tellus 55B: 391–415.

Sirois, L. 1992. The transition between boreal forest and tundra. pp. 196–215. *In*: H. H. Shugart, R. Leemans and G. B. Bonan (eds.). A System Analysis of the Global Boreal Forest. Cambridge University Press, Cambridge, UK.

Strack, M. (ed.). 2008. Peatlands and Climate Change. International Peat Society, Saarijärven Offset Oy, Saarijärvi, Finland.

Waide, R. B., M. R. Willig, C. F. Steiner, G. Mittelbach, L. Gough, S. I. Dodson et al. 1999. The relationship between productivity and species richness. Annual Review of Ecology and Systematics 30: 257–300.

Walter, H. 1979. Vegetation of the Earth and Ecological System of the Geo-Biosphere. 2nd edition. Springer-Verlag, New York, USA.

Wulder, M. A., S. Campbell, J. C. White, M. Flannigan and I. D. Campbell. 2007. National circumstances in the international circumboreal community. The Forestry Chronical 83(4): 539–556.

Managed Boreal Forests in Local Context

ABSTRACT

Boreal forests in northern Europe are widely used to produce timber and biomass for wood-based materials, products, and energy. They extend from the northern parts of the hemi-boreal vegetation zone (N 60°) through the boreal zone to the tundra (N 70°). In the boreal forests Scots pine (*Pinus sylvestris* L.) dominates mainly on poor sites and Norway spruce [*Picea abies* (L.) Karst.] mainly on fertile sites. Birch (*Betula pendula* Roth and *B. pubescens* Ehrh.) grows commonly with conifers. These forests are in the high latitudes, where the future climate will probably be much warmer than currently.

Keywords: Finland, boreal forests, forest resources, diversity, carbon sequestration, managed boreal forest, natural boreal forests, climatic limits, bedrock, soils, climate change

3.1 Bioclimatic Limits

Temperature and precipitation

Boreal forests in northern Europe, including Finland, are widely used to produce timber for wood-based products and energy. The territory of Finland extends from the upper edge of the hemi-boreal vegetation zone (N 60°) through the boreal zone to the tundra in the northern part of the country (N 70°). The climate is characterized by clear south-north gradients in temperature and precipitation across the country. In the southernmost areas, the mean annual temperature is from +4 to +5°C and the mean annual precipitation 600–700 mm, whereas in the northernmost areas the mean annual temperature is from –1 to –2°C and the mean annual precipitation 400–500 mm (Fig. 3.1). The west-east gradient in climate is not as pronounced but in the western part of country precipitation remains lower than in the eastern part. The Atlantic Ocean has a strong effect on the annual and seasonal variability of climate and weather in northern Europe.

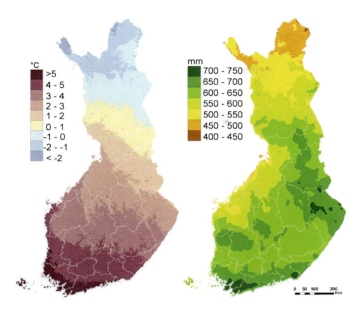

Fig. 3.1 Left: Mean annual temperature, and Right: Mean annual precipitation over the period 2001–2010 (http://ilmatieteenlaitos.fi/vuositilastot, visited 21.1.2014). Courtesy of the Finnish Meteorology Institute.

Length of growing season

The south-north gradient in temperature affects the length of growing season. The length of the growing season refers to the time in days from the beginning to the end of the thermal growing season: from the day with a diurnal mean temperature ≥ + 5°C to the day with a diurnal temperature < + 5°C (Fig. 3.2). According to this range, the length of the thermal growing season in the southernmost part of the country is 160–180 days, whereas in the northernmost part 100–110 days. Temperature sum (TS, d.d. degree days) is another way to indicate the length of the growing season and temperature:

$$TS = \sum_{d=1}^{n} (T_d - 5), \qquad if\ T_d \geq + 5°C \tag{3.1}$$

where T_d [°C] is the daily mean temperature minus the threshold temperature of +5°C. In the south, the temperature sum is 1,300–1,400 d.d., whereas in the north the temperature sum remains 600–700 d.d. At the northern timberline, the temperature sum is less than 600 d.d., corresponding a mean annual temperature of –2 to –1°C.

Bedrock and soil

The bedrock in Finland is very old (Pre-Cambric) and consists mainly of various granites, gneisses and quartzites. They cover large areas in the middle boreal zone (i.e., central and eastern Finland), where the bedrock is acid and erodes and decomposes very slowly. They include only small amounts of calcium, phosphate, potassium and other minerals important for plant growth. In a few places, the bedrock is comprised of dolomites or other alkaline stones but this is less than 1% of the total area of bedrock. During the last glaciation, the bedrock surface was broken and the movement of ice ground the stone into pieces of varying sizes, from large boulders and small stones (size > 20 mm) to clay particles (size < 0.002 mm). A large area of the bedrock is covered by moraines: glacial drift consisting of a mixture of clay, sand, pebbles and boulders. The moraines may be further classified,

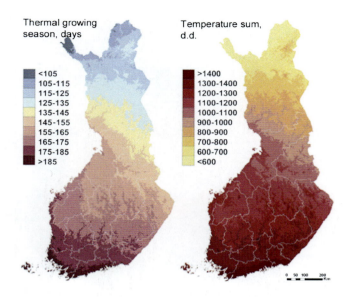

Fig. 3.2 Left: Mean annual length of thermal growing season (days), and Right: Mean annual temperature sum (degree days, d.d.) over the period 2001–2010 (http://ilmatieteenlaitos.fi/vuositilastot, visited 21.1.2014). Courtesy of the Finnish Meteorology Institute.

based on the dominant soil type, for example as sand or clay moraines. Various moraines cover more than 80% of the soil in Finland. Water running from melting ice sorted the moraines and formed soils dominated by gravel, sand or clay. In many places, gravel and sand were formed as ridges at the terminal edges of continental ice, and clay soils represent the sediments in waters outside the melting ice.

3.2 Forest Vegetation and Forest Resources

Main tree species

The forests in Finland are divided into the hemi-boreal, and the southern, central and northern boreal forests (Fig. 3.3). The hemi-boreal forests represent the transition from the Central-European temperate forests to boreal forests, dominated by Scots pine (*Pinus sylvestris* L.), mainly on poor sites, and Norway spruce [*Picea abies* (L.) Karst.], mainly on fertile sites. Birch (*Betula pendula* Roth and *B. pubescens* Ehrh.) grows commonly with conifers if the site is fertile enough. Several other deciduous species are common throughout the country, including aspen (*Populus tremula* L.), gray alder (*Alnus incana* (Linnaeus) Moench), black alder [*Alnus glutinosa* (L.) Gaertn.] and mountain ash (*Sorbus aucuparia* L.), but their percentage in total timber resources is marginal. Many deciduous species common in Central Europe (e.g., oak (*Quercus robur* L.), lime (*Tilia cordata* L.), maple (*Acer platanoides* L.), ash (*Fraxinus excelsior* L.) and elm (*Ulmus glabra* L.)) have their northern limits in Finland, but these species are very rare, with the main populations in the southernmost areas of the country. Tree species richness declines northwards, and the northern boreal forests represent the transition, with the Scots pine dominating up to the arctic tundra. Here closed Scots pine forests turn open forests or forests dominated by timberline deciduous forests (*Betula pubescens var. tortuosa*). The timber line for Norway spruce and Pendula birch is more southern than the timber line for Scots pine and Pubescent birch. There are 31 native tree or woody species in Finland, of which four are coniferous and others deciduous.

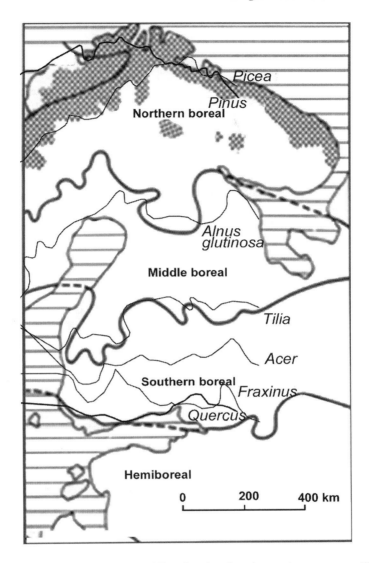

Fig. 3.3 Distribution of hemi-boreal and southern, middle and northern boreal vegetation zones across Finland and northern distribution limit of selected tree species growing in Finland, based on Kalliola (1973). In the hemi-boreal zone, the mean length of the growing season (the number of days with mean temperature $\geq +5°C$) is >175 days, in the southern boreal zone 160–175 days, in the middle boreal zone 140–160 days, and in the northern boreal zone 100–140 days.

Ground vegetation and site types

Ground vegetation is dominated by the same taxa as elsewhere in boreal forests. On poor sites, lichens (e.g., *Cladionia* sp.) often dominate the ground cover, especially in the northern parts of the boreal forests with open canopy structure. On more fertile sites, the ground cover is dominated by mosses (e.g., *Hylocomium*, *Pleurozium* and *Dicranum* sp.). This is especially the case in mature coniferous forests on sites of intermediate fertility. In these conditions, the ground cover communities include further dwarf shrubs (e.g., *Vaccinium*, *Empetrum* sp.), grasses (e.g., *Descampsia*, *Poa*, *Calamagrostis* sp.) and herbs (e.g., *Epilobium* sp.). Peatlands in the boreal north support a flora that typically includes a large number of *Sphagnum* mosses and vascular species such as *Ledum*,

Eriophorum and *Carex* species. In the northern parts of boreal forests and even on poor sites in the southern boreal areas, the growth of ground vegetation represents a large part of the productivity of forested ecosystems (Bonan and Shugart 1989).

On upland sites, the species composition of ground vegetation is used to indicate the fertility of sites in terms of site types under the assumption that the properties of ground vegetation indicate productivity (P) as a function of the environment (climatic and edaphic factors):

$$P = f(Environment) = f(Climatic\ and\ edaphic\ factors) \Rightarrow P = f(Ground\ vegetation) \qquad (3.2)$$

By definition, all the sites with similar ground vegetation belong to the same forest type (Cajander 1909). This applies most clearly to sites occupied by old-growth tree stands with closed canopies, where the ground vegetation is stable (i.e., in subclimax or climax phase). The site types also apply to sites occupied by tree stands that are still developing, with the successive ground vegetation changing towards the climax phase, indicating a given forest type. For example, the grass- and herb-rich ground vegetation typical on sites of medium fertility will be dominated by dwarf shrubs (e.g., *Vaccinium myrtillius*) and mosses (e.g., *Hylocomium splendens*), when a tree stand matures and the canopy closes. Figure 3.4 shows the dependence of ground vegetation and the subsequent sites types on the climatic and edaphic properties of sites over Finland. The dominance of lichens and dwarf shrubs increases northwards and herbs and grasses southwards. The capital letters refer to the main dominant species or taxa used to label the site types.

Site fertility	Site type in southern boreal	Southern boreal	Middle boreal	Northern boreal
Extremely poor	*Cladonia* type (CIT)	Lichen rich, scattered dwarf shrubs		
Very poor	*Calluna* type (CT)	Dwarf shrub rich, scattered lichens and mosses		
Poor	*Vaccinium* type (VT)	Dwarf shrub rich, moss rich, scattered lichens		
Medium fertile	*Myrtillus* type (MT)	Dwarf shrub rich, moss rich, scattered grasses and herbs		
Fertile	*Oxalis-Myrtillus* type (OMT)	Moss rich, grass rich, scattered dwarf shrubs and herbs		
Very fertile	*Oxalis-Maianthemum* type (OMaT)	Grass and herb rich, scattered mosses and dwarf shrubs		

Soil: Increasing grain size, reducing soil moisture and available nitrogen ↑

Climate: Declining radiation, temperature, precipitation and length of growing season →

Fig. 3.4 Schematic presentation of the distribution of site fertility in site types (as indicated by some main ground cover plant species) for southern boreal forests in relation to selected climatic and edaphic factors. The parallel site types apply for middle and northern boreal forests but the species composition in ground cover is dependent on regions. The same site fertility class in the south indicates higher productivity than that in the north regardless of tree species. The capital letters refer to specific site types indicated (labeled) by selected ground cover species.

Forest resources and ownership

In Finland, the total land area for forestry is 26 million hectares including the productive (the annual growth ≥ 1 m^3 ha^{-1} yr^{-1}), low productive (the annual growth 0.1–1 m^3 ha^{-1} yr^{-1}) and non-productive (the annual growth < 0.1 m^3 ha^{-1} yr^{-1}) land given in the terms of stem wood growth per area and year. The forests are dominated by Scots pine and Norway spruce: 50% of total stocking represents Scots pine, 30% Norway spruce, and 17% birches. The share of other deciduous species (including aspen, alders, mountain ash, oak, etc.) is only 3% of the total stocking (Finnish Forest Research Institute

2011, 2012). The total growing stock of trees is 2,306 million m³ which make 111 m³ ha⁻¹. Totally, 71% of the entire stocking is owned by private persons while the state owns 18%, industry 7% and others 4%.

The total growth of forests is 104 million m³ yr⁻¹, with the mean growth 4.6 m³ ha⁻¹ yr⁻¹ but growth varies substantially throughout the country (Fig. 3.5). In general, growth decreases towards the north following the south-north temperature gradient but varies substantially between regions following the variability of climate and the fertility of sites. The time scale in forestry is long, with a rotation of 60–160 years according to species and region. Currently (2012), total annual growth exceeds annual total removal (71 million m³ yr⁻¹) (cuttings and natural mortality), which is 68% of the total growth (Finnish Forest Research Institute 2011, 2012).

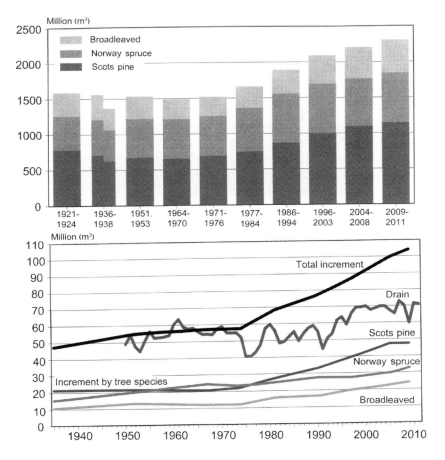

Fig. 3.5 Above: Growing stock of main tree species over the period 1921–2011, and Below: Total growth of forests and the growth of main tree species over the period 1935–2010 (Finnish Forest Research Institute 2011, 2012). Courtesy of the Finnish Forest Research Institute.

Carbon sequestration

Liski and Westman (1997a,b) studied carbon density in the organic layer on mineral soils, and down to one meter in soil profiles throughout Finland, in the temperature sum range of 800–1,300 d.d. (i.e., within 66°–60° N). The carbon density was 2.2 kg C m⁻² in the organic soil layer at the most fertile sites (*Oxalis-Myrtillus* type and sites with higher fertility). At the medium-fertile and medium-poor (*Myrtilllus* and *Vaccinium* site types) sites, carbon density was 1.9 kg C m^{-2}. At the poor sites (*Calluna* site type and the site types of lower fertility), carbon density was 1.2 kg C m^{-2}. When considering soil organic layers and soil profile deeper than one meter, they found that the organic layer contained 28% of total carbon in soil, while the carbon in the mineral layer down to one meter was 68% and below that 4%.

According to Liski and Westman (1997a,b), the amount of carbon in soil was more than double that in trees. Forest type explained 36 and 70% of the variability of carbon in organic and mineral layers. On the other hand, the carbon density in the mineral soil layer increased 0.266 kg C m^{-2} per the increase of 100 d.d. in the temperature sum. Carbon density in soil thus increased along with the increase in productivity indicated by the increasing temperature sum, the largest density values being in the middle and southern boreal forests (see also Olsson et al. 2008, Stendahl et al. 2010). Based on the National Forest Inventory data, Liski and Westman (1997b) estimated that the total carbon storage in the upland sites throughout Finland falls in the range 1,109–1,315 Tg.

In managed forests, carbon density and the total amount of carbon are affected by management and harvest. Based on the inventory data, Liski et al. (2006) analyzed the long-term dynamics of carbon sequestration at the national scale (Fig. 3.6). Over an 80-year period, Net Primary Production (NPP) increased from 0.29 to 0.39 kg C m^{-2} yr^{-1} (35% increase) and the heterotrophic respiration (Rh, decay of soil organic matter) increased from 0.25 to 0.29 kg C m^{-2} yr^{-1} (16% increase). This implied that the Net Ecosystem Production (NEP) (or Net Ecosystem Exchange (NEE)) increased from 0.04 to 0.10 kg C m^{-2} yr^{-1} (150% increase). The Net Biome Production (NBP) (carbon in NEE plus carbon in harvests) fell below zero in the 1920s and 1930s, and occasionally in the 1960s, but thereafter remained positive: the harvest remained smaller than growth. In the 1990s, the mean net ecosystem production was 0.38 kg C m^{-2} yr^{-1} but 0.28 kg C m^{-2} yr^{-1} (70% of NEE) of this was lost in heterotrophic respiration. At the same time, a half of the net ecosystem production (0.060 kg C m^{-2} yr^{-1}) was harvested in timber (0.039 kg C m^{-2} yr^{-1}). At the same time, 72% (0.028 kg C m^{-2} yr^{-1}) of the net biome production was accumulated in the tree biomass and the rest (0.011 kg C m^{-2} yr^{-1}) in soil.

3.3 Biodiversity

In general, the number of species per unit area is low at northern latitudes, and this is also true for Finland, where the total number of known species is about 45,000 (Rassi et al. 2010, Finnish Forest Research Institute 2011). This low total number is mainly due to the short time elapsed since the last glaciation (11,000 years), with the consequence that migration is still on the way. The subsequent biodiversity involves a large number of species, which are native of the eastern taiga or continental boreal forests in Eurasia. Currently, about 2,250 of these species are classified as threatened, which is about 10% of the species that are known well enough for their classification to be identified as threatened or not (Rassi et al. 2010, Finnish Forest Research Institute 2011).

The most recent estimates show that forest habitats are preferred by 800 out of the total number of threatened species (Rassi et al. 2010). This is especially the case for fungi, which comprise 70% of all the threatened species in forest habitats. The main threat is the change in forest habitats due to forestry operations, such as a reduction of decaying wood, reduction of old-growth forest, and changes in tree species composition, with an increasing dominance of coniferous species. The impact of forestry on biodiversity is closely related to the high proportion of living forest species (20–25%) that are dependent on dead wood (800 coleopterans, 1,000 dipterans, 1,000 fungi, 200 lichens, etc.). Many of these species are further specialized to live on burnt dead wood, which makes these species especially vulnerable to changes in the properties of forest environment under regular management (e.g., Kellomäki et al. 2001). The importance of dead wood for species richness is demonstrated in Fig. 3.7. The number of saproxylic species is positively correlated with the amount of dead wood (Siitonen et al. 2005), and decaying birch wood in particular seems to support a wide variety of species. The positive correlation probably indicates that there are more specialized microhabitats available in forests with a high amount of dead wood habitat, providing dead wood for the species most affected by management and timber harvest.

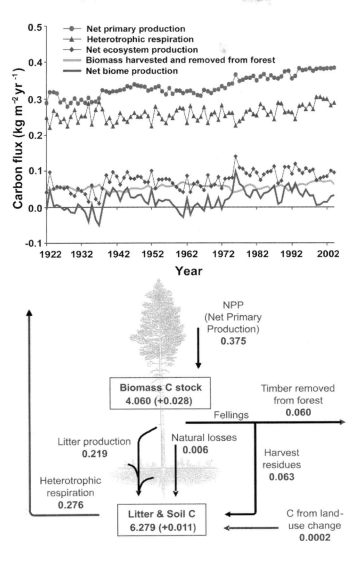

Fig. 3.6 Above: Annual carbon fluxes over Finland in the period 1922–2002, and Below: Mean carbon budget over Finland in the 1990s (Liski et al. 2006). Permission of EDP Sciences. Fluxes in upper Figure are in kg C m^{-2} yr^{-1}, and carbon mass in lower Figure is kg C m^{-2}.

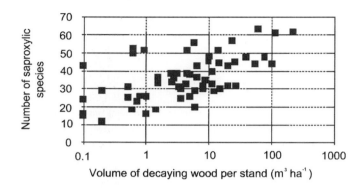

Fig. 3.7 Correlation between the number of saproxylic species and the volume of decaying wood after logarithmic transformation of the x-axis (Siitonen et al. 2005). Courtesy of the Finnish Forest Research Institute.

3.4 Unmanaged vs. Managed Boreal Forests

Management operations in controlling the growth and development of forest

Dynamics of trees in unmanaged and managed forests involve regeneration, growth and mortality of trees, which changes the structure of tree populations/communities over time. Regeneration and growth link the dynamics of forest ecosystems with the climatic and edaphic properties of site through the energy flow and cycles of nutrients and water. Where there is no management, these processes may create an uneven-aged forest structure, or a forest structure that is a mosaic of even-aged stands. In the former case, succession is driven by autogenic disturbances (e.g., the death of single trees), whereas in the latter case succession is driven by allogenic disturbances (e.g., fire, strong wind killing trees over smaller or larger areas) (Box 3.1). In this context, management operations can be divided into autogenic measures and allogenic measures in making the resources available for regeneration and growth of trees. In the former, trees are managed individually and in the latter they are managed stand by stand. Tree-by-tree management leads to an uneven-aged forest structure and stand-by-stand management to a mosaic-type forest structure, where each stand is evenly aged, but in a different phase of development compared to neighboring stands. Between these two extremes, the forest structure of boreal forests is normally a mixture of both, created concurrently due to allogenic and autogenic factors (Kellomäki 2009).

In natural forests, abiotic and biotic disturbances (e.g., fire, strong wind, insects and pest attack) control the ecosystem dynamics and produce landscapes of varying structures. The same factors effect in managed forests, but their impacts are exceeded by management operations. The emerging functioning and structure of ecosystem make it possible to produce the goods and services, which are aimed at management (Fig. 3.8). For example, soil preparation to enhance the availability of nutrients and subsequent restocking can take place before regeneration. The tending of seedling stands means that fewer trees are productive, thus providing faster growth of individual trees than that would be possible without increased spacing. Regular thinning enhances the nutrient cycle, and subsequent growth in the intermediate phase of stand development and regeneration in more mature stands. The further enhancement of natural regeneration can occur through regeneration cuttings (e.g., shelter wood or seed tree methods) and soil preparation. Timber, biomass and other goods and services are concurrently supplied throughout the production cycle (Kellomäki 2009).

Box 3.1 Measures available for management

Table 3.1 lists some management measures, with their impact on the properties of sites and the trees occupying sites. Allogenic measures primarily control the properties of the soil system, such as the nutrient cycle and soil moisture, with the largest enhancement of growth and regeneration of pioneer species (a species most successful under an ample supply of resources like birch) through the disturbance of soil surface. Site preparation involves mechanical measures to reduce the effects of ground vegetation on seedlings and to control the physical and chemical properties of the soil. Ditching lowers the ground water level and reduces excess soil moisture. Prescribed burning influences the chemical and physical properties of soil and reduces the effects of ground vegetation on seedlings. As described burning, fertilization increases the availability of nutrients for growth.

Autogenic measures primarily represent spacing of tree stands and enhance the growth and regeneration of both pioneer and climax species (a species successful also under a short supply of resources like Norway spruce). Regenerative cutting involves terminal cuts for reforesting through natural seeding or artificial seeding or planting. In tending of seedling stands, tree species not preferred for management are removed, while pre-commercial thinning aims at proper spacing for further growth of remaining trees for commercial thinning. Pruning is employed to remove dead and living branches from the lower crown to increase the amount of stem wood without the knots.

Table 3.1 Schematic presentation of the effects of some management measures on the properties of site and tree populations (Kellomäki 2009).

Measure	Effect on site properties and population/communities				
	Nutrients	Moisture	Temperature	Soil physics	Species
Allogenic measures (subjected mainly to environment), with effects on site and trees					
• Site preparation	Strong	Moderate	Strong	Strong	Strong
• Ditching	Moderate	Strong	Moderate	Moderate	Strong
• Prescribed burning	Strong	Small	Moderate	Moderate	Strong
• Fertilizing	Strong				
Autogenic measures (subjected mainly to trees), with effects on site and trees					
• Regenerative cuttings	Moderate	Moderate	Moderate	Small	Strong
• Pre-commercial thinning	Small	Small	Small	Small	Strong
• Commercial thinning	Moderate	Moderate	Small	Small	Moderate
• Pruning	Small	Small	Small		Small

Structure and function of unmanaged vs. managed forest ecosystems

Similä et al. (2012) compared the structural and functional features of unmanaged and managed boreal forests as listed in Table 3.2.

Regardless of disturbances, the amount of standing and lying dead trees in a naturally developing forest is substantially larger than in managed forests, where dying and dead trees are regularly harvested in order to avoid economic losses and to reduce the risk of insect attacks on living trees (Esseen et al. 1992, 1997). Furthermore, the biogeochemical cycles are more closed in unmanaged than managed forests, where nutrients are made purposefully available by enhancing the decomposition of litter and humus in soil management. This implies that the carbon stored in soil is reduced compared to that in unmanaged forests. Management also makes the hydrological cycle more open, which enhances nutrient leaks outside sites (Tamm et al. 1974). These structural and functional differences between unmanaged and managed forests are further reflected in the

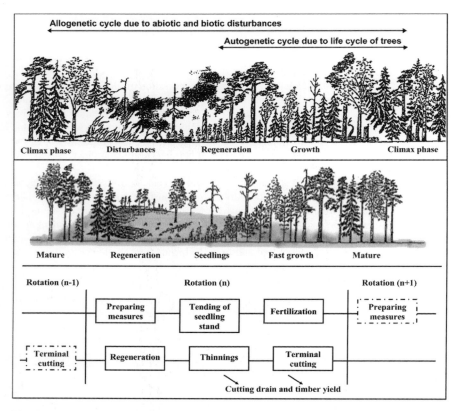

Fig. 3.8 Schematic presentation of how growth and development (i.e., successional cycle) proceeds in naturally growing and developing forest (above), and in the forest managed in different phases of the production cycle (middle), and the timing of selected management measurers in controlling the cycle (below), assuming stand-by-stand management (Kellomäki 2009).

diversity; i.e., species specialized to natural forest habitats created by varying disturbances are no more successful in managed forests. The short supply of dead wood, for example, leads to the replacement of specialized species by generalist species, which are successful in habitats typical of managed forests (Siitonen and Martikainen 1994, Siitonen et al. 1995).

3.5 Concluding Remarks

Boreal forests in Finland have been utilized intensively for more than 100 years, which has inevitably modified the structure and the subsequent functioning of the ecosystem. For example, management to maximize the timber yield has led to the fragmentation of the forest landscape, which deviates from that emerging under natural disturbances. Fragmented structure, with clear stand boundaries is likely to make managed boreal forests vulnerable to wind damage even under the current climate. Regular thinning, together with a preference for coniferous species have reduced the proportion of deciduous. Effective firefighting and measures increasing forest resistance to wind damages have also reduced the formation of dead wood and the occurrence of burnt wood, for which many species are specialized. Under warming climate, such structural features of boreal forest ecosystems are still important for sustaining the production of ecosystem and goods and services in the future.

Table 3.2 Comparison of the structures and dynamics of unmanaged (natural) and managed boreal forests, as adopted from Similä et al. (2012) with slight modifications. Management is assumed to be aimed at the structure of the forest landscape, with mosaics of even-aged stands in varying phases of development.

Characteristics	Natural forests	Managed forests
Living trees	• Trees of various age and size. • Trees spatially aggregated, with variations in the canopy cover.	• Stands dominated mainly by a single tree species, with the same age growing rapidly and uniformly.
Dead trees	• Dead wood present at any phase of succession. Many kinds of dead wood by tree species, diameter and degree of decay. • Stumps and snags of different heights. • Continuity of deadwood.	• Small amount of dead wood. • Dead trees only of limited diameter, no continuity. • Stumps may even be absent in sites harvested for energy.
Soil conditions	• Variable soil structure varied due to fire and wind throws. • Varying degree of upheavals due to disturbances of varying intensity. • Decaying stems for seedlings to establish.	• Soil treated uniformly for regeneration. • Few areas of exposed mineral soil in mature forests due to uprooted trees.
Hydrological conditions	• Natural streams, springs, groundwater seepage areas, moist hollows creating variety in forest microclimate.	• Springs, groundwater seepage areas and moist hollows often drained; streams cleared to improve drainage.
Species	• Species adapted to natural forest habitats and disturbance regimes. • Species assemblages change radically following major disturbances. • Species utilizing dead wood and are fire-dependent.	• Generalist species. • Many species depend on dead wood species are absent. • Species assemblages change greatly after clear-cutting.
Nutrient and soil carbon	• Nutrients and carbon stored in living wood, dead wood and soil. • No or negligible leakage of nutrients to ground and surface waters.	• Thinning, site preparation and harvest of energy biomass enhance nutrient cycle and reduce carbon in ecosystem. • Use of fertilizers may result in nutrient leakages and result in eutrophication of ground and surface waters.
Disturbance dynamics	• Disturbances occur irregularly. • Major disturbances such as forest fires and serious storm damage are rare. • Small and moderate disturbances such as death of single and groups of trees.	• Regular disturbances, e.g., logging and site preparation. • Most trees killed in natural disturbances are removed.
Structure of forest landscape	• Mature and forests with closed canopy dominate landscapes. • Forest stands merge into each other with no clear boundaries. • Connectivity of habitats preserved due to natural disturbance dynamics.	• Stands of different age-classes, with high proportions of young stands. • Clear boundaries between stands with trees of different species and ages. • Habitats for species dependent on decaying wood are fragmented.

References

Bonan, G. B. and H. H. Shugart. 1989. Environmental factors and ecological processes in boreal forests. Annual Review of Ecology and Systematics 20: 1–28.

Cajander, A. K. 1909. Über Waldtypen. Fennia 28(2): 1–175.

Esseen, P., A. Ehnström, L. Eriksson and K. Sjöberg. 1992. Boreal forest—the focal habitats of Fennoscandia. pp. 252–325. *In*: L. Hansson (ed.). Ecological Principles of Nature Conservation. Applications in Temperate and Boreal Environments. Elsevier Applied Science, London, UK.

Essen, P. A., B. Ehnström, L. Ericson and K. Sjöberg. 1997. Boreal forests. Ecological Bulletins 46: 16–47.

Finnish Forest Research Institute. 2011. Finnish Statistical Yearbook of Forestry 2011. Sastamala, Finland.
Finnish Forest Research Institute. 2012. Finnish Statistical Yearbook of Forestry 2012. Sastamala, Finland.
Kalliola, R. 1973. Suomen kasvimaantiede. Werner Söderström Oy, Porvoo, Finland.
Kellomäki, S., J. Kouki, P. Niemelä and H. Peltola. 2001. Timber industry. Encyclopedia of Biodiversity. Academic Press, London, UK 5: 655–666.
Kellomäki, S., V. Koski, P. Niemelä, H. Peltola and P. Pulkkinen. 2009. Management of forest ecosystems. pp. 252–373. *In*: S. Kellomäki (ed.). Forest Resources and Sustainable Management. Papermaking Science and Technology. Book 2. Paper Engineers' Association/Paperi ja Puu Oy. Gummerus Oy, Jyväskylä, Finland.
Liski, J. and C. J. Westman. 1997a. Carbon storage in forest soil of Finland. 1. Effect of thermoclimate. Biogeochemistry 36: 239–260.
Liski, J. and C. J. Westman. 1997b. Carbon storage in forest soil of Finland. 2. Size and regional patterns. Biogeochemistry 36: 261–274.
Liski, J., A. Lehtonen, T. Palosuo, M. Peltoniemi, T. Eggers, P. Muukkonen et al. 2006. Carbon accumulation in Finland's forests 1922–2004—an estimate obtained by combination of forest inventory data with modelling of biomass, litter and soil. Annals of Forest Science 63: 687–697.
Olsson, M. T., M. Erlandersson, L. Lundin, T. Nilsson, Å. Nilsson and J. Stendahl. 2008. Organic carbon stocks in Swedish podzol soils in relation to soil hydrology and other site characteristics. Silva Fennica 43(2): 2009–222.
Rassi, P., E. Hyvärinen, A. Juslén and I. Mannerkoski (eds.). 2010. Suomen lajien uhanalaisuus—Punainen kirja 2010. Ympäristöministeriö ja Suomen ympäristökeskus, Helsinki, Finland. www.ymparisto.fi.
Siitonen, J. and P. Martikainen. 1994. Occurrence of rare and threatened insects living on decaying *Populus tremula*: A comparison between Finnish and Russian Karelia. Scandinavian Journal of Forest Research 9(2): 185–191.
Siitonen, J., P. Martikainen, L. Kaila, A. Nikula and P. Punttila. 1995. Kovakuoriaslajiston monimuotoisuus eri tavoin käsitellyillä metsäalueilla Suomessa ja Karjalan Tasavallassa. Metsäntutkimuslaitoksen Tiedonantoja 564: 43–63.
Similä, M., K. Junninen, E. Hyvärinen and J. Kouki. 2012. Restoring heathland forest habitats. pp. 6–8. *In*: M. Similä and K. Junninen (eds.). Ecological Restoration and Management in Boreal Forests—Best Practices from Finland. Metsähallitus Natural Heritage Services. Erweko Painotuote, Helsinki, Finland.
Stendahl, J., M. -B. Johansson, E. Eriksson, E. Nilsson and O. Langvall. 2010. Soil organic matter in Swedish spruce and pine forests—differences in stock levels and regional patterns. Silva Fennica 44(1): 5–21.
Tamm, L. O., H. Holmen, B. Popovic and G. Wiklander. 1974. Leaching of plant nutrients from soils as a consequence of forestry operations. Ambio 3(6): 211–221.

PART II
Climate Change and Impact on Growing Conditions

Global Climate Change

ABSTRACT

Climate change is any change in the climate over time, whether due to natural variability or as a result of human activities. Climate change is most related to the increase in atmospheric concentrations of carbon dioxide (CO_2), and the increased concentration of other greenhouse gases. The warming is probably most pronounced at the high latitudes in the northern hemisphere, where the mean annual temperature is likely to be 6–8°C higher by the end of this century. The annual precipitation in high latitudes may increase by 20–30%, especially in the maritime areas.

Keywords: Global warming, climate change, greenhouse gas, radiative forcing, temperature, precipitation, greenhouse gas emission, IPCC

4.1 Basic Mechanisms

Increase of greenhouse gases (GHG) in atmosphere

According to the Framework Convention on Climate Change, climate change refers to a change in climate which is directly or indirectly due to human activity altering the composition of the atmosphere and which can be observed over a given time period. This definition differs from that of the International Panel for Climate Change (IPCC), which defines climate change to mean any change in climate over time, whether due to natural variability or a result of human activities. This book uses the IPCC definition, based mainly on the findings of the fourth assessment published in 2007 (IPCC 2007).

Climate change is generally related to an increase in atmospheric concentrations of carbon dioxide (CO_2), and the increased concentration of other greenhouse gases (GHG) like methane (NH_4) and nitrous oxide (N_2O). From the beginning of the industrial age (since 1750), CO_2 concentration in the atmosphere has increased from 280 ppm to 395 ppm (IPCC 2007), and it is currently increasing at about 2 ppm per year (Fig. 4.1). The main source of increased CO_2 is the use of fossil fuels in energy production, but CO_2 emissions are also increased by the conversion of forests into agricultural fields (i.e., land use change, Box 4.1). At the same time, methane (NH_4) concentration has increased from 715 ppb to 1,774 ppb, and the concentration of N_2O from 270 ppb

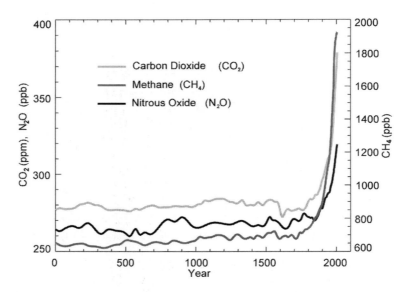

Fig. 4.1 Concentrations of carbon dioxide (CO_2), methane (CH_4) and nitrous oxide (N_2O) in the atmosphere over the last 2000 years (IPCC 2007). Permission of Oxford University Press.

Box 4.1 Carbon emissions from the conversion of forests to crop and pasture lands

One of the main drivers increasing atmospheric CO_2 is the conversion of forest to croplands and pastures. Houghton (1999) estimated that in the period 1850–1990 124 Pg C has been emitted due to changes in land use. This amount was half that emitted in the use of fossil fuels in the same period. Houghton (1999) further estimated that 108 Pg C of loss is due to the conversion of forests to croplands and pastures, i.e., two thirds were lost from tropical forests, and one third from boreal forests. At the same time, about 20 Pg C was lost in the harvest of wood. This estimate included carbon stored in wood products (17 Pg C) and woody debris (4 Pg C) used for energy and oxidized to CO_2. Houghton (1999) estimated that the annual net flux was 2.0 Pg C yr^{-1} in the 1980s, mainly due to tropical deforestation.

---

to 318 ppb. In the last century (1900–1999), the amount of CO_2 and methane has increased by 20 and 70%, respectively (IPCC 2007).

Increase of radiative forcing due to the increase of GHG and land use

Increasing concentrations of CO_2, other GHGs and water vapor allow short-wave solar radiation (wave length < 900 nm) to enter the atmosphere, but reduce the release of long-wave radiation (wave length > 900 nm) into space (Fig. 4.2). Changes in the properties of the atmosphere disturb the radiative balance in the earth/atmosphere system, affecting the prevailing climate. IPCC uses the concept of radiative forcing when referring to changes in the radiative balance resulting from the increase of GHGs in the atmosphere (IPCC 2007): radiative forcing is a measure of rate of flow of excess energy entering the earth system, altering the balance of incoming and outgoing energy in the Earth-Atmosphere system and is an index of a potential climate change mechanism. In the IPCC report (2007), "the values of radiative forcing are for changes relative to pre-industrial conditions defined at 1750 and are expressed in W m^{-2}."

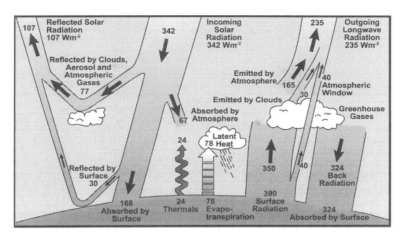

Fig. 4.2 Schematic presentations of how changes in the gaseous content of the atmosphere may affect the energy balance of the lower atmosphere. Water vapor, CO_2, CH_4, N_2O and other GHGs and clouds control how solar radiation passes in and out of the atmosphere. The energy in incoming solar radiation is 343 Wm^{-2}, of which 103 Wm^{-2} is reflected to space and 240 Wm^{-2} is emitted in long-wave radiation (heat). The atmosphere acts like a greenhouse, letting a large part of the short-wave radiation in but allowing the long-wave radiation partly to escape to space (IPCC 2007). Permission of Oxford University Press.

Since 1750, changes in the atmosphere's properties have increased radiative forcing by +2.30 Wm^{-2}, which is further increased by +0.35 Wm^{-2} due to the changes in tropospheric ozone, and by +0.12 Wm^{-2} due to the natural changes in solar irradiance (IPCC 2007). On the other hand, the increases in the radiative forcing are counterbalanced by aerosols from anthropogenic sources such as sulfates, organic carbons and dust in the atmosphere. This has a direct cooling effect of –0.5 Wm^{-2} and an indirect cooling effect of –0.7 Wm^{-2} due to increased cloudiness. Nevertheless, the temperature of soils, waters and the lower atmosphere will increase along with the increasing concentrations of GHGs, which further affect atmospheric humidity and precipitation.

4.2 Climate Change so Far and Future Expectations

Observed warming and changes in precipitation

Globally, the mean annual temperature has increased by 0.76°C during the period 1850–2005, especially in the last 50 years (Fig. 4.3). The mean annual temperature is expected to rise 1.4–5.8°C by 2100 (IPCC 2007). Warming has been highest at high latitudes in the northern hemisphere, including the boreal zone, but the increase in temperature has been identified throughout the world. The growing season in the northern hemisphere has lengthened by about 10 days since the early 1960s. This implies shorter snow duration, and particularly earlier snow melt in spring, i.e., the mean annual snow cover has reduced by 5% per year since 1966. At the same time, precipitation has increased significantly in eastern areas of North and South America, northern Europe and Central Asia, and reduced in the Sahel, the Mediterranean and southern Africa, and in parts of South Asia. Higher temperatures and a reduction in precipitation has resulted in more frequent droughts, but the frequency of heavy rain episodes has also increased with the warming and the increase of water vapor in the atmosphere. Westerly winds at middle latitudes have increased since the 1960s (IPCC 2007).

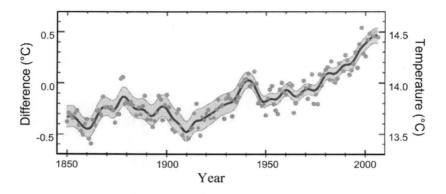

Fig. 4.3 Observed changes (left axis) in the global average surface temperature (right axis) in the period 1850–2005. The changes are relative to the average for the period 1961–1990. The smoothing is for the decades based on the annual values. The shaded area represents the range of uncertainty (IPCC 2007). Permission of Oxford University Press.

Future expectations

The global projections of future climate are based on models, which simulate the variability of the climate in the interaction between atmosphere, oceans and continents. Emissions of GHGs and particles change the properties of the atmosphere, and this is used in modeling the warming of the climate. Emission scenarios are based on projections of how human population, economic activity and consumption grow and drive GHG emissions. Several story lines (SRES) are used to describe how the drivers behind the emission scenarios develop over time (Naki´cenovi´c 2000, IPCC 2007):

- "The A1 story line and scenario family describe a future world of very rapid economic growth, global population that peaks in mid-century and declines thereafter, and the rapid introduction of new and more efficient technologies. Major underlying themes are convergence among regions, capacity building and increased cultural and social interactions, with a substantial reduction in regional differences in per capita income. The A1 scenario family develops into three groups that describe alternative directions of technological change in the energy system. The three A1 groups are distinguished by their technological emphasis: fossil-intensive (A1FI), non-fossil energy sources (A1T) or a balance across all sources (A1B) (where balanced is defined as not relying too heavily on one particular energy source, on the assumption that similar improvement rates apply to all energy supply and end use technologies).
- The A2 storyline and scenario family describe a very heterogeneous world. The underlying theme is self-reliance and preservation of local identities. Fertility patterns across regions converge very slowly, which results in continuously increasing population. Economic development is primarily regionally oriented and per capita economic growth and technological change more fragmented and slower than other storylines.
- The B1 storyline and scenario family describe a convergent world with the same global population, that peaks in mid-century and declines thereafter, as in the A1 storyline, but with rapid change in economic structures toward a service and information economy, with reductions in material intensity and the introduction of clean and resource-efficient technologies. The emphasis is on global solutions to economic, social and environmental sustainability, including improved equity, but without additional climate initiatives.
- The B2 storyline and scenario family describes a world in which the emphasis is on local solutions to economic, social and environmental sustainability. It is a world with continuously increasing global population, at a rate lower than A2, intermediate levels of economic development, and

Global Climate Change   41

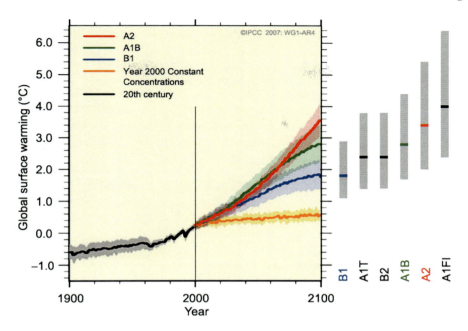

Fig. 4.4 Global warming under the emission scenarios SRES A2, A1B and B1 for 2000–2100, relative to the mean temperature in 1980–1999. Shading indicates one standard deviation around the annual averages generated by several models used in calculations. The orange line is for the calculations, where GHG concentrations in the atmosphere have been kept constant, representing the values in 2000. The bars on the right give the estimate of warming by 2100 with the likely range based on different emissions scenarios (IPCC 2007). Permission of Oxford University Press.

less rapid and more diverse technological change than in the B1 and A1 storylines. While the scenario is also oriented towards environmental protection and social equity, it focuses on local and regional levels."

Future temperature conditions are specific to each emission scenario (and the climate model), but a clear increase in the global temperature is expected regardless of the emission scenarios, as shown in Fig. 4.4 (IPCC 2007). Under the SRES B1 scenario, with low emissions, the global climate is likely to be 1.8°C warmer (the range 1.1°C–2.9°C) than currently, whereas with high emissions in the SRES A1FI scenario, the global mean temperature is likely to increase by 4.0°C (the range 2.4°C–6.4°C). The warming under SRES A1FI is thus nearly double that under SRES B1. Regardless of the GHG scenario, the range of expected warming is wide, which makes the predictions uncertain, especially from 2050 onwards.

Globally, the warming is uneven, as shown in Fig. 4.5. By 2100, warming is probably most pronounced at high latitudes in the northern hemisphere, where the mean annual temperature is likely to be 6–8°C higher than currently (IPCC 2007). Warming is likely to increase most during winter (December, January, February). For example, in northern Europe (above N 50°), the mean winter temperature may be 10°C higher than that over the last 30-year period (1980–2010). This is also true for summer temperatures (June, July, August), which may be 2–4°C higher than currently (IPCC 2007). The predictions given by single climate models are variable, however, as indicated by the probability distribution on the left side of Fig. 4.5. At the middle latitudes, warming is moderate, even in the continental areas, which include temperate forests. In the subtropics and tropics, with dry seasonal forests and moist evergreen forests, warming is also moderate in the range 2–3°C. Even in the near future, by 2020, clear warming may take place at high latitudes, but warming in this scenario is still lower than year-to-year variability.

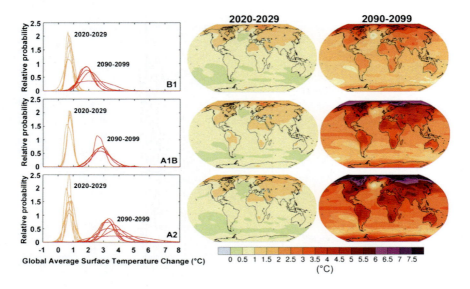

Fig. 4.5 Left: Relative probabilities of global warming in 2000–2020 and 2090–2099, and Right: Changes in surface temperature in the early and late 21st century relative to the period 1980–1999 under the selected emission scenarios (SRES B1, A1B, A2) (IPCC 2007). Permission of Oxford University Press.

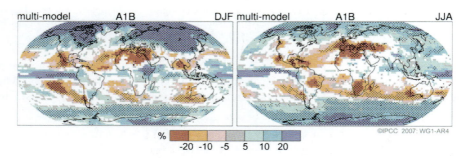

Fig. 4.6 Relative changes (%) in precipitation for the period 2090–2099 compared to precipitation in the period 1980–1999. Values are multi-model averages based on the SRES A1B scenario for the period from December to February (left, DJF) and from June to August (right, JJA). White areas indicate the areas where less than 66% of the model estimates agree. The shading indicates the areas where more than 90% of the models agree the direction of change (IPCC 2007). Permission of Oxford University Press.

Global warming implies changes in precipitation. Annual precipitation in high latitudes, for example, may increase by 20–30% by the end of this century, especially in the maritime areas of the northern hemisphere in North America and Eurasia (Fig. 4.6). The increase is likely to be especially pronounced during autumn (September, October, November) and winter (December, January, February); whereas the increase in spring (March, April, May) and summer (June, July, August) may remain less than 10% or even remain locally smaller than currently. At the middle latitudes, with seasonal moist and/or dry forests, precipitation may clearly reduce (up to 20%); especially in summer time (IPCC 2007). For example, precipitation is likely to be reduced in southern Europe, Mediterranean Africa and Central America, where summer drying is probably common due to reduced precipitation and enhanced evaporation. In the tropical zone, precipitation is likely to increase in the summer monsoon season of South and Southeast Asia and East Africa, whereas in the dry tropics, such as in Sahelian area in Africa, precipitation is likely to reduce.

4.3 Concluding Remarks

Climate change scenarios provide the basis for assessing the impact of climate change on forests, globally and locally. They are based on assumptions about the future demographic and socio-economic context and technical development, which all affect the anthropogenic emissions of GHGs and aerosols, and further the global radiation balance and climate. In this respect, the future climate is uncertain due to problems to predict future demographics, socio-economics and technological development. This is also true for Finland, where likely warming is clearly demonstrated in the recent Coupled Model Intercomparison Project (CMIP5) (Taylor 2012) using several Representative Concentration Pathways (RCP) for the period 1980–2099. According to high-resolution grid over Finland, summer warming under the pathway RCP4.5 was close to that under the SRES A2 scenario, with warming of 3°C by the end of this century. Under RCP8.5 warming is 5°C, which represents a more extreme warming scenario than any provided using the SRES emission scenarios (Lehtonen et al. 2015). On the other hand, global biogeochemical cycles affect the content of GHGs in the atmosphere, further increasing the uncertainties in climate scenarios. The climate system thus varies naturally, which generates the main uncertainties in short-term climate scenarios. However, projections of future climate change, even uncertain, are needed in order to address the potential anthropogenic influence on the climate, and the necessary mitigation and adaptation measures to avoid/reduce the harmful impact of climate change on societies (Fischlin et al. 2009).

References

Fischlin, A., M. Ayres, D. Karnosky, S. Kellomäki, S. Louman, C. Ong et al. 2009. Future environmental impacts and vulnerabilities. *In*: R. Seppälä, A. Buck and P. Katila (eds.). Adaptation of Forests and People to Climate Change—A Global Assessment Report. IUFRO World Series 22: 53–100.

Houghton, R. A. 1999. The annual net flux of carbon to the atmosphere from changes in land use 1850–1990. Tellus 51B: 298–313.

IPCC. 2007. Climate Change 2007: The Physical Science Basis. Contribution of Working Group I to the Fourth Assessment Report of the Intergovernmental Panel on Climate Change. Cambridge University Press, Cambridge, UK.

Lehtonen, I., A. Venäläinen, M. Kämäräinen, H. Peltola and H. Gregow. 2015. Risk for large-scale fires in boreal forests of Finland under changing climate. Natural Hazads and Earth System Sciences 3: 4753–4795.

Naki´cenovi´c, N., A. D. Joseph, G. B. de Vries, J. Fenhann, S. Gran et al. 2000. Special Report on Emission Scenarios, A Special Report of Working Group III of the Intergovernmental Panel on Climate Change. Cambridge University Press, Cambridge, UK.

Taylor, K. E., R. J. Stouer and G. A. Meehl. 2012. An overview of CMIP5 and the experimental design, B. American Meteorological Society 93: 485–498.

Climate Change in the Local Context, with Changes in Growing Conditions

ABSTRACT

In boreal Europe, the climate is subarctic, with a reducing south-north trend in temperature and precipitation. In southern Finland (60° N), the annual mean temperature is currently 5–6°C, and in the north (70° N) from –3 to –2°C. By the end of this century, warming is clear: temperature is expected to be 9–10°C in the south and up to 0°C in the north, while the mean annual precipitation may increase by 15–40%. Scaling global scenarios to the local scale used in forestry includes uncertainties for identifying the occurrence of extreme weather episodes detrimental to forest ecosystems and forests.

Keywords: boreal forest, temperature elevation, changes in temperature sum, changes in precipitation, evaporation in relation to precipitation, soil moisture, extreme weather episodes

5.1 Climate Change so Far

Temperature and temperature sum

The whole of Finland comprises high northern latitudes (60°–70° N) with the subarctic climate, where climate change is likely to greatly affect temperature, precipitation, snow fall and the duration of snow cover. The magnitude and timing of changes in climate are dependent on the global context, but the local conditions are likely to generate large spatial and temporal variability in future environmental conditions.

During the period 1908–2010, the mean annual temperature has increased by 0.92°C over the whole country. The increase has been greatest in spring (1.72°C) but also recognizable in summer (0.84°C), autumn (0.69°C) and winter (0.52°C) (http://ilmasto-opas.fi/fi/ilmastonmuutos/suomen-muuttuva-ilmasto/-/artikkeli/, visited 28.1.2014). Figure 5.1 shows that warming has been clearest in the southern (Helsinki, 60° N) and central (Jyväskylä, 61° N) parts of country but unclear in the north (Sodankylä, 67° N). Regardless of region, the year-to-year variability is large, and still exceeds the mean increase indicated by the trends.

Currently, the mean annual temperature is 5–6°C in the southern (60° N), about 0°C central (63° N), and –3 to –2°C northern (70° N) parts of country. Under the climate warming, the temperature increase implies an elongation of the growing season and an increase in the temperature sum. The phenological time series show that the growing season has become 8–10 days longer during the period from the 1960s to the 2000s (Linkosalo et al. 2009). Consequently, the temperature sum has increased most in central parts of the country (latitudes 62–63° N) (Fig. 5.2), where the temperature sum in 1981–2010 in large areas was the same as that in the south during the period 1961–1990 (latitudes < 62° N). Until now, the increase of temperature sum in the north (latitudes > 63° N) has been moderate as may be expected on the basis of the increase in the mean annual temperature during the period 1981–2010 (Fig. 5.1).

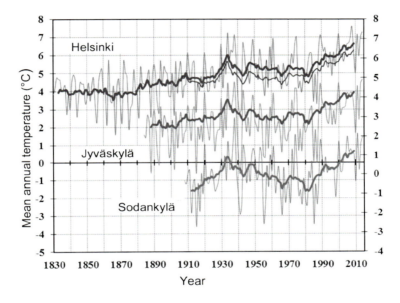

Fig. 5.1 Mean annual temperature in 1830–2010 in Helsinki (Kaisaniemi, 60° N), Jyväskylä (61° N) and Sodankylä (67° N). The annual values are indicated by thin and the ten-year moving averages by thick lines. In the case of Helsinki, the semi-thick lines indicates temperature with the effect of urbanization removed (http://ilmasto-opas.fi/fi/ilmastonmuutos/suomen-muuttuva-ilmasto/-/artikkeli/, visited 28.1.2014). Courtesy of the Finnish Meteorological Institute.

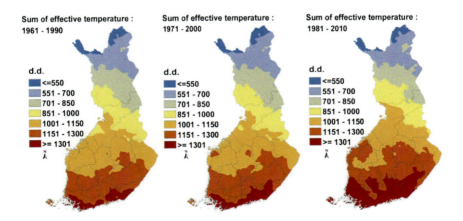

Fig. 5.2 Mean annual temperature sum for the periods 1961–1990, 1971–2000 and 1981–2010 (Finnish Meteorological Institute 2015, unpublished). Courtesy of the Finnish Meteorological Institute.

Precipitation and wind, extreme weather episodes

So far, there has been no evidence that climate has changed in other respects than temperature (https://ilmasto-opas.fi/fi/ilmastonmuutos/suomen-muuttuva-ilmasto/-/artikkeli/, visited 28.1.2014). On the contrary, the precipitation in May through September has varied, with no apparent trend, over the past 100 years. This is also the case for the frequency and length of periods with no precipitation. The frequency of strong winds (velocity > 14 m s^{-1}) has reduced rather than increased, except in the last 50 years, but the latter increase is not statistically significant. Cloud cover has varied randomly, with no systematic change during the past 100 years (Jylhä et al. 2009, https://ilmasto-opas.fi/fi/ilmastonmuutos/suomen-muuttuva-ilmasto/-/artikkeli/, visited 28.1.2014).

Extreme weather episodes involve high wind velocity, heavy rainfall or snow fall, and exceptional high or low temperatures. Even under the current climate, the occurrence of extreme weather episodes is poorly known due to short time series. However, Gregow et al. (2008) found that in 1961–2000 the wind speed exceeded 14 m s^{-1} at least 155 times, and five times 17 m s^{-1} at inland sites. At the same period, the average snow load on trees exceeded 20 kg m^{-2} 65 times. The past weather records show further that in the north very cold winter weather (daily minimum temperature below –20°C, lasting two to three weeks) occurs once every 20 years, while in the south the length of this cold weather seldom lasts longer than a week. On the other hand, summer temperatures exceed 31–32°C once every 20 years, on average, but hot weather with a daily maximum temperature above 25°C lasting longer than one week occurs once in every 20 years (Jylhä et al. 2009).

5.2 Expected Changes in Temperature and Precipitation

Temperature and precipitation under selected scenarios

In the global scale, Fig. 5.3 shows the likely changes in the mean annual temperature and precipitation at high latitudes (above 60° N) as adopted from the Arctic Climate Impacts Assessment (ACIA 2005). Up to 2040, both climate scenarios suggest a fairly similar increase in temperature and precipitation: 2–3°C in temperature and 7–8% in precipitation. Thereafter, the expected changes deviate substantially, depending on the emission scenario. By 2100, the temperature increase could be 5°C under the SRES B2 scenario, and 7°C under the SRES A2 scenario, whereas the corresponding increase in precipitation could be 17 and 24%.

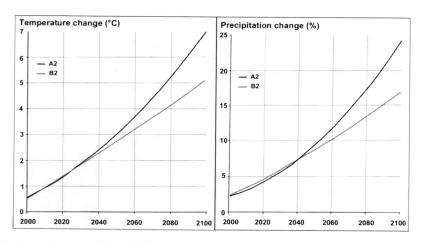

Fig. 5.3 Global change in mean annual temperature and precipitation in the Arctic area (above 60° N) under SRES A2 and B2 climate scenarios (ACIA 2005). Permission of Cambridge University Press.

The differences between long-term climate scenarios are further demonstrated in Table 5.1, which shows the mean annual temperature and precipitation over the whole of Finland for the selected time perspectives. For example, the atmospheric concentration CO_2 increases most rapidly in the SRES A1 scenario based on a heterogeneous world and increasing population in societies oriented to economic growth. By the end of this century, the atmospheric CO_2 concentration will be more than 800 ppm, with the mean annual temperature up 7°C and precipitation 37% higher than currently. Under the SRES B2 scenario, the changes are much smaller than under the SRES A1: the CO_2 concentration in 2080 is about 570 ppm, while the mean annual temperature is 5°C and precipitation 28% higher than currently. The SRES B2 scenario is based on an assumption that societies with moderate population growth prefer local solutions for economic and social sustainability (Carter et al. 2004, Carter et al. 2005).

Table 5.1 Annual mean values of temperature and precipitation in the future based on varying story lines (SRES) throughout the territory of Finland (Carter et al. 2004, Carter et al. 2005).

Year and climatic factor	Climate scenarios			
	A1	A2	B1	B2
2020				
• CO_2, ppm	432	429	421	414
• Change in temperature, °C	1.5–3.1	1.3–2.8	1.5–2.4	1.5–2.8
• Change in precipitation, %	4–14	2–13	3–14	3–16
2050				
• CO_2, ppm	590	545	492	486
• Change in temperature, °C	3.8–5.2	2.9–4.0	1.8–3.5	2.1–3.7
• Change in precipitation, %	9–28	7–21	4–17	1–20
2080				
• CO_2, ppm	829	718	534	567
• Change in temperature, °C	5.6–7.4	4.4–5.9	2.4–4.4	3.0–5.0
• Change in precipitation, %	14–37	8–29	8–23	6–28

Changes in temperature and temperature sum

Changes in climate vary from region to region depending on the current climate, topography and the area of water bodies, as shown in Fig. 5.4 for mean annual temperature. The current climate is represented by the period 1979–2000 in the 10 x 10 km grid, and the changing climate is calculated in 50 x 50 km grid for the periods 1991–2020, 2021–2050 and 2070–2099 based on the emission scenario of SRES A2 (Ruosteenoja et al. 2005).

Under climate change, it is very likely (with more than 95% probability) that the mean annual temperature in the period 2011–2020 will be 1°C higher than the mean in the period 1971–2000 (Jylhä et al. 2009). Temperature zones will shift northwards in such a way that in the last simulation period (2070–2100), the annual mean temperature in the south will be 9–10°C. At the same time, the mean annual temperature of 0°C currently in the central country shifts northwards some 500 km beyond the current northern timberline (Kellomäki et al. 2005). These changes suggest that even in the first simulation period (1991–2020) the temperature sum will increase 10–15% throughout the country. During the second period (2021–2050), the increase will be 30–50% and further 50–90% during the third period (2070–2099). By the end of this century, the temperature sum could be 800–900 d.d. (currently 600 d.d.) in the northernmost and 2,000–2,300 d.d. (currently 1,400 d.d.) in the southernmost parts of country.

Climate warming will also effect on the frequency distributions of mean, maximum and minimum temperatures, all shifting towards higher values. The greatest changes will occur in the lowest values of daily minimum winter temperatures (Jylhä et al. 2009), thus reducing the variability

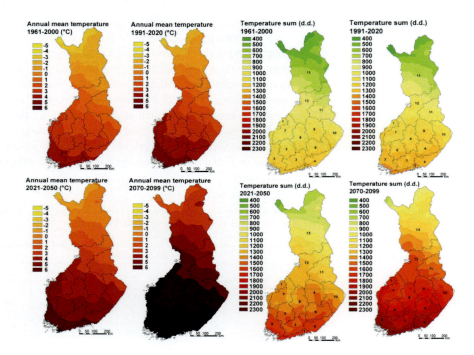

Fig. 5.4 Left: Distribution of mean annual temperature currently and under climate change, and Right: Distribution of mean annual temperature sum, currently and under climate change in the periods 1991–2020, 2021–2050 and 2070–2100 (Kellomäki et al. 2005). Simulations are based on SRES A2 emission scenario (Ruosteenoja et al. 2005). Lines and numbers in the Figures refer to the forest centers.

in winter temperatures. Winter temperatures may increase to such an extent that the thermal winter (daily mean temperature is permanently < 0°C) would shorten by one and a half months. Both thermal summer (daily mean temperature is permanently above 10°C) and the thermal growing season (daily mean temperature is permanently > 5°C) would elongate by one to one and a half month. The lengthening is likely to be greatest in the southern parts of the country. By the end of this century, the length of the thermal growing season in the north might be the same as currently in the south (Jylhä et al. 2009).

Changes in precipitation and evaporation, changes in occurrence of drought

Figure 5.5 shows the distribution of annual precipitation and potential evaporation over Finland, both currently and under climate change (Kellomäki et al. 2005). In the southernmost areas (60° N), the mean annual precipitation is currently 600–800 mm, in the central area (63° N) 500–600 mm and in the northernmost country (70° N) 300–400 mm. Under climate change, precipitation increases most in the northern areas: up to 15% in the period 2021–2050 and up to 40% in the period 2071–2100, while the increase in the south will be less than 10%. At the same time, precipitation will increase most in winter (10–40%), and in summer the increase will be 0–20% by the end of this century. The average snow depth (in terms of snow water equivalent) may decrease in the southern and central parts by 70–80% or even more in the north (Jylhä et al. 2009). The average annual maximum snow water content and the duration of snow cover (number of days with snow cover) will reduce, most in early and late winter.

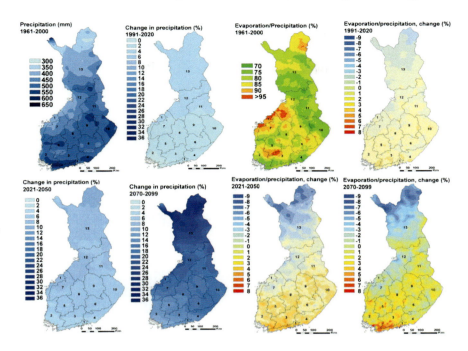

Fig. 5.5 Left: Distribution of mean annual precipitation, and Right: Percentage of potential evaporation from precipitation, currently and under climate change in the periods 1991–2020, 2021–2050 and 2070–2100 (Kellomäki et al. 2005). Potential evaporation is related only to temperature, here based on SRES A2 emission scenario (Ruosteenoja et al. 2005). Lines and numbers in the Figures refer to the forest centers.

Even during the first calculation period (1991–2020), the elevating temperature may increase potential evaporation: up to 5%, more in the south than in the north. During the second period (2021–2050), potential evaporation may increase further up to 8–13%, again more in the south than in the north. During the last period (2070–2099), potential evaporation may increase up to 25% in the south and 10–15% in the north. In the south, potential evaporative demand seems locally to increase more than precipitation, whereas in the north the situation tends to be opposite (Kellomäki et al. 2005).

Changes in precipitation and evaporation are reflected in soil moisture, which is the water held in the spaces between soil particles. The water in the rooting zone is especially important, as it affects the availability of water for growth. The rooting zone of Scots pine, Norway spruce and birch extends down to 100 cm, but most of the fine roots important in water uptake are located in the upper mineral soil (thickness < 10 cm) and the boundary between the upper mineral soil and organic layer on soil. Water in the rooting zone available for growth is defined by the field capacity and the wilting point specific to different soil types (Fig. 5.6). Field capacity indicates the maximum amount of water (m^3 m^{-3}) held in the spaces between soil particles, whereas the wilting point indicates the minimum soil moisture (m^3 m^{-3}) required by plants not to wilt. The water available for plants (extractable by roots for transpiration) is the water between these boundaries, but plant capacity to extract water is species-specific. Soil moisture below the wilting point shows drought episodes but higher limit may be set to indicate water shortage and droughts (Granier et al. 1999).

The frequency and duration of drought episodes may be indicated by the number of dry days as used Kellomäki et al. (2005) in assessing the availability of soil moisture in boreal conditions. Dry day refers to a day, with soil moisture below wilting point. For example, under the warming climate scenario, by 2070–2099, the mean annual temperatures are expected to increase by 4°C

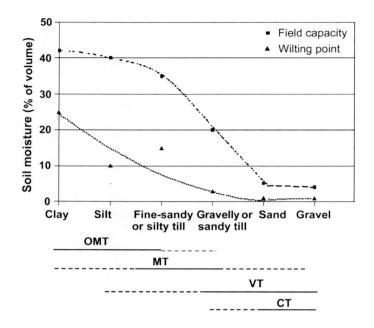

Fig. 5.6 Field capacity and wilting point of soil for sites of different forest types, based on Urvas and Erviö (1977), Talkkari and Hypén (1996) and Kellomäki et al. (2008). Legend: OMT = fertile site, MT = medium fertile site, VT = poor site and CT = very poor site.

in the summer and by 6°C in the winter. At the same time, wintertime precipitation may increase by 20%, while in summer the precipitation will remain nearly unchanged. In this case, Kellomäki et al. (2005) found that in the period 1991–2020 the number of dry days may reduce in some places, but clearly increase in southern Finland (Fig. 5.7). This tendency was further increased during the second calculation period (2021–2050). During the third calculation period (2070–2099), the number of dry days may increase up to 90% in the south (62–63° N). Even in northern sites, the number of dry days increased several tens of percentage in response to climate warming.

5.3 Expected Changes in Soil Moisture in Forest Sites

Changes in water dynamics and soil moisture at stand scale

Kellomäki and Väisänen (1996) analyzed how temperature elevation alone may affect the interception, transpiration, evaporation and drainage of water at the stand scale, and how they were affecting the soil moisture on two sites occupied by Scots pine in the south (61° N) and north (66° N). The mean annual temperature was assumed to increase gradually by 5°C over a 100-year simulation period. Precipitation was created using a weather generator based on changes in temperature and cloud cover (Strandman et al. 1993). In both sites, the moisture content of the top soil was 0.21 $m^3 m^{-3}$ at wilting point and 0.54 $m^3 m^{-3}$ at field capacity. The computations were made with and without a tree stand. In the former case, the density of the pure Scots pine stand at the onset of simulation was 2,500 trees per hectare, with an initial diameter (at 1.3 m above ground level) of one centimeter.

On both sites, the share of water intercepted by the canopy increased under rising temperatures, but the percentage increase of the proportion was greater in the north (14%) than in the south (6%) (Fig. 5.8). Contrary to the share of interception, the share of transpiration decreased in both sites

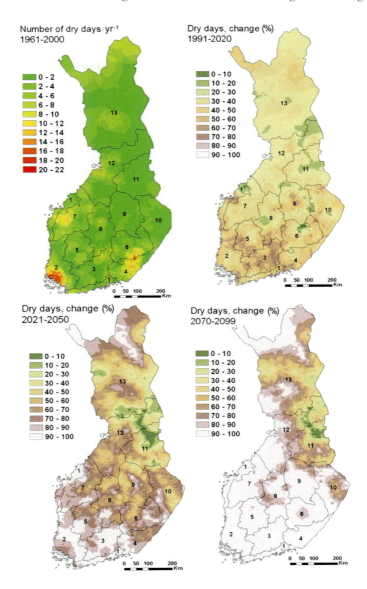

Fig. 5.7 Distribution of mean annual number of dry days currently and under climate change in the periods 1991–2020, 2021–2050 and 2051–2100 (Kellomäki et al. 2005) based on the SRES A2 emission scenario (Ruosteenoja et al. 2005). Lines and numbers in the Figures refer to the forest centers.

when the rising temperature was assumed: the reduction was 11% in the south and 29% in the north. The share of evaporation from the soil surface increased in both sites, but the temperature elevation caused a larger increase in the northern (about 39%) than in the southern (about 10%) site. At the same time, the share of drainage was more reduced in the south (13%) than in the north (11%).

Under the current climate, the water content in soil decreased towards the end of the simulation (from 0.45 m³ m⁻³ to 0.39 m³ m⁻³) if no trees were on the site (Fig. 5.9). This was related to the reduction of organic layer (litter, humus) on the soil surface, which enhanced heat transfer through soil increasing evaporation and lengthening of soil frost in autumn and early spring, thus reducing infiltration of water into the soil. The role of the organic layer was also important under tree cover.

52 Managing Boreal Forests in the Context of Climate Change

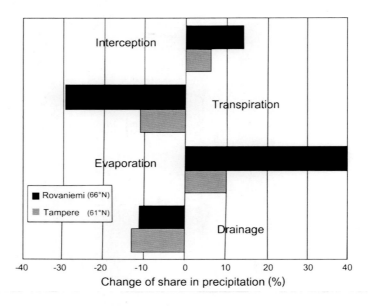

Fig. 5.8 Percentage change under warming in the share of interception, transpiration, evaporation and drainage in a Scots pine stand as related to the prevailing precipitation in southern (61° N, Tampere) and northern (66° N, Rovaniemi) boreal conditions (Kellomäki and Väisänen 1996). Permission of Springer.

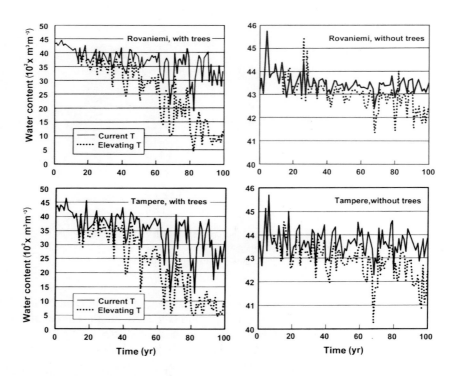

Fig. 5.9 Mean water content of the uppermost layer of soil (upper 10 cm) for May–August under the current and elevating temperatures on the sites with (left) and without (right) tree cover in southern (61° N, Tampere) and northern (66° N, Rovaniemi) boreal forests (Kellomäki and Väisänen 1996). Permission of Springer.

The moisture content of mineral soil decreased at first due to the reducing organic layer but stabilized later in the simulation, when litter fall balanced its reduction. In this phase, the mean soil moisture was 0.37–0.42 $m^3 m^{-3}$ in the north and 0.35–0.40 $m^3 m^{-3}$ in the south, the lower limit representing the rising temperature and the upper limit the current temperature.

Soil moisture became more diversified, if only the growing season (May through August) was considered (Fig. 5.9). When no trees were present, the moisture content decreased substantially towards the end of the simulation, particularly under the increasing temperatures. However, the moisture content remained clearly higher than under tree cover. Regardless of the site, the soil moisture was 50–70% of that under the current climate. By the end of the simulation, the moisture content under tree cover frequently dropped below the wilting point (0.21 $m^3 m^{-3}$) even in the north.

Changes in water dynamics and soil moisture at a regional scale

In the regional scale, the interception, transpiration, evapotranspiration and drainage depend on the climate and properties of sites and tree stands. Ge et al. (2013a) studied how warming may affect evapotranspiration (combined water depletion in evaporation and transpiration) and water infiltration across the south-north temperature gradient over Finland. Over a 100-year period, the mean annual evapotranspiration under climate warming was 19–25% higher than that under the current climate (Table 5.2); i.e., the share of cumulative evapotranspiration of total precipitation increased up to 7–18%, more in the south than in the north. These findings suggest again that in the south the evaporative demand may increase more than precipitation, while in the north the situation may be the opposite. At the same time, water infiltration to the soil profile decreased up to 6% in the south, while infiltration increased up to 7% in the north. Consequently, the soil water content

Table 5.2 Mean annual evaporation under the current and changing climate and the percentage of evaporation to precipitation, with the annual infiltration of water for a Norway spruce stand letting to growth with no management in the selected temperature sum zones (Ge et al. 2013a). Numbers in the parethesis indicate the percentage change of that under the current climate.

Bioclimatic zone (XX°–XX° N), climate scenario	Evapotranspiration, mm	% of evapotranspiration to precipitation	Water infiltration, mm
Z_I: 60–62			
• Current	431	81	348
• Changing	534 (+24)	92 (+14)	327 (–6)
Z_{II}: 62–64			
• Current	455	76	376
• Changing	544 (+19)	90 (+18)	362 (–4)
Z_{III}: 64–66			
• Current	388	76	290
• Changing	463 (+20)	83 (+10)	281 (–3)
Z_{IV}: 66–68			
• Current	358	65	256
• Changing	442 (+23)	71 (+9)	272 (+7)
Z_V: 68–70			
• Current	302	68	190
• Changing	375 (+25)	73 (+7)	202 (+6)

The simulations were done over a period (2000–2100) based on the SRES A2 scenario (Ruosteenoja et al. 2005). Under climate warming, the mean annual temperature increased by 4°C in the summer and 6°C in the winter, while precipitation increased by 10–15% in the south and 20–30% in the north, mainly in winter time. The atmospheric CO_2 concentration increased from 352 to 841 ppm.

reduced more in the south than in the north, where the increase in precipitation partly compensated for water depletion (Ge et al. 2013a,b).

5.4 Expected Changes in Snow Cover and Soil Frost

Changes in the duration and depth of snow cover

Under boreal conditions, soil temperature and moisture vary substantially, following the current annual weather cycle. In both cases, climate change also modifies soil conditions outside the growing season as demonstrated by Kellomäki et al. (2010). They studied how climate change may affect the accumulation of snow and soil frost at sites without tree cover in boreal conditions (60°–70° N) by using a process-based ecosystem model (FinnFor).

Under the current climate (1971–2000), the duration of snow cover was 160–180 days in the southern parts of country (Fig. 5.10). Under climate warming, the snow duration reduced slightly in the first calculation period (2001–2020), but substantially (30%) in the third period (2071–2100). In the northern sites, the duration of snow cover reduced to a lesser extent: the duration may be less than 180 days per year at the end of this century, implying a 15–20% reduction compared to that under the current climate. The thickness of snow cover was related to the duration of snow. Consequently, the current mean snow depth in January was 60–100 cm in the northern, 40–50 cm in central and 20–30 cm in southern parts of the country. Under a warming climate, the mean snow depth reduced, mainly in southern and central parts of country, even in the first calculation period

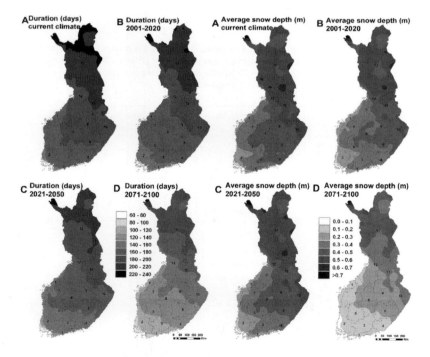

Fig. 5.10 Left: Mean annual duration of snow cover (the days per year when snow cover ≥ 1 cm), and Right: Mean annual depth of snow cover (m) in January under the current (Figures A) and changing climate in the periods 2001–2100 (Figures B), 2021–2050 (Figures C) and 2071–2100 (Figures D) based on the SRES A2 emission scenario (Ruosteenoja et al. 2005). By 2070–2100, the mean annual temperatures increased by 4°C in the summer and 6°C in the winter, while the winter (December, January, February) precipitation increased by 20% (Kellomäki et al. 2010). Lines and numbers in the Figures refer to forest centers. Courtesy of and the Finnish Society of Forest Science and the Finnish Forest Research Institute.

(2001–2020). This was also true for the second period (2021–2050), but in the third period (2071–2100) the depth of snow cover reduced substantially, even in the north. In general, the decreasing duration and depth of snow cover towards the end of this century were mostly related to the higher temperatures in early and late winters (Venäläinen et al. 2001, Kellomäki et al. 2010).

Changes in duration and depth soil frost

The duration and depth of soil frost substantially affect the carrying capacity of soil, and thus have a direct effect on the mechanized timber harvest. The duration of soil frost refers to the number of days per month with the temperature < 0°C in the uppermost 10 cm in the soil profile. Under the current climate, the soil freezes in late October, except in the north where the soil froze in late September

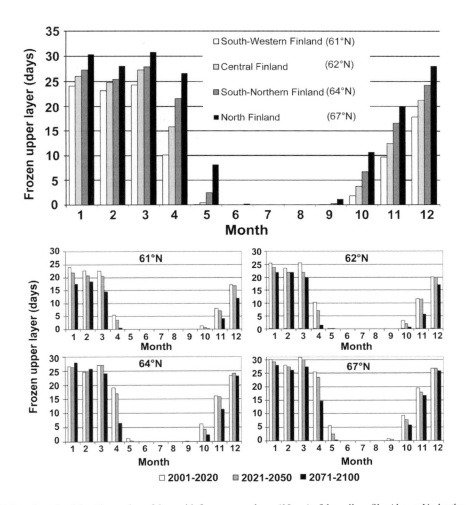

Fig. 5.11 Duration of soil frost in number of days with frozen upper layer (10 cm) of the soil profile. Above: Under the current climate, and Below: Under a warming climate in selected provinces in Finland based on the SRES A2 emission scenario (Ruosteenoja et al. 2005). The duration of soil frost means the sum of days with a soil temperature < 0°C in the uppermost layer (Kellomäki et al. 2010). Courtesy of and the Finnish Society of Forest Science and the Finnish Forest Research Institute.

(Fig. 5.11). The soil frost disappears in April in the southern and central parts of the country, but there is still soil frost in May in the north. The current pattern of soil frost may hold under a warming climate, but the period with no soil frost may increase even in the north, representing a lengthening of the no-frost period both in spring and autumn. The duration of soil frost may decrease from 4–5 months to 2–3 months in the south and from 5–6 months to 4–5 months in the north, if a temperature elevation of 4°C is assumed (Peltola et al. 1999, Venäläinen et al. 2001).

Following the current temperature pattern, the average depth of soil frost may exceed 100 cm across wide areas in the northern and north-eastern parts of country (Kellomäki et al. 2010). In southern areas, the mean depth of soil frost is less than elsewhere in the country, and in the most southern areas it is only a few tens of centimeters. Under climate change, this spatial pattern may hold for the first (2001–2020) and second (2021–2050) calculation periods, but the depth of soil frost may reduce by several tens of centimeters throughout the country. In the third calculation period (2071–2100), soil frost nearly disappears in the southern and central parts of the country.

5.5 Expected Changes in Soil Organic Matter and Changes in Available Nitrogen

In general, the growth of boreal forests is greatly limited by the short supply of available nitrogen (Jarvis and Linder et al. 2000), even though several Mg of nitrogen per hectare is bound in Soil Organic Matter (SOM, litter and humus on soil). Based on the thickness of SOM layer, the current mean amount of SOM on the mineral soil in southern boreal forests may exceed 60–90 Mg ha^{-1} on fertile sites (*Myrtillus* type and more fertile), while on poor sites (*Vaccinium* type and less fertile) the amount of SOM may remain less than a half of this (Fig. 5.12). In the northern boreal forests, the amount of SOM is 80–90% of that in the southern part, regardless of the site type (Kellomäki et al. 2005, Kellomäki et al. 2008). The declining amount of SOM towards the north is related to the declining forest growth across the boreal forests. However, the spatial variability SOM is large following, e.g., the variability in site fertility and tree species, and management and disturbance history.

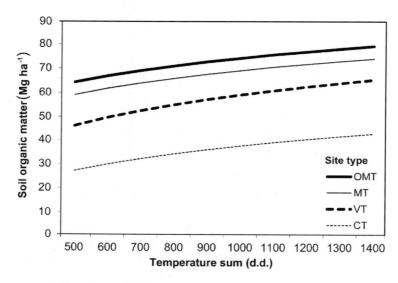

Fig. 5.12 Mean amount of soil organic matter (litter, humus) in forest sites on mineral soil as a function of temperature sum and site fertility (site type), based on the Finnish National Forest Inventory providing the thickness of soil organic layer (Kellomäki et al. 2005).

Table 5.3 Mean amount of soil organic matter (litter, humus) and available nitrogen in soil currently and in the period 1990–2099 under climate change divided between southern and northern Finland (Kellomäki et al. 2005).

Region and period	Organic matter, Mg ha^{-1} (% of that under current climate)	Available nitrogen, kg ha^{-1} (% of that under current climate)
Southern Finland		
• Current	80.3	43.2
• 1990–2020	80.3 (0)	44.5 (+3.0)
• 2021–2050	104.0 (+29.5)	55.4 (+28.3)
• 2071–2099	133.5 (+66.3)	63.7 (+47.6)
Northern Finland		
• Current	63.2	26.2
• 1990–2020	63.1 (–0.3)	27.1 (+3.7)
• 2021–2050	74.2 (+17.4)	43.4 (+65.9)
• 2071–2099	88.6 (+40.1)	46.7 (+78.2)
Total		
• Current	73.4	36.3
• 1990–2020	73.3 (–0.1)	37.4 (+3.2)
• 2021–2050	91.9 (+25.3)	50.4 (+39.3)
• 2071–2099	115 (+57.1)	56.8 (+56.6)

The percentage change in relation to the current values is in parentheses. Southern Finland refers to the forests below 63° N and northern Finland to the forests above 63° N. Note that the results only represent upland sites on mineral soils excluding peatlands. The values are based on the thickness of litter and humus on mineral soil calculated from the National Forest Inventory data.

The growth increase under a warming climate suggests the higher accumulation of SOM. Based on model simulations, Kellomäki et al. (2005) showed that climatic warming may increase SOM in southern and middle boreal forests by 5% even in the short term (the period 2001–2020). The same takes place later in the north, where the amount of SOM increases in the period 2001–2050, by as much as 15% in some places. During the period 2071–2100, the amount of SOM is further increased in the north, where the increase exceeds 30% occasionally. In the south, the amount of SOM also increases locally, but in many places SOM reduces by 20–30%. By the end of this century, climatic warming may, however, increase the mean amount of SOM by 40% in the north and 60% in the south (Table 5.3). This difference between the southern and northern sites follows the difference in growth and the consequent litter fall, controlling much the accumulation of organic matter. In the north, low temperature increases further the accumulation of SOM due to low decay.

In boreal forests, the nitrogen cycle is very temperature-dependent, and only seldom does a short supply of water limit the decay of SOM. A warming climate is thus likely to enhance the decay of litter and humus, making the nitrogen cycle faster and increasing the amount of available nitrogen as shown Table 5.3. The increase was several percentages even in the period 2001–2020, with a further increase alongside the warming of up to 30–60% in the period 2021–2050, and up to 50–80% in the period 2070–2100. This is especially pronounced in the north, where the percent changes clearly exceeded those in the south. At the end of the simulation period, the amount of available nitrogen is still greater in the south than in the north.

5.6 Concluding Remarks

Climate models show that greenhouse gas emissions are likely to warm the climate in boreal Europe in some 50 years. At the same time, climate change is likely to slightly increase precipitation. From a perspective of 100 years, the mean annual temperature may increase by up to 6–8°C, with an increase of 10–15% in temperature sum and 10–40% in precipitation. The changes will probably be most pronounced in the northern boreal forest. By the end of this century, soil frost will likely be disappearing in the southern and central but still remaining in the northern parts of boreal forests. Warming outside the proper growing season is likely to increase soil moisture due to increasing rainfall and snowmelt events, thus further reducing the carrying capacity of soil and hindering and the year-round harvest of timber and biomass.

Currently, low temperatures and short nitrogen supply imply low productivity of boreal forests. Warming is likely to enhance the nitrogen cycle including the uptake and binding of nitrogen in growth, the return of nitrogen to the soil in litter, and finally the release of nitrogen in the decay of soil organic matter for reuse in growth. Magnani et al. (2007), for example, showed that the carbon balance in temperate and boreal forests is tightly controlled by the availability of nitrogen including nitrogen deposition related from anthropogenic activities (e.g., use of fossil fuels, agriculture, traffic), affecting greenhouse emissions and climate warming. Currently, the nitrogen deposition from anthropogenic sources, including ammonium (NH_4) and nitrate (NO_3) compounds, is greatest (4–10 kg N ha^{-1} yr^{-1}) in the most urbanized and industrialized parts of southern boreal areas. In the sparsely-populated northern boreal areas, the mean nitrogen deposition is < 4 kg N ha^{-1} yr^{-1}, and in many places even about 1 kg N ha^{-1} yr^{-1}, mainly originated from natural sources, as biological nitrogen fixation and lighting (http://www.ymparisto.fi, visited 9.9.2014).

References

ACIA. 2005. Arctic Climate Impacts Assessment. Cambridge University Press, Cambridge, UK.
Carter, T. R., S. Fronzek and I. Bärlund. 2004. FINSKEN: a framework for developing consistent global change scenarios for Finland in 21th century. Boreal Environmental Research 9: 91–107.
Carter, T. R., K. Jylhä, A. Perrels, S. Fronzek and S. Kankaanpää. 2005. FINADAPT scenarios for the 21st century. Alternative futures for considering adaptation to climate change in Finland. Finnish Environment Institute, FinAdapt Working Paper 2: 1–42.
Ge, Z. -M., S. Kellomäki, H. Peltola, X. Zhou and H. Väisänen. 2013a. Adaptive management to climate change for Norway spruce forests along a regional gradient in Finland. Climatic Change 118: 275–289.
Ge, Z. -M., S. Kellomäki, H. Peltola, X. Zhou, H. Väisänen and H. Strandman. 2013b. Impact of climate change on primary production and carbon sequestration of boreal Norway spruce forests: Finland as a model. Climatic Change 118: 259–273.
Granier, A., N. Bréda, P. Biron and S. Villette. 1999. A lumped balance model to evaluate duration and intensity of drought constraints in forest stands. Ecological Modelling 116: 269–283.
Gregow, H., U. Puranen, A. Venäläinen, H. Peltola, S. Kellomäki and D. Schulz. 2008. Temporal and spatial occurrence of strong winds and large snow load amounts in Finland during 1961–2000. Silva Fennica 42(2): 515–534.
http://ilmasto-opas.fi/fi/ilmastonmuutos/suomen-muuttuva-ilmasto/-/artikkeli/, visited 28.1.2014.
http://www.ymparisto.fi, visited 9.9.2014.
Jarvis, P. G. and S. Linder. 2000. Constraints to growth of boreal forests. Nature 405: 904–905.
Jylhä, K., K. Ruoteenoja, J. Räisänen, A. Venäläinen, H. Tuomenvirta, L. Ruokolainen et al. 2009. Arvioita Suomen muuttuvasta ilmastosta sopeutumistutkimuksia varten. ACCLIM-hankkeen raportti 2009. Raportteja 2009: 4–102.
Kellomäki, S. and H. Väisänen. 1996. Model computations on the effect of elevating temperature on soil moisture and water availability in Scots pine dominated ecosystems in the boreal zone in Finland. Climatic Change 32: 423–445.
Kellomäki, S., H. Strandman, T. Nuutinen, H. Peltola, K. T. Korhonen and H. Väisänen. 2005. Adaptation of forest ecosystems, forests and forestry to climate change. Finnish Environment Institute, FinAdapt Working Paper 4: 1–50.
Kellomäki, S., H. Peltola, T. Nuutinen, K. T. Korhonen and H. Strandman. 2008. Sensitivity of managed boreal forests in Finland to climate change, with implications for adaptive management. Philosophical Transactions of the Royal Society B363: 2341–2351.
Kellomäki, S., M. Maajärvi, H. Strandman, A. Kilpeläinen and H. Peltola. 2010. Model computations on the climate change effects on snow cover, soil moisture and soil frost in the boreal conditions over Finland. Silva Fennica 44(2): 213–233.
Linkosalo, T., R. Häkkinen, J. Terhivuo, H. Tuomenvirta and P. Hari. 2009. The time series of flowering and leaf bud burst of boreal trees (1846–2005) support the direct temperature observations of climatic warming. Agricultural and Forest Meteorology 149: 453–461.
Magnani, F., M. Mencuccini, M. Borghetti, P. Berbigier, F. Berninger, S. Delzon et al. 2007. The human footprint in the carbon cycle of temperate and boreal forests. Nature 447: 848–850.
Peltola, H., S. Kellomäki, H. Väisänen and V. -P. Ikonen. 1999. A mechanistic model for assessing the risk of wind and snow damage to single trees and stands of Scots pine, Norway spruce and birch. Canadian Journal of Forest Research 29: 647–661.

Ruosteenoja, K., K. Jylhä and H. Tuomenvirta. 2005. Climate scenarios for FINADAPT studies of climate change adaptation. Finnish Environment Institute, FinAdapt Working Paper 15: 1–15.

Strandman, H., H. Väisänen and S. Kellomäki. 1993. A procedure for generating synthetic weather records in conjunction of climatic scenario for modelling ecological impacts of changing climate in boreal conditions. Ecological Modelling 70: 195–220.

Talkkari, A. and H. Hypén. 1996. Development and assessment of a gap-type model to predict the effects of climate change on forests based on spatial forest data. Forest Ecology and Management 83: 217–228.

Urvas, L. and R. Erviö. 1974. Metsätyypin määrittäminen maalajin ja maaperän kemiallisten ominaisuuksien perusteella. Influence of the soil type and the chemical properties of soil on the determining of the site type. Maataloustieteellinen Aikakauskirja 46: 307–319.

Venäläinen, A., H. Tuomenvirta, M. Heikinheimo, S. Kellomäki, H. Peltola, H. Strandman et al. 2001. Impacts of climate change on soil frost and snow cover in a forested landscape. Climate Research 17: 63–72.

PART III
Impact of Climate Change on the Eco-physiological Performance of Selected Boreal Tree Species

6

Carbon Uptake and Climate Change

ABSTRACT

In general, the response in carbon uptake of trees to the elevated CO_2 and temperature at the level of leaves/needles is immediate, but in the long term the adjustment of biochemical processes and morphology of leaves/needles to elevated CO_2 and temperature modifies the response. The CO_2 assimilation rate is enhanced by increasing nitrogen concentration in leaves/needles. The temperature increase in spring and autumn may reduce the period of depressed photosynthetic capacity, allowing boreal coniferous trees to utilize radiation longer during spring and autumn, with greater efficiency than possible under the current temperature conditions.

Keywords: carbon uptake, photosynthesis, nitrogen, biochemical model for photosynthesis, effect of nitrogen on photosynthesis, seasonality of photosynthesis

6.1 Processes Linking the Growth of Trees to Climate Change

Under global warming, the increasing temperature is linked to increasing CO_2, vapor pressure deficit, soil moisture and nutrient availability, which drive carbon uptake in trees. Carbon uptake is further dependent on the biochemical and eco-physiological processes which link the growth and development of trees to variability and changes in climate. Biochemical responses refer to the changes in biochemical reactions behind carbon assimilation in photosynthesis, whereas eco-physiological responses involve the interrelationship between the eco-physiological functioning of trees and their environment. Eco-physiological responses include, e.g., photosynthesis, respiration, transpiration and the uptake of water and nutrients controlled by the physical and chemical properties of the environment and the structural properties of trees.

In general, photosynthesis is limited by the availability of radiation and CO_2, while temperature, nitrogen and water drive the biochemical processes behind carbon uptake. In this context, stomatal conductance controls the flow of CO_2 into the stomatal cavity, where it is available for the photosynthetic processes occurring in the chloroplasts in surrounding tissues (Fig. 6.1; Box 6.1). The higher atmospheric CO_2 behind climate change increases the CO_2 concentration in the stomatal cavity and enhances the primary fixation of carbon in C_3-plants, thus increasing resources for regeneration and growth. The uptake of CO_2 through stomata allows the H_2O to evaporate into the atmosphere, with consequent effects on the water status of trees. The increase in atmospheric CO_2 may change stomatal conductance in leaves/needles (e.g., changes in the number of stomata per unit area, degree of stomatal aperture), thus affecting the CO_2 uptake rate. Under high CO_2, the use of H_2O per unit production of photosynthates is reduced with the increase in the water use efficiency of plants (Jarvis 1993).

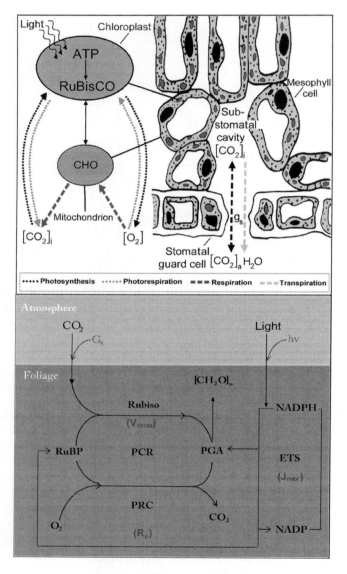

Fig. 6.1 Above: Schematic presentation of the stomatal functions in leaves/needles controlling photosynthetic and transpiration rates with the links to biochemical processes converting CO_2 to carbohydrates in photosynthesis. Below: Outline of the links between the light and dark reactions controlling carbon fixation. Legend: PCR is the Calvin cycle (process converting CO_2 to sugars and carbohydrates (dark reactions)). PRC is the photosynthetic reaction center capturing energy (light reactions), and ETS is the electron transfer system. PGA refers to three-carbon intermediates that further reduce to carbohydrates using photochemically generated ATP and $NADPH_2$. V_{cmax} is the maximum photosynthetic rate under saturating CO_2 and J_{max} is the maximum photosynthetic rate under saturating radiation. R_d refers to photorespiration, which occurs at low carbon dioxide levels. RuBP (ribulose bisphosphate) are the five-carbon acceptor molecules in generating PGA. The reaction between carbon and RuBP is catalyzed by the Rubisco enzyme (ribulose bisphosphate carboxylase/oxygenase). g_s is stomatal conductance and hv is quantum of light (Box 6.1).

Box 6.1 Outlines of photosynthetic processes

In photosynthesis, chlorophyll pigments in the mesophyll tissue of leaves/needles absorb light to oxidize water (H_2O) to molecular oxygen and reduce CO_2 into primary sugars (Fig. 6.2). Light drives the transfer of electrons between compounds that act as the donors and acceptors of electrons. Light reactions produce ATP (adenosine triphosphate) and $NADPH_2$ (Nicotinamide Adenine Dinucleotide Phosphate) for the synthesis of sugars (CH_2O) in light-independent reactions, where carbon and water are combined with the five-carbon acceptor molecules (RuBP) to generate two molecules of three-carbon intermediates (PGA, 3-phosphoglycerate). They are further reduced to carbohydrates using the photochemically generated ATP and $NADPH_2$. The reaction between carbon and RuBP (ribulose bisphosphate) (carboxylation) is catalyzed by the Rubisco enzyme (ribulose bisphosphate carboxylase/oxygenase). Rubisco favors carbon dioxide, but in C_3-plants it is replaced by oxygen (photorespiration) when the oxygen concentration is high relative to the carbon dioxide. C_3-plants (e.g., boreal trees) cannot directly recycle CO_2 released in the daytime as do C_4-plants (e.g., some tropical plants). On the other hand, C_3-plants can use the increase in atmospheric CO_2, unlike C_4-plants which are only slightly responsive to the increase in atmospheric CO_2 (Salisbury and Ross 1992). https://en.wikipedia.org/wiki/Chloroplast.

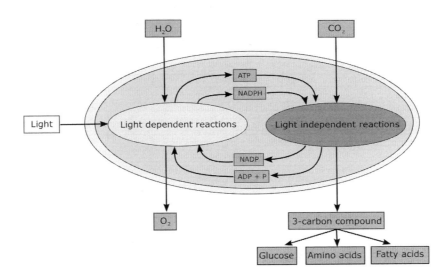

Fig. 6.2 Light-dependent and light-independent photosynthetic reactions in chlorophyll pigment.

Trees grow under variable environmental conditions representing energy (radiation, heat, mechanical) and resources (CO_2, H_2O, nutrients). A warming climate may change both the variability and availability of energy (e.g., heat) and resources (e.g., CO_2 and H_2O). At the level of leaves/needles, the responses in carbon uptake to these changes are immediate. Immediate or short-term responses (i.e., ≤ day) include the variability of CO_2 uptake and H_2O emissions related to the short-term variability in environmental factors controlling photosynthesis and stomatal conductance. The carbon uptake mechanism (stomatal conductance, biochemical photosynthetic processes and morphology of leaves/needles) may further adjust to elevated CO_2 and temperature, modifying short-term responses (Jarvis 1993) if the elevation extends over several days or weeks. Over longer periods (i.e., decades and/or centuries), the elevation of CO_2 and temperature is likely to change the genotypic properties of carbon uptake in trees adapting to the trend-like changes in environmental conditions (e.g., gradual warming (Box 6.2)).

> **Box 6.2 Response, acclimation and adaptation**
>
> Response refers to the reactions of processes (e.g., photosynthesis, transpiration) that a tree has to environmental factors (e.g., CO_2, temperature, etc.). The response may be on various time scales (e.g., from seconds to days), but disappears when the stimulus is removed.
>
> Acclimation refers to the adjustment of the reactions of processes (e.g., photosynthesis, transpiration) to a gradual or permanent change in the environment (e.g., the elevated CO_2, temperature, etc.). Acclimation may involve varying time scales (e.g., days to weeks), but removing the change in environment reverses (redirects) the adjustment.
>
> Adaptation refers to evolutionary changes in genotypes (i.e., natural selection) induced by long-term/permanent alteration of environmental conditions. Adaptation (adaptive responses) may include changes in phenology, growth and development, morphology, biochemistry, etc., representing different ecotypes of the same species with varying responses to the same environmental factors, such as tolerance of low temperatures.

6.2 Biochemical Responses of Carbon uptake to Elevated CO_2 and Temperature

Infrastructure used to study climate change impact on carbon uptake

Many experimental findings on the impact of climate change (the elevated CO_2 and temperature), on photosynthesis and stomatal functions, are based on experiments in growth chambers and greenhouses. Branch bags and enclosures of part of a mature tree or an ecosystem are further used to assess the impact of climate change on trees and ecosystems (Pontailler et al. 1998). In any of this infrastructure, the growing conditions also change in other respects, such as excluding wind force and/or natural falls of rain and snow, increasing air humidity, reducing the radiation on leaf/needle surfaces, etc. Nevertheless, the use of such infrastructure with replicates in the proper factorial design will facilitate the study of how the deliberate changes in atmospheric CO_2 and temperature may modify the physiological and ecological responses of plants to climate change. Many of these problems are avoided in the FACE (Free-Air CO_2 Enrichment) experiment, where trees are grown in the field under elevated CO_2 in an environment only slightly changed in other respects (Ainsworth and Long 2005). Only seldom are the effects of experimental infrastructure missing, as is the case in studies using trees growing around natural CO_2 wells (Hättenschwiler et al. 1997).

Chamber experiments are among the most popular methods used to study how climate change may affect physiological processes in forest trees. Key questions have included how the change in climate immediately affects carbon fixation through the stomatal functions and the photosynthetic and respiration processes. An important question is whether the key physiological processes are acclimating to climate change with impact on the growth of trees and finally the long-term dynamics of forest ecosystem. The research findings from the chamber experiments represent mainly short-term (physiology) or medium long (growth) responses of up to few years at the most. This information is widely used in modeling, which makes it possible to expand short-term findings over longer periods, for example those applicable to forestry and forest management, in a time scale of decades or even longer (Kellomäki and Väisänen 1997, Matala et al. 2006).

Growth chambers may be in laboratory conditions or field conditions. In field conditions, open and closed chambers are used mainly in seedling phase to expose trees to the elevated CO_2 and/or elevated temperature. A continuous and costly supply of CO_2 is needed when using closed chambers (Pontailler et al. 1998), where the excessive elevation of temperature is a further problem. Proper ventilation with the necessary addition of a CO_2 supply is needed to maintain the target CO_2 concentration in the chambers (e.g., 800 µmol mol^{-1} vs. 400 µmol mol^{-1}). Special problems are found in conditions, where the seasonality of weather is pronounced, as in the boreal regions, with below zero ambient temperatures under clear skies in late winter. Under these conditions, the heat load is high, which makes it difficult to maintain the target temperature in the chambers without using an excessive amount of CO_2 due to the need for effective ventilation (Kellomäki et al. 2000).

Chamber experiments in boreal conditions

The Mekrijärvi chamber experiment, the University of Joensuu (the University of Eastern Finland from 2010 onwards), the Mekrijärvi Research Station, Finland, was used to study climate change impact on boreal trees. The focus was on studies asking how the elevation of CO_2 and temperature alone or concurrently may affect the physiological and ecological performance of Scots pine [*Pinus sylvestris* L.] (Kellomäki et al. 2000). In this experiment, the exposure to the elevation of temperature and CO_2 covered two separate periods: 1990–1995 and 1996–2002, the former period using an open top chambers and the latter period closed chambers. In both cases, the experimental treatments simulated the expected changes in atmospheric CO_2 and temperature conditions (Box 6.3).

The University of Kuopio (the University of Eastern Finland from 2010 onwards), Department of Environmental Science, Finland, carried out a chamber experiment (1994–1996) in collaboration with the University of Joensuu, with the focus on the interactive effects of carbon dioxide (CO_2) and ozone (O_3) on the physiology of Scots pine (Palomäki et al. 1998, Kellomäki and Wang 1998a,b). Open-top chambers were also used in the Suonenjoki Research Station, the Finnish Forest Research Institute (Metla) (the Finnish Natural Resource Institute from 2015 onwards), to study how the elevation of CO_2 and O_3 may affect the physiological and ecological performance of boreal birch (*Betula pendula* Roth) during the period 1999–2001 (Vapaavuori et al. 2002). At the same time (1997–2004), a large-scale chamber experiment was carried out in Sweden, the Swedish University of Agricultural Sciences, close to Umeå, in order to study how the elevation in CO_2 and temperature combined with nitrogen addition and irrigation may affect the physiological and ecological performance of boreal Norway spruce [*Picea abies* (L.) Karst.] (Medhurst et al. 2006). This experiment also provided much information about how the physiology and growth of Norway spruce responds to the availability of nitrogen and soil water under boreal conditions.

Box 6.3 The Mekrijärvi chamber experiments

The open-top (1991–1995) and closed chamber (1996–2002) experiments were established in a naturally regenerated Scots pine forest near the Mekrijärvi Research Station (62°47' N, 30°58' E, 145 m a.s.l.), University of Joensuu, Finland. The site is sandy with low supply of nitrogen (Fig. 6.3). Scots pines were 15–20 years old, past their juvenility.

Both experiments followed the factorial design: the CO_2 and temperature were elevated alone or concurrently, each of the treatments being replicated four times. In the control chambers, the CO_2 and temperature followed the values outside the chambers. Measurements were also taken for trees growing outside the chambers in order to quantify the effects of closing trees in the chambers. All the chambers in both experiments were irrigated in summer so that the total sum of natural rainfall and irrigation was at least 10 mm per week, representing the long-term mean precipitation for the area in June. In winter, snow was added to compensate for evaporative water losses.

In the open-top chambers, the temperature elevation was designed to cause premature growth onset during winter (Hänninen et al. 1993), and therefore, a year-round minimum target temperature of $\geq 0°C$ was used. The mean temperature was 2°C above the mean outside the chambers in summer (from April 15 to September 15) and 5–20°C above (September 15 to April 15) in winter. CO_2 was elevated on April 15 (the onset of growth) and terminated on September 15 (the fall in dormancy). The elevated CO_2 was kept between 550–600 µmol mol^{-1}, with the mean of 580 µmol mol^{-1} in daytime (i.e., between 6 a.m. and 6 p.m.) (Hänninen et al. 1993, Repo et al. 1996, Wang 1996b, Wang et al. 1996). In the closed chambers, the temperature followed the seasonal pattern outside, with the increase depending on the season: 6°C in winter (from December to February), 4°C in spring and autumn (from March to May and from September to November) and 2°C in summer (from June to August). The fixed CO_2 concentration of 700 µmol mol^{-1} was used 24 hours a day throughout the year (Kellomäki et al. 2000).

Box 6.3 contd....

Box 6.3 contd....

Fig. 6.3 Open-top (on the right) and closed chambers (on the left) in the Mekrijärvi experiment. Photo and permission of Topi Ylä-Mononen.

Carbon uptake under elevated CO_2 and temperature

In general, varying responses of stomatal conductance and carbon uptake to climate change (the elevated CO_2 and temperature) may be expected (Fig. 6.4). According to Jarvis (1993), elevated CO_2 may result in a higher assimilation rate per unit area, partially offset by lower Rubisco activity per unit area. At the same time, the stomatal conductance may remain the same as in the ambient CO_2 or reduce per unit area. A change of energy balance in leaf/needle may further lead to a higher leaf temperature, which will partially offset the reduction in transpiration rate due to reduced stomatal conductance under the CO_2 elevation. The overall enhancement of assimilation rate under the elevated CO_2 may vary widely depending on the reduction of the A_1/C function and the direction and magnitude of stomatal acclimation (Fig. 6.4) (Box 6.4).

Short- and long-term responses of carbon uptake to elevated CO_2 and temperature

In the short term, elevated CO_2 enhances photosynthesis by increasing carboxylation. Stitt (1992) found that doubling the ambient CO_2 from 350 to 700 µmol mol^{-1} increased the carboxylation by 78% (at 25°C and 21% O_2). Similarly, Long 1991 found CO_2 elevation to increase both the carboxylation and oxygenation rates. However, the oxygenation increased more relative to carboxylation under the elevated temperature, resulting in greater loss of CO_2 in photorespiration, thus reducing the efficiency of CO_2 uptake at higher temperatures. In the long term, the carboxylation rate may be smaller than the short-term rate due to the acclimation of photosynthetic processes to higher CO_2 (Sage et al. 1989, Dang et al. 2008). For example, Cure and Accock (1986) estimated that the net CO_2 exchange rate in crop plants may increase, on average, up to 52% following the first exposure to the doubling of CO_2, but the increase remained about 30% after the plants were acclimated to higher CO_2. Similarly, Curtis (1996), demonstrated that the relative CO_2 effect (i.e., the ratio between net assimilate (NA) at CO_2 of 750 µmol mol^{-1} to NA at CO_2 of 350 µmol mol^{-1}) on selected deciduous and coniferous seedlings was greatest, if the length of exposure to high CO_2 was < 50 days.

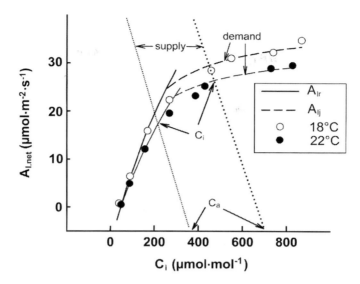

Fig. 6.4 Schematic presentation of how the CO_2 uptake depends on the ambient (C_a) and doubled CO_2 outside the leaf/needle and CO_2 in the stomatal cavity (C_i). CO_2 in the stomatal cavity is assumed to be a linear function of CO_2 outside leaf/needle with the slope defined by the stomatal conductance (Wang and Kellomäki 2003). The CO_2 response of photosynthesis is given as a function of two leaf temperatures, with A_{lr} indicates the Rubisco activity and A_{lj} the regeneration of RuBP controlling the photosynthesis. The rate of substrate supply is derived from the gradient of CO_2 for current and elevated CO_2 outside the leaf.

Cao et al. (2007) showed that the photosynthesis of birch (*Betula papyrifera* Marsh.) was regulated down by 57% when grown 80 days under elevated CO_2 (720 μmol mol^{-1} vs. 360 μmol mol^{-1}). Curtis (1996) found that the net assimilation response to doubling CO_2 was greater in unstressed than in stressed seedlings.

Experimental findings about the effects of elevated CO_2 on carbon uptake

Light saturated photosynthesis is widely used to indicate the long-term effects of elevated CO_2 on photosynthetic production. Wang et al. (1995) found that CO_2 treatment alone increased maximum photosynthetic rate of boreal Scots pines grown in two years in open-top chambers in elevated temperature or CO_2 concentration, or both. In contrast, the temperature elevation decreased maximum photosynthetic rate and photosynthetic efficiency. The photosynthetic responses of one- and two-year-old needles to varying photon flux densities (0–1,500 μmol m^{-2} s^{-1}) and CO_2 concentrations (350, 700 and 1,400 μmol mol^{-1}) during measurements showed that the CO_2 treatment decreased and the temperature treatment enhanced the reduction of maximum photosynthesis due to needle aging.

FACE experiments also show that the elevation of CO_2 increases light saturated photosynthesis. For example, in loblolly pine (*Pinus taeda* L.) and sweetgum (*Liquidambar styraciflue* L.) in subtropical conditions may increase 30–60% under elevated CO_2 (the ambient + 200 μmol mol^{-1}) (Ellsworth et al. 2012). Ainsworth and Long (2005) also found that the elevated CO_2 in the FACE experiments increased light-saturated carbon uptake by about 30%, while stomatal conductance was reduced 20% with no acclimation of stomatal conductance to the elevated CO_2. This was also the case for selected European deciduous and coniferous species, as demonstrated by Medlyn et al. (2001). Based on a meta-analysis of data from 13 long-term (> 1 year) field-based studies, they found that stomatal conductance was reduced by 21% in response to elevated CO_2. The response was stronger in young trees than in old ones, and stronger in deciduous trees than in coniferous

Box 6.4 Biochemical photosynthetic model to analyze the impact of climate change on photosynthetic production

The model developed by Farquhar et al. (1980) and von Caemmerer and Farquhar (1981) is widely used to relate gas exchange measurements for the biochemical processes controlling photosynthesis. The model is based on the assumption that the photosynthetic rate is related to the amount of activated Rubisco, the regeneration rate of RuBP and the partial pressures of CO_2 and O_2 in the CO_2 fixation site. The carbon assimilation rate (A_n, µmol m^{-2} s^{-1}) is:

$$A_n = \min(A_c, A_q) \qquad (6.1)$$

where A_c (µmol m^{-2} s^{-1}) is the net photosynthesis rate controlled by the availability of CO_2 (i.e., the amount of activated Rubisco), and A_q (µmol m^{-2} s^{-1}) is the net photosynthesis rate controlled by the availability of radiation (i.e., the regeneration rate of RuBP). The values of A_c are:

$$A_c = \frac{V_{cmax} \times (C_{in} - 0.5 \times O_i / S_R)}{C_{in} + K_c \times (1 + O_i / K_o)} - R_d \qquad (6.2)$$

where V_{cmax} [µmol m^{-2} s^{-1}] is the maximum rate of carbon fixation (carboxylation) under an ample supply of Rubisco and CO_2. The variable R_d [µmol m^{-2} s^{-1}] indicates the release of CO_2 in the photosynthesis (day respiration) and C_{in} is the concentration of CO_2 [µmol mol^{-1}] in the stomatal cavity. The variable O_i is the concentration of O_2 in the stroma and K_c and K_o are parameters, and S_R is the CO_2/O_2 specificity factor of Rubisco. If radiation is limiting photosynthesis, the photosynthetic rate is:

$$A_q = \frac{\delta \times Q \times (C_{in} - 0.5 \times O_i / S_R)}{4(C_{in} - O_i / S_R)} \times \left[1 + \left(\frac{\delta \times Q}{J_{max}}\right)\right]^{-0.5} - R_d \qquad (6.3)$$

where Q photon flux density [µmol m^{-2} s^{-1}], J_{max} is the maximum rate of electron transfer [µmol m^{-2} s^{-1}] and σ is the efficiency of light energy conversion [electron quanta^{-1}] (quantum yield).

The values of K_c, K_o and R_d are depending on temperature and R_d and σ further on the nitrogen content of leaves/needles (N_l) (Wang et al. 1996, Kellomäki and Wang 1997b):

$$K_c = \exp(c - H_a / (R \times T_a)) \qquad (6.4)$$

$$K_o = 1000 \times \exp(c - H_a / (R \times T_a)) \qquad (6.5)$$

$$R_d = (c_1 \times N_l + c_2) \times \exp((T_a - T_o) \times H_a / (R \times T_a \times T_o)) \qquad (6.6)$$

$$\delta = c_{\delta 1} \times N_l + c_{\delta 2} \qquad (6.7)$$

where R is the universal gas constant (8.314 J K^{-1} mol^{-1}), c, c_1, c_2, $c_{\delta 1}$ and $c_{\delta 2}$ are parameters, H_a is the activation energy, T_a [K] is the leaf/needle temperature and T_o [K] is the optimum temperature for photosynthesis. The values of J_{max} and V_{cmax} are further dependent on the seasonality of photosynthesis, leaf/needle temperature and the optimum temperature for photosynthesis, and leaf/needle nitrogen content (Harley et al. 1992, Wang et al. 1996, Kellomäki and Wang 1997b):

$$J_{max}, V_{cmax} = K \times c_N \times (N_l - c_{min}) \times \frac{\exp[(T_a - T_o) \times H_a / [R \times T_a \times T_o]]}{1 + \exp[(S_a \times T_a - H_d) / (R \times T_a)]} \qquad (6.8)$$

where K is the seasonality of photosynthesis capacity [0,1], c_N and c_{min} are parameters, S_a is an entropy term and H_d is the energy of deactivation. Figure 6.5 shows that at low temperature but high CO_2, the quantum yield (the slope of curve relating CO_2 uptake to absorbed light) was smaller, but light compensation (the light intensity on the light curve where the rate of photosynthesis meets the rate of respiration) was higher than under other combinations. At low temperature and CO_2, photosynthesis saturated at low radiation with the lowest maximum photosynthesis, while saturating light was high at high temperature and CO_2 with the highest maximum photosynthesis.

Box 6.4 contd....

Box 6.4 contd....

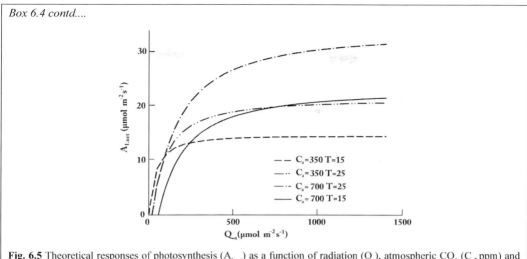

Fig. 6.5 Theoretical responses of photosynthesis ($A_{l,net}$) as a function of radiation (Q_a), atmospheric CO_2 (C_a, ppm) and temperature (T, °C) (Wang and Kellomäki 2003).

ones and stronger in water stressed than in nutrient stressed trees. No evidence of the acclimation of stomatal conductance to the elevated CO_2 was found. However, the stomatal conductance and carbon assimilation responded in parallel to the elevated CO_2 except if water was limiting.

Effect of nitrogen on carbon uptake under elevated CO_2 and temperature

The photosynthetic activity of many tree species including boreal and temperate coniferous and deciduous species is linearly proportional to the nitrogen content in foliage, due to the limited availability of nitrogen (Reich et al. 1998, Cao et al. 2007, Hyvönen et al. 2007). The effects of nitrogen on carbon uptake are associated with the biochemical processes that drive photosynthesis related to the amount and activity of Rubisco enzyme, which accounts for 50% of soluble leaf protein in C_3 plants (20–30% of total leaf nitrogen). In this context, the acclimation of photosynthesis to elevated CO_2 is most likely related to the accumulation of nonstructural carbohydrates in leaves/needles, thus reducing the nitrogen concentration in foliage. A sufficient/increasing availability of nitrogen is needed to balance the increased availability of carbon, otherwise the carbon uptake may be reduced under the elevated CO_2 (Jarvis 1993, Ainsworth and Long 2005, Hyvönen et al. 2007) related to the reduction of the amount of Rubisco and/or Rubisco activity (Rogers and Humphries 2000) (Box 6.5).

Box 6.5 Role of nitrogen in carbon uptake

In general, the CO_2 assimilation rate is enhanced by increasing nitrogen concentration in leaves/needles. Nitrogen is involved in the enzymes of the PCR cycle, and the nitrogen in thylakoid proteins includes the main part of nitrogen in the leaves/needles involved in light reactions with light harvesting, electron transport and photophosphorylation (see Box 6.1). The other half is involved in dark reactions in PCR. They include proteins involved in CO_2 assimilation, photorespiration, RuBP regeneration and the synthesis of carbohydrates (Evans 1989). This implies that the short supply of nitrogen from soil and/or from translocation from plant tissues may affect both carboxylation efficiency and CO_2 saturated photosynthesis. Nitrogen limitation tends to affect more Rubisco activity (i.e., the slope of A_n/C_i curve) than RuBP regeneration (i.e., the plateau part of the curve). In general, the Rubisco activity and the electron transport are closely related to each other, thus representing the co-limitation of photosynthesis under a short supply of nitrogen (Medlyn et al. 1999).

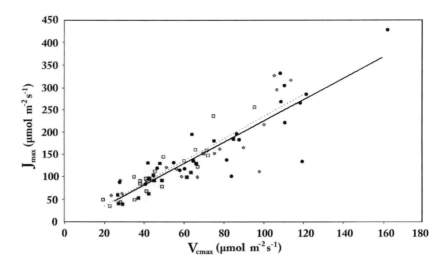

Fig. 6.6 Relationship between the mean value of J_{max} and V_{cmax} for selected European deciduous and coniferous trees exposed to the long-term elevation of CO_2 (Medlyn et al. 1999). Permission of Wiley & Sons. Legend: filled symbols represent the trees grown under ambient CO_2 (350 µmol mol^{-1}) and the open symbols trees grown under elevated CO_2 (700 µmol mol^{-1}). The squares are deciduous and the squares standing on the edge are coniferous species. The solid regression line is for the ambient CO_2: $J_{max} = 2.39 V_{cmax} - 14.2$, $R^2 = 0.80$, and the dotted line for the elevated CO_2: $J_{max} = 2.25 V_{cmax} - 14.3$, $R^2 = 0.78$.

Medlyn et al. (1999) found that long-term exposure to elevated CO_2 reduced photosynthesis 10–20% in selected European deciduous and coniferous trees if measured at the same CO_2 concentration. They also found that long-term exposure to elevated CO_2 affects the biochemical processes as indicated by the down-regulations of J_{max} and V_{cmax} by 10% in such a way that the values of both parameters were correlated closely (Fig. 6.6). The reduction was linked to the effects of elevated CO_2 on the reduction of nitrogen concentration in leaves/needles. The reduction was 15% on a mass basis. Medlyn et al. (1999) concluded that: "the general reduction in leaf nitrogen concentration may thus be thought of as a dilution effect, caused by increased leaf mass per unit area." It was evident that a part of the increase of mass per unit area was due to the accumulation of starch in elevated CO_2. A dilution effect with down-regulation of photosynthesis was not found on an area basis, because nitrogen content per unit area was not changed by elevated CO_2. It was evident that V_{cmax} tended to reduce alongside reducing nitrogen per leaf area.

The nitrogen content in leaves and needles is dependent, e.g., on the species, site fertility (i.e., edaphic and climatic conditions), maturity of trees, canopy position, etc. as demonstrated in Table 6.1 for boreal Scots pine, Norway spruce and birch. The nitrogen content of foliage varies from 0.5 to 4% of dry weight as a function of species and site fertility. The maximum photosynthesis at the upper limit is doubled compared to that at the lower limit, provided no other factors limit photosynthesis (Cao et al. 2007). The foliage mass of trees will increase simultaneously with a further increase in the total photosynthesis of trees (Wang 1996a,b). This links the growth of trees and the productivity of sites (site fertility) to the nitrogen cycle, i.e., the nitrogen content of foliage is related to the availability of nitrogen in the site.

The links between the effects of elevated CO_2 on leaf/needle nitrogen and photosynthesis in boreal Scots pines are demonstrated in Fig. 6.7. It shows how the CO_2 saturated (V_{cmax}) and the radiation saturated (J_{max}) photosynthesis, and the values of other parameters of the photosynthetic model in Box 6.4 depend on the nitrogen content of needles exposed to elevated CO_2 and temperature, and their combination. The values of V_{cmax} and J_{max} increased along with the increasing nitrogen content, regardless the climate treatment (Kellomäki and Wang 1997b, Kellomäki and Wang 1997d). Similarly, the elevated CO_2 alone or combined with the elevated temperature tended

Table 6.1 Mean nitrogen content in the foliage of selected boreal tree species in Finland as a function of site type and site fertility based on several sources (Kellomäki et al. 1992, Matala et al. 2006).

Species	Site type	Site index, H_{100} (Scots pine, Norway spruce) or H_{50} (birch)	Nitrogen content in foliage, %
Scots pine	OMT	30	2.43
	MT	27	1.90
	VT	24	1.30
	CT	18	0.54
Norway spruce	OMT	30	3.74
	MT	27	2.92
Birch	OMT	26	4.68
	MT	24	3.65

Site type refers to fertility defined based on ground vegetation while the site index refers to the dominant height (e.g., H_{100} dominant height at the age of 100 years) of species at a given age.

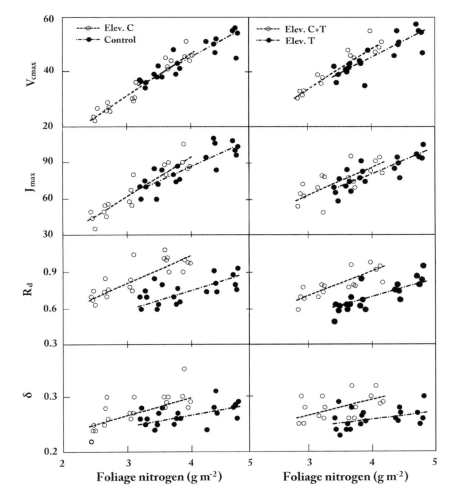

Fig. 6.7 Dependence of CO_2 (V_{cmax}, µmol m^{-2} s^{-1}) and light (J_{max}, µmol m^{-2} s^{-1}) saturated photosynthetic rates, day respiration (R_d, µmol m^{-2} s^{-1}) and quantum yield (δ, electron quanta^{-1}) on the nitrogen content in needles of Scots pine grown under ambient conditions (Control) and under elevated temperature (Elev. T) and CO_2 (Elev. C) and their combination (Elev. C + T) (Kellomäki and Wang 1997b). Permission of Elsevier.

to increase the values of other parameters of the photosynthetic model. The increase was especially pronounced in the case of day respiration (R_d) and the efficiency of light use (quantum yield, δ) (Wang 1996a). In conifers like Scots pine, the structure of shoots and needles is further related to the nitrogen content, which may increase the photosynthetic capacity of shoots by increasing the needle level photosynthetic rate and reducing the internal shading in the shoot. For example, Smolander et al. (1990) found the reduction of internal shading to substantially increase the photosynthetic capacity of shoots of Scots pine grown on fertile site (*Oxalis-Myrtillus* site type, OMT) compared to those grown on poor sites (*Vaccinium* site type, VT).

Effect of ozone on carbon uptake under elevated CO_2

The wide-scale use of fossil fuels increases atmospheric CO_2 and other air pollutants, including ozone (O_3) in the lower atmosphere. Ozone may reduce photosynthesis by reducing stomatal conductance, and damaging photochemical processes in light harvest and carbon fixation (Reich 1987). At the same time, the increase of atmospheric CO_2 may increase the Rubisco activity and reduce stomatal conductance. Increasing CO_2 may thus partly compensate for the detrimental effects of ozone, but chronic ozone exposure is likely to lead to a decline in the activity and quantity of Rubisco, the capacity of electron transport and stomatal conductance.

Kellomäki and Wang (1997a,c, 1998b) exposed naturally regenerated, 30-year-old Scots pines in open-top chambers *in situ* to doubled ambient O_3, doubled ambient CO_2 and a combination of elevated O_3 and CO_2 from April 15 to September 15, for three growing seasons (1994–1996). Doubled ambient O_3 significantly decreased the rate of photosynthesis, regardless of photon flux density. This implied a significant decrease in the photochemical efficiency of photosystem II and the rate of whole electron transport, rather than a decrease in stomatal conductance. When measurements were made at the doubled ambient concentration of CO_2 (700 μmol mol^{-1}), the doubled ambient CO_2 treatment had no clear effect on the intrinsic capacity of photosynthesis, but the elevated CO_2 increased the sensitivity of stomatal conductance to light and decreased maximal stomatal conductance. When O_3 and CO_2 were combined, the O_3-induced decrease in photosynthesis rate was reduced significantly at a high concentration of CO_2. This may be partly related to the decrease in stomatal conductance caused by the high concentration of CO_2.

Kellomäki and Wang (1998a) found further that elevated O_3 decreased photosynthetic capacity and stomatal conductance in Scots pine shoots during the whole photosynthetic period in the main growing season. On the other hand, elevated O_3 delayed the springtime onset of photosynthetic recovery, while the combined elevation of O_3 and CO_2 decreased the photosynthetic capacity and stomatal conductance. The introduction of O_3 and CO_2 impact in a model simulation showed that elevated O_3 alone, and a combination of elevated O_3 and CO_2 decreased the annual total of net photosynthesis per unit leaf area by 55 and 38%, while elevated CO_2 alone increased the annual total net photosynthesis by 13% compared to the ambient conditions.

6.3 Eco-physiological Responses of Carbon uptake to Elevated CO_2 and Temperature

Carbon uptake during the active period

Under boreal conditions, trees are adapted to a strong seasonal cycle (ontogenetic cycle) of dormant (rest period) and active periods (growing period), with an autumn transition from the active period to winter dormancy, and a spring transition from winter dormancy to the active period.

In the active period, climate change (the elevation CO_2 and temperature) directly affects the photosynthetic rate. Figure 6.8 shows the net photosynthetic rate of Scots pine as a function of photon flux density and temperature (above), and the CO_2 concentration in the stomatal cavity and temperature (below) in shoots grown under elevated CO_2, and temperature alone or combined. The net photosynthetic rate increases to saturation if radiation and CO_2 concentrations increase. Regardless of climate treatment, the photosynthetic rate is highest when the temperature is 22°C, if the intercellular CO_2 concentration was > 250 ppm. This pattern implies that climate change may affect net photosynthetic production through the elevation of temperature and CO_2. However, it also limits the net photosynthetic rate due to the concurrent increase in respiration.

Exposure to elevated CO_2 and temperature may further modify photosynthetic responses to single environmental factors (Fig. 6.9). When the measuring temperature was < 25°C, the values of V_{cmax} in response to temperature drifted upwards and at temperature > 25°C the values of V_{cmax} reduced. The drift was especially pronounced if the growing temperature and CO_2 were elevated concurrently. The temperature response of V_{cmax} culminated in higher temperatures if the trees were grown under elevated temperature or combined with increased CO_2. The values of J_{max} were fairly similar regardless of climatic treatment, when the measuring temperature was < 22°C. At higher measuring temperature, there were clear differences in temperature response, the lowest values being from the trees grown under elevated CO_2 alone or ambient conditions. The temperature response of J_{max} culminated at higher measuring temperatures if trees were grown under elevated temperature alone or combined with elevated CO_2.

Figure 6.9 also shows that in Scots pine the increased CO_2 supply (the ambient CO_2 concentration) from 230 to 540 ppm enhanced the maximal light saturated assimilation 28–34%. The increase was larger in the shoots grown under the elevation of temperature or CO_2 alone than in the shoots grown under the combined elevation of temperature and CO_2 (Wang et al. 1996). When using variable air temperature, atmospheric CO_2 and needle nitrogen as an input in a process-based model (FinnFor), it appeared that the elevation of temperature alone increased the optimum temperature for net photosynthesis but elevated CO_2 alone reduced the optimum temperature (Fig. 6.10). When the elevation of temperature and CO_2 were combined, the optimum temperature increased as did under the elevation of temperature alone (Wang et al. 1996). Nitrogen content also affected the photosynthesis responses to temperature and CO_2 as was also found by Hyvönen et al. (2006). The elevation of CO_2 increased the maximum photosynthesis, but it also shifted the culmination of the photosynthetic rate to a higher temperature. The photosynthetic rate was higher under high CO_2 even under low temperatures, than it was under low CO_2, and the increasing effect of high nitrogen content was still clear under low temperatures.

During growing seasons, dry spells may limit photosynthesis of conifers even in boreal conditions. In general, the reducing carbon uptake is attributable to the reducing stomatal conductance in response to the reducing water potential of foliage, and reducing intercellular CO_2 pressure (Fig. 6.11). The consequent suppression of carbon uptake may further be related to the non-stomatal limitation caused by reducing water potential in the intrinsic photosynthetic capacity. When using the biochemical photosynthetic model developed by Farquhar et al. (1980), it may be assumed that its parameters are affected by the water potential in leaves/needles (Kellomäki and Wang 1996):

$$Parameters\ (V_{c\max}, J_{\max}, C_i, g_s) = K_{\max} / \left[1 + (\Psi_l / K_t)^{-K_s}\right] \tag{6.9}$$

$$R_d = K_m \times \left[1 + (\Psi_l / K_t)^{-K_s}\right] \tag{6.10}$$

where V_{cmax} is the maximum rate of carbon fixation, J_{max} is the maximum rate of electron transfer, R_d is day respiration, K_m is the maximum value of the parameter under well-watered conditions, K_t is the threshold value of leaf/needle water potential ψ_l and K_s is the coefficient connecting the sensitivity of the parameter to the decreasing water potential.

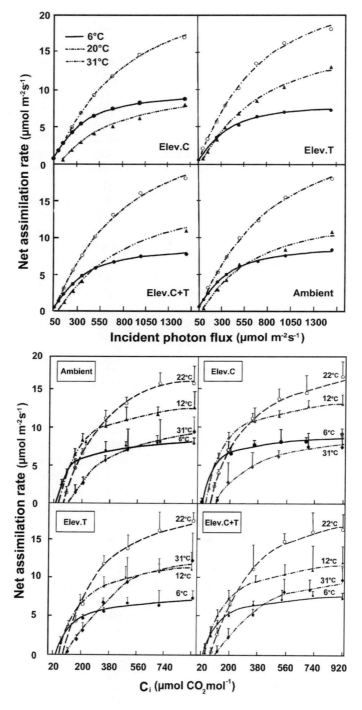

Fig. 6.8 Net photosynthesis of Scots pine shoots grown under the current (Ambient), and elevated CO_2 (Elev. C) and temperature (Elev. T) and combined both (Elev. C + T) as a function of incident photon flux and temperature. Above: under a constant CO_2 concentration of 1,400 μmol mol^{-1}, and intercellular concentration of CO_2 and temperature. Below: under a saturating photon flux 1,500 μmol m^{-2} s^{-1} (Wang et al. 1996). Permission of Elsevier.

Carbon Uptake and Climate Change 77

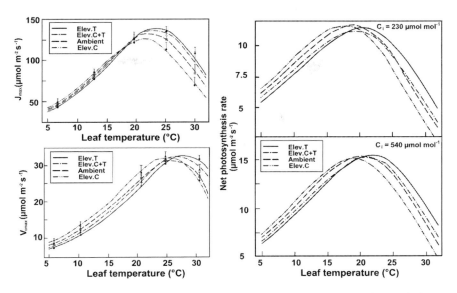

Fig. 6.9 Temperature dependence of J_{max} (upper left), V_{cmax} (lower left), and light saturated net photosynthetic rate under two concentrations of CO_2 in stomatal cavity (right) as a function of leaf temperature in Scots pines grown under the current conditions (Ambient), elevated temperature (Elev. T) and CO_2 (Elev. C) alone, and their combination (Elev. C + T) (Wang et al. 1996). Permission of Elsevier.

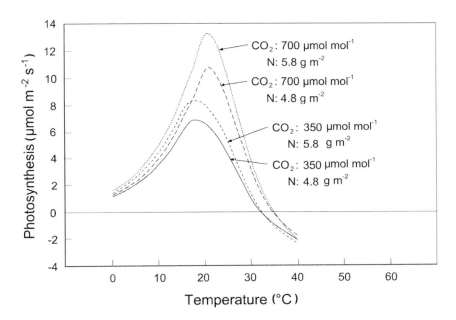

Fig. 6.10 Dependence of light saturated photosynthetic response of Scots pine on the nitrogen content of needles, temperature, and atmospheric CO_2. The graphs are based on calculations through a process-based simulation model (FinnFor) (Kellomäki and Väisänen 1997).

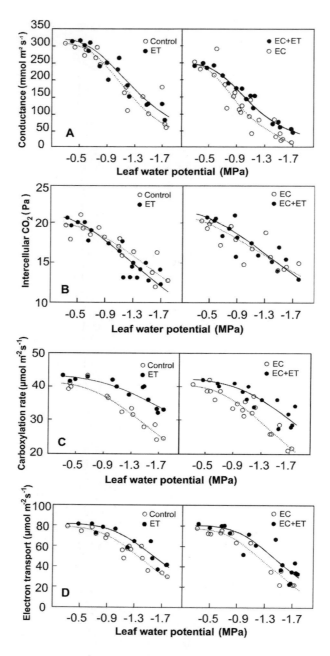

Fig. 6.11 Effect of needle water potential on stomatal conductance (A, upper), intercellular CO_2 (B, lower upper), carboxylation rate (V_{cmax}) (C, upper lower) and electron transport rate (J_{max}) (D, lower) in Scots pines grown four years in open-top chambers with elevated temperature (ET), elevated CO_2 (EC) or a combination of elevated temperature and CO_2 (EC + ET). Data used in the analysis was measured at a high photon flux density (1,500 µmol m^{-2} s^{-1}), saturated leaf to air water vapor pressure deficit (VPD < 0.4 kPa) and optimal temperature (18 ± 0.5°C) in the Mekrijärvi chamber experiment (Kellomäki and Wang 1996). Permission of Oxford University Press. The lines are the fit of Equation (6.9).

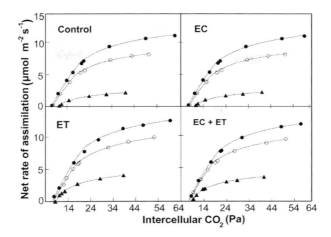

Fig. 6.12 Net photosynthesis as a function of intercellular partial pressure of CO_2 in Scots pine needles grown under elevated CO_2 (EC), elevated temperature (ET) and a combination of both (EC + ET) with varying water potential values indicated by filled dots for potential −0.4 MPa, open dots for −1.0 MPa, and filled triangles for −1.8 MPa (Kellomäki and Wang 1996). Permission of Oxford University Press.

Kellomäki and Wang (1996) found that in Scots pine the suppression of the CO_2 assimilation rate at low needle water potential was related to both stomatal and non-stomatal limitations (Fig. 6.11). The stomatal limitation was greatest under the early phase of water stress ($\psi_1 < -1.2$ MPa), with the immediate adjustment of stomatal conductance to the decreasing water potential. Under long-term water stress, non-stomatal limitations dominated, with a slow response to water stress. In this case, photosynthesis was reduced more by the decline in J_{max} than that in V_{cmax}. Elevated CO_2 decreased the values of V_{cmax}, J_{max} and R_d but significantly increased the sensitivity of V_{cmax}, J_{max} and R_d to decreasing ψ_1. The effects were opposite under elevated temperature alone and under combined elevated CO_2 and temperature, which further decreased the sensitivity of stomatal conductance to the decreasing ψ_1. This may increase the tolerance of Scots pine to short supply of water under climatic change (Lebourgeois et al. 1998, Grassi et al. 2005). In this context, Fig. 6.12 combines the effects of drought and intercellular CO_2 on the net assimilation rate in boreal Scots pines grown four years under elevated CO_2 and temperature. Under the needle water potential of −0.4 MPa, the light saturated net photosynthesis was one and a half times larger than under the water potential −1.0 MPa and five times larger than under the water potential −1.8 MPa under high intercellular CO_2.

Seasonality of carbon uptake

In northern Europe, tree species such as Scots pine and Norway spruce are exposed to a pronounced fluctuation in temperature, varying from −50°C in winter to +40°C in summer. In winter, photosynthetic processes are inactive and partly damaged, with a reduction in quantum yield and photochemical efficiency. The time required for photosynthetic capacity to recover is 60–85 days, depending on the spring weather, with faster recovery during warm springs without severe frost nights (Troeng and Linder 1982). The recovery of photosynthetic capacity with the increase of quantum yield and light compensation (Wang 1996a) is mainly controlled by mean air temperature and the frequency of severe night frosts as found in soil heating experiments (Bergh and Linder 1999). Troeng and Linder (1982), estimated that 95% of the annual carbon gain in Scots pine took place from May to October in central Sweden (62–63° N). Thus, climatic warming in spring and autumn may reduce the period of suppressed photosynthetic capacity, allowing boreal conifers to utilize radiation during the spring and autumn with greater efficiency than possible under the current temperature conditions.

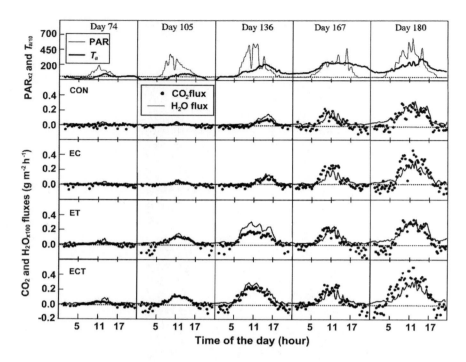

Fig. 6.13 Seasonal course of photosynthetic and transpiration rates of Scots pine under climate change before the beginning of the growing season (D74 and D105 indicating the number of day from the beginning of the year), at the beginning of the growing season (D136 and 167) and during the growing season (D180). The time series show the hourly photosynthetic and transpiration rates in the control chambers (CON, ambient conditions), under elevated CO_2 (EC) alone, elevated temperature (ET) alone and under the combined elevation of CO_2 and temperature (ECT). The uppermost time series show temperature and irradiation in the control chambers (ambient conditions) (Wang et al. 2003). Permission of Oxford University Press.

Based on Wang et al. (2003), Fig. 6.13 shows the net photosynthetic rate of Scots pine for the selected days (D84, D118, etc. from the beginning of the year) from late March to the middle of July in the Mekrijärvi chamber experiment. Under ambient conditions, the continuous uptake of CO_2 began in late April, but occasional uptake was recorded on warm days even in March (D74). The CO_2 uptake rate increased rapidly in May, and maximum photosynthesis took place in late June. Under elevated CO_2, the onset of CO_2 uptake took place at the same time as under ambient conditions, as it did in Norway spruce (Wallin et al. 2013). In contrast to elevated CO_2, photosynthesis in Scots pine recovered 15–20 days earlier under higher temperatures (elevated temperature alone or combined with elevated CO_2) with a four week longer growing period, as is also the case for Norway spruce (Hall et al. 2013, Walling et al. 2013). The maximum photosynthetic rate was achieved in the middle of June for both boreal coniferous species.

Regardless of the climatic treatment, H_2O lost in transpiration of Scots pine were closely coupled with CO_2 uptake during spring (Wang et al. 2003). For example, on D136, the total daily H_2O flux increased by 104% under elevated temperature and by 110% under elevated temperature combined with elevated CO_2, but decreased by 6% under elevated CO_2 alone. On D180, the temperature elevation alone increased H_2O flux by 38%. The temperature elevation combined with CO_2 increased H_2O flux by 41%. Regardless of the treatment, the H_2O flux increased in spring more than expected on the basis of the increase in the CO_2 flux.

In autumn, the physiological functioning of trees falls into dormancy, as indicated by the photosynthetic rate for Scots pine in Fig. 6.14. In the middle of September, the CO_2 uptake rate is reduced, even though radiation is high enough to support a higher photosynthetic rate (Wang

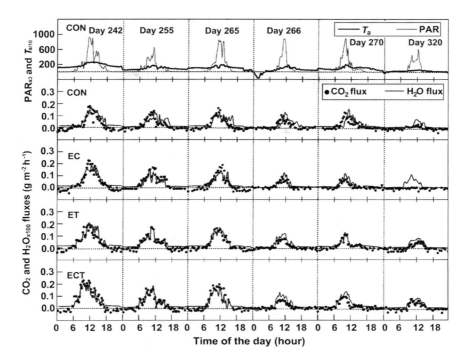

Fig. 6.14 Seasonal course of the photosynthetic and transpiration rates of Scots pine in the autumn (D242, D255, D265, D270 and D320 from the beginning of the year). The time series show the hourly photosynthetic and transpiration rates in the control chambers (CON, ambient conditions), under elevated CO_2 (EC), elevated temperature (ET) and combined elevation of CO_2 and temperature (ECT). The uppermost time series show temperature and irradiation in the control chambers (ambient conditions) (Wang et al. 2003). Permission of Oxford University Press.

et al. 2003). Under ambient conditions, the CO_2 uptake virtually stopped by the end of October, but positive values were still recorded under warm spells, with the daytime temperature above 0°C over several days. The increases in CO_2 uptake caused by CO_2 elevation declined with the acclimation to low temperatures, and the reduction in H_2O fluxes showed a smaller change over the acclimation period relative to CO_2 fluxes. Higher temperatures prolonged the duration of CO_2 uptake until the end of November, and led to a greater increase in H_2O flux over the autumn. The freezing of soil seems to finally end the photosynthesis of boreal Scots pine in the autumn (see also Troeng and Linder 1982). In spring, frozen soil curbs water uptake and limits stomatal conductance and the subsequent carbon uptake.

Wang et al. (2003) found further that temperature and CO_2 elevation either separately or combined, increased the sensitivity of CO_2 uptake in the day followed by a night with frost. This is exemplified in Fig. 6.14 for D265, D266 and D270, when the incident radiation, daytime air temperature and vapor pressure deficit were similar. On D265, the mean night temperature was −9°C, with the consequence that on D266 the CO_2 uptake under ambient conditions was 24% less than on the previous day (D265). Under elevated CO_2, temperature and their combination, the reduction was 35, 41 and 39%, respectively. On D270, the CO_2 uptake had recovered regardless of the treatment: under ambient conditions recovery was 16%, while under elevated CO_2 alone 10% and elevated temperature alone 12% and under the combined elevation of CO_2 and temperature 13%. Figure 6.14 shows further that H_2O flux was also reduced substantially on the day after the frost on D265. In both spring and autumn, the CO_2 uptake rate under elevated CO_2 seemed to be affected by night frosts more than under ambient conditions. This may delay the recovery of photosynthesis in spring, and reduce and even finish it earlier than under ambient conditions (Wang et al. 2003).

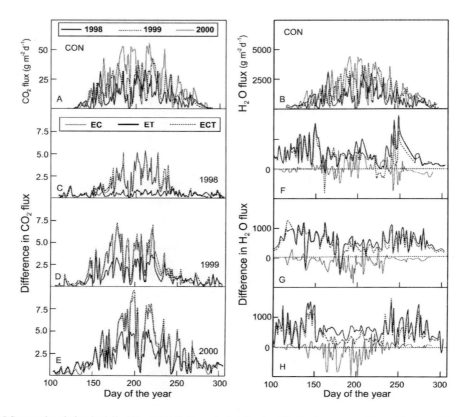

Fig. 6.15 Seasonal variation in daily CO_2 and H_2O fluxes in the shoots of Scots pines exposed to ambient conditions (CON, A and B), elevated temperature and CO_2, and their combination over three years, and the differences between the fluxes for the shoots grown under elevated CO_2 (EC), elevated temperature (ET) or the combined elevation of CO_2 and temperature (ECT) minus the fluxes from shoots grown under ambient conditions (CON) (Wang et al. 2003). Permission of Oxford University Press.

Seasonal and annual variation and differences in CO_2 and H_2O fluxes are demonstrated in Fig. 6.15 on a daily basis over three years under the temperature and CO_2 treatments (treatment–ambient conditions). Under ambient conditions (control), the maximum value for CO_2 uptake was in the range 27–53 g m^{-2} d^{-1}, and the maximum H_2O flux varied in the range 2,600–4,800 g m^{-2} d^{-1}. In both cases, the maximum mean values were in July, which is the warmest month in these conditions. Regarding the treatment-induced effects in relation to the ambient conditions (treatment–control), elevated CO_2 and temperature interacted in 1999 and 2000, and elevated temperature alone increased CO_2 uptake from 1998 to 2000 (Wang et al. 2003).

Wang et al. (2003) analyzed further the results in Fig. 6.15 by dividing the observations into those representing the growing season (D150–239) and those outside the growing season (D1–149 and D240–364). Table 6.2 shows that elevated CO_2 alone increased the CO_2 flux both in the growing season and outside the growing season, but reduced the H_2O flux in all years except 1998. Similarly, elevated temperature increased CO_2 flux in both cases, but the increase was greater outside the growing season than in the growing season. Elevated temperature also increased the H_2O flux, especially outside the growing season. When the elevation of CO_2 and temperature were combined, the CO_2 flux increased. The H_2O flux also increased as it did under elevated temperature alone. The increase was similar both in the growing season and outside the growing season.

Table 6.2 Cumulative flux of CO_2 (net photosynthesis) and H_2O for Scots pine shoots growing under elevated CO_2 and temperature (Wang et al. 2003).

Year and treatment	CO_2 flux, g m^{-2} d^{-1} (increase/decrease, % of ambient)		H_2O flux, g m^{-2} d^{-1} (increase/decrease, % of ambient)	
	Growing season	Non-growing season	Growing season	Non-growing season
1998				
• Ambient	15.10	2.51	1,503	704
• CO_2	19.31 (+28)	2.86 (+14)	1,615 (+7)	755 (+7)
• T	15.85 (+5)	2.92 (+16)	1,996 (+33)	1,524 (+116)
• CO_2 + T	19.15 (+27)	2.98 (+19)	1,953 (+30)	1,456 (+107)
1999				
• Ambient	20.19	4.89	1,750	783
• CO_2	25.21 (+25)	5.40 (+10)	1,529 (–13)	751 (–4)
• T	22.61 (+12)	5.78 (+18)	2,204 (+26)	1,523 (+95)
• CO_2 + T	26.37 (+31)	6.04 (+24)	2,152 (+23)	1,404 (+79)
2000				
• Ambient	24.80	5.62	1,981	1,010
• CO_2	31.24 (+26)	6.76 (+20)	1,728 (–13)	909 (–10)
• T	28.30 (+14)	7.16 (+27)	2,433 (+23)	1,764 (+75)
• CO_2 + T	31.88 (+29)	7.37 (+31)	2,405 (+21)	1,713 (+70)
Mean over 1998–2000				
• Ambient	20.0	4.3	1,745	832
• CO_2	25.3 (+27)	5.0 (+16)	1,624 (–7)	805 (–2)
• T	22.3 (+6.0)	5.1 (+23)	2,211 (+27)	1,604 (+93)
• CO_2 + T	25.8 (+29)	5.5 (+28)	2,170 (+24)	1,524 (+83)

The growing season was D150–240 and the non-growing season was D1–149 and D239–364 from the star of a year. Elevated CO_2 is ambient CO_2 + 350 ppm and elevated T is ambient T + 2°C in the growing season and ambient T + 6°C in the non-growing season.

Modeling seasonality in carbon uptake

In general, the seasonality of carbon uptake is related to the annual temperature cycle as found by Pelkonen and Hari (1980). Based on these findings, Kellomäki and Väisänen (1997) assumed that the maximum values of radiation and carbon dioxide limited rates of photosynthesis (J_{max} and V_{cmax}) are specific to the seasons: the sensitivity of J_{max} and V_{cmax} to temperature is controlled by the annual temperature cycle. By replacing $V_{cmax} = V_{omax}$ and $J_{max} = J_{omax}$, the effect of the annual temperature cycle was introduced in calculating the photosynthetic rate at any time of the year using the photosynthetic model in Box 6.3:

$$J_{omax} = K \times J_{max} \qquad (6.11)$$

$$V_{omax} = K \times V_{cmax} \qquad (6.12)$$

where K (0–1) is the multiplier introducing the effect of seasonality on photosynthetic capacity. The value of multiple K = 0 in full dormancy and K = 1 in full activity (full recovery), whereas 0 < K < 1 for the transition in autumn (falling in dormancy) and in spring (recovering from dormancy). K was modeled as proposed by Pelkonen and Hari (1980):

$$K(t) = \begin{vmatrix} 0 & \text{if } S_p(t) \le 0 \\ \dfrac{S_p}{S_{max}} & \text{if } 0 \prec S_p(t) \prec S_{max} \\ 1 & \text{if } S_p(t) \ge S_{max} \end{vmatrix} \qquad (6.13)$$

where $S_p(t)$ is the stage of development of the annual cycle at the moment t calculated on the hourly basis and S_{max} is the parameter with a value of 6,500:

$$S_p(t) = S_p(t-1) + \sum_{t=1}^{24} m(T(t), S_p(t)) \qquad (6.14)$$

where $m(T(t), S_p(t))$ is the development rate of annual cycle dependent on the temperature and the stage of development:

$$m(T(t), S_p(t)) = \frac{100}{1+100a^{-(T(t)-S_p(t-1/c))}} - \frac{100}{1+100a^{(T(t)-S_p(t-1)/c)}} \qquad (6.15)$$

where T(t) is the hourly temperature [°C] and a and c are parameters [dimensionless] with the values 2 and 600, respectively. Consequently, the photosynthetic capacity at a given moment is acclimated to prevailing temperature, with varying response to temperature dependent on the developmental stage of annual cycle of photosynthetic activity (Pelkonen and Hari 1980).

The effect of seasonality on photosynthetic capacity is demonstrated in Fig. 6.16, based on calculations using a process-based model (FinnFor) (Kellomäki and Väisänen 1997). Assuming a changing climate, canopy photosynthesis gained full capacity earlier and lost it later than under the current climate, regardless of the site. In the south, however, photosynthesis recovered, in winter even under the current climate, if the temperature was > 0°C for several days. In the north, too, the temperature elevation triggered photosynthesis earlier than under the current climate, but recovery occurred in the north substantially later than in the south. In the active period, the photosynthetic capacity was fairly similar in both sites, but winter dormancy was triggered much earlier in the north than in the south.

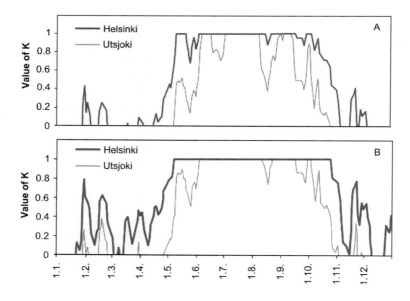

Fig. 6.16 Seasonal course of photosynthetic capacity in terms of K values in southern (60° N, Helsinki) and northern Finland (70° N, Utsjoki) under current (A) and elevated temperatures (B). Current temperature was the mean daily one for the period 1961–1990, and the elevated temperature was the same as the current one but with 4°C daily addition throughout the year.

Photosynthesis mainly increased in early spring at both sites under a warming climate; i.e., during the time when it was not limited by the availability of radiation and soil water, whereas the increase was small in autumn due to the limited radiation (Fig. 6.17). On the other hand, temperature elevation reduced photosynthesis in late July due to the limiting effect of soil moisture. The extended period of photosynthesis was associated with an increase in maximum photosynthesis if a combined elevation of temperature and CO_2 was assumed. The seasonal course of net photosynthesis under elevated CO_2 was the same as under the current climate, but the maximum values for photosynthesis were 25–30% larger. The increase was highest in late June and early July, but the values for photosynthesis during the rest of the growing season still exceeded those for the current climate, except in late July (difference between photosynthesis under elevated temperature + CO_2 and elevated CO_2).

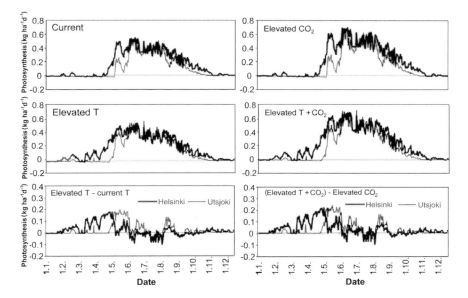

Fig. 6.17 Seasonal course of the daily values of net canopy photosynthesis in southern (Helsinki, 60° N) and northern boreal conditions (Utsjoki, 70° N) under climate change (Beuker et al. 1996). Computations were run for Scots pine stands (density 2,500 seedling per hectare, mean height 1.3 m, *Myrtillus* site type) using a process-based model (FinnFor) (Kellomäki and Väisänen 1997). The simulations were run assuming the current climate and climatic conditions with elevated temperature or CO_2 or both. Under the elevated temperature, 4°C was added to current temperatures. The elevated CO_2 was 700 ppm, and the current CO_2 was 350 ppm. In both sites, precipitation was that currently, even under elevated temperature and CO_2.

6.4 Concluding Remarks

Under forest management, tree populations/communities and their sites are manipulated in order to optimize yields (e.g., biomass, timber) and sustain the productive capacity of sites. In this context, it is important to understand how the extent and rate of carbon uptake depends on light, temperature, water and nutrients, and the maturity of trees, and how they may be influenced by varying management operations/treatments. In this respect, climate change affects production capacity with a need to assess: (i) how trees respond to climate change (biochemical/physiological mechanisms) and (ii) how to scale (integrate) the biochemical/physiological responses over single trees and populations/communities of trees exposed to varying management. Knowledge of the response of carbon uptake of trees to climate change is mainly based on seedlings in the juvenile phase. There are still many questions about whether these findings are applicable over the life span of trees, including mature trees, and the maturity of different tree organs.

The primary productivity of tree populations may be different from that expected on the basis of the immediate response of trees to changes in climate. Reynolds and Acock (1985) have suggested that the growth models used to simulate long-term responses of plants to the changing climate should be mechanistic or semi-mechanistic in order to link the plant responses in a realistic way to the elevating temperature and CO_2. Long (1991) also pointed out that such models should include the direct effects of CO_2 and temperature, and the basic interactions of temperature and CO_2 affecting photosynthesis. This is especially important under boreal conditions, as shown by Troeng and Linder (1982). They found that nearly all annual photosynthetic gain for Scots pine occurs from late March through late October. In spring, photosynthesis recovered fully after soil thawing, varying from year to year as related to the differences in spring air temperatures. In autumn, lowering air temperature reduced photosynthetic capacity, but low photosynthesis in autumn is further related to the reduced radiation. Bergh et al. (1998) estimated that a 40% error in estimating the annual photosynthetic gain is possible if the annual cycle of photosynthetic capacity is excluded from model calculations. The elongation of the growing season, especially that in spring time, seems to be the most important factor in increasing photosynthetic production of conifers like Norway spruce under a changing climate (Hall et al. 2012).

References

Ainsworth, E. A. and S. P. Long. 2005. What have we learned from 15 years of free-air CO_2 enrichment (FACE)? A meta-analytic review of the responses of photosynthesis, canopy properties and plant production to rising CO_2. New Phytologist 165: 351–372.

Bergh, J., R. E. McMurtrie and S. Linder. 1998. Climatic factors controlling the productivity of Norway spruce: a model-based analysis. Forest Ecology and Management 110: 127–139.

Bergh, J. and S. Linder. 1999. Effects of soil warming during spring on photosynthetic recovery in boreal Norway spruce stands. Global Change Biology 5: 245–453.

Beuker, E., P. Hari, J. Holopainen, T. Holopainen, H. Hypén, H. Hänninen et al. 1996. Metsät. pp. 71–106. In: E. Kuusisto, L. Kauppi and P. Heikinheimo (eds.). Ilmastonmuutos ja Suomi. Helsinki University Press, Helsinki, Finland.

Cao, B., Q. -L. Dang and S. Zhang. 2007. Relationship between photosynthesis and leaf nitrogen concentration in ambient and elevated [CO_2] in white birch seedlings. Tree Physiology 27: 891–899.

Cure, J. D. and B. Accock. 1986. Crop responses to carbon dioxide doubling: a literature survey. Agricultural and Forest Meteorology 38: 127–145.

Curtis, P. S. 1996. A meta-analysis of leaf gas exchange and nitrogen in trees grown under elevated carbon dioxide. Plant, Cell and Environment 19: 127–137.

Dang, Q. -L., J. M. Maepea and W. H. Parker. 2008. Genetic variation of ecophysiological responses to CO_2 in *Picea glauca* seedlings. The Open Forest Science Journal 1: 68–79.

Ellsworth, D. S., R. Thomas, K. Y. Crous, S. Palmroth, E. Ward, C. Maier et al. 2012. Elevated CO_2 affects photosynthetic responses in canopy pine and subcanopy deciduous trees over 10 years: a synthesis from Duke FACE. Global Change Biology 18: 223–242.

Evans, J. R. 1989. Photosynthesis and nitrogen relationships in leaves of C_3 plants. Oecologia 78: 9–19.

Farquhar, G. D., S. von Caemmerer and J. A. Berry. 1980. A biochemical model of photosynthetic assimilation in leaves of C_3 species. Planta 149: 67–90.

Grassi, G., E. Vicinelli, F. Ponti, L. Cantoni and F. Magnani. 2005. Seasonal and interannual variability of photosynthetic capacity in relation to leaf nitrogen in a deciduous forest plantation in northern Italy. Tree Physiology 25: 349–360.

Hall, M., B. E. Medlyn, G. Abramowitz, O. Franklin, M. Räntfors, S. Linder et al. 2013. Which are the most important parameters for modeling carbon assimilation in boreal Norway spruce under elevated [CO_2] and temperature conditions? Tree Physiology 33: 1156–1176.

Hänninen, H., S. Kellomäki, K. Laitinen, B. Pajari and T. Repo. 1993. Effect of increased winter temperature on the onset of height growth of Scots pine: a field test of a phenological model. Silva Fennica 27: 251–257.

Harley, P. C., R. B. Thomas, J. F. Reynolds and B. R. Strain. 1992. Modelling photosynthesis of cotton grown in elevated CO_2. Plant, Cell and Environment 15: 271–282.

Hättenschwiler, S., F. Miglietta, A. Raschi and C. Körner. 1997. Thirty years of *in situ* tree growth under elevated CO_2: a model for future forest growth. Global Change Biology 3: 463–471.

Hyvönen, R., G. I. Ågren, S. Linder, T. Persson, M. F. Cotrufo, A. Ekblad et al. 2007. The likely impact of elevated [CO_2], nitrogen deposition, increased temperature and management on carbon sequestration in temperate and boreal forest ecosystems: a literature review. New Phytologist 173: 463–480.

Jarvis, P. G. 1993. Global change and plant water relations. pp. 1–13. In: M. Borgetti, J. Grace and A. Raschi (eds.). Water Transport in Plants under Climatic Stress. Cambridge University Press, Cambridge, UK.
Kellomäki, S., H. Väisänen, H. Hänninen, T. Kolström, R. Lauhanen, U. Mattila et al. 1992. Sima: a model for forest succession based on the carbon and nitrogen cycles with application to silvicultural management of the forest ecosystem. Silva Carelica 22: 1–85.
Kellomäki, S. and K. -Y. Wang. 1996. Photosynthetic responses to needle water potentials in Scots pine after a four-year exposure to elevated CO_2 and temperature. Tree Physiology 16: 765–772.
Kellomäki, S. and H. Väisänen. 1997. Modelling the dynamics of the boreal forest ecosystems for climate change studies in the boreal conditions. Ecological Modelling 97(1,2): 121–140.
Kellomäki, S. and K. -Y. Wang. 1997a. Effects of elevated O_3 and CO_2 on chlorophyll fluorescence and gas exchange in Scots pine during the third growing season. Environmental Pollution 97(1–2): 17–27.
Kellomäki, S. and K. -Y. Wang. 1997b. Effects of long-term CO_2 and temperature elevation on crown nitrogen distribution and daily photosynthetic performance of Scots pine. Forest Ecology and Management 99: 309–326.
Kellomäki, S. and K. -Y. Wang. 1997c. Effects of O_3 and CO_2 concentrations on photosynthesis and stomatal conductance in Scots pine. Plant, Cell and Environment 20: 995–1006.
Kellomäki, S. and K. -Y. Wang. 1997d. Photosynthetic responses of Scots pine to elevated CO_2 and nitrogen supply: results of a branch-in-bag experiment. Tree Physiology 17: 231–240.
Kellomäki, S. and K. -Y. Wang. 1998a. Daily and seasonal CO_2 exchange in Scots pine grown under elevated O_3 and CO_2: experiment and simulation. Plant Ecology 136: 229–248.
Kellomäki, S. and K. -Y. Wang. 1998b. Growth, respiration and nitrogen content in needles of Scots pine exposed to elevated ozone and carbon dioxide in the field. Environmental Pollution 101: 263–274.
Kellomäki, S., K. -Y. Wang and M. Lemettinen. 2000. Controlled environment chambers for investigating tree response to elevated CO_2 and temperature under boreal conditions. Photosynthetica 38: 69–81.
Lebourgeois, F., G. Lévy, G. Aussenac, B. Clerc and F. Willm. 1998. Influence of soil dying on leaf water potential, photosynthesis, stomatal conductance and growth of two black pine varieties. Annals of Forest Science 55: 287–299.
Long, S. P. 1991. Modification of the response of photosynthetic productivity to rising temperature by atmospheric CO_2 concentrations: has its importance been underestimated? Plant, Cell and Environment 14: 729–739.
Matala, J., R. Ojansuu, H. Peltola, H. Raitio and S. Kellomäki. 2006. Modelling the response of tree growth to temperature and CO_2 elevation as related to the fertility of current temperature sum of a site. Ecological Modelling 199: 39–52.
Medhurst, J., J. Parsby, S. Linder, G. Wallin, E. Ceschia and M. Slaney. 2006. A whole-tree chamber system for examining tree-level physiological responses of field-grown trees to environmental variation and climate change. Plant, Cell and Environment 29(9): 1853–1869.
Medlyn, B., P. G. Jarvis, F. Badeck, D. Pury, C. Barton, M. Broadmeadow et al. 1999. Effects of elevated CO_2 on photosynthesis in European forest species: a meta-analysis of model parameters. Plant, Cell and Environment 22: 1475–1495.
Medlyn, B. E., C. V. M. Barton, M. S. J. Broadmeadow, R. Ceulemans, P. De Angelis, M. Forstreuter et al. 2001. Stomatal conductance of forest species after long-term exposure to elevated CO_2 concentration: a synthesis. New Phytologist 149: 247–264.
Palomäki, V., A. Hassinen, M. Lemettinen, T. Oksanen, H. -S. Helmisaari, H. Holopainen et al. 1998. Open-top chamber fumigation system for exposure of field grown Pinus sylvestris to elevated carbon dioxide and ozone concentration. Silva Fennica 32(3): 205–214.
Pelkonen, P. and P. Hari. 1980. The dependence of the spring recovery of CO_2 uptake in Scots pine on temperature and internal factors. Flora 169: 398–404.
Pontailler, J. -Y., G. V. M. Barton, D. Durrent and M. Forstreuter. 1998. How can we study CO_2 impacts on trees and forests. pp. 1–28. In: P. G. Jarvis (ed.). European Forests and Global Change. The Likely Impacts of Rising CO_2 and Temperature. Cambridge University Press, Cambridge, UK.
Reich, P. B. 1987. Quantifying plant response to ozone: a unifying theory. Tree Physiology 3: 63–91.
Reich, P. B., M. B. Walters, M. G. Tjoelker, D. Van Der Klein and C. Buschena. 1998. Photosynthesis and respiration rates depend on leaf and root morphology and nitrogen concentration in nine boreal species differing in relative growth rate. Functional Ecology 12: 395–405.
Repo, T., H. Hänninen and S. Kellomäki. 1996. The effect of long-term elevation of air temperature and CO_2 on frost hardiness of Scots pine. Plant, Cell and Environment 19: 209–216.
Reynolds, J. F. and B. Accock. 1985. Predicting the response of plants to increasing carbon dioxide: a critique of plant growth models. Ecological Modelling 29: 107–129.
Rogers, A. and S. W. Humphries. 2000. A mechanistic evaluation of photosynthetic acclimation at elevated CO_2. Global Change Biology 6: 1005–1011.
Sage, R. F., T. D. Sharkey and J. R. Seemann. 1989. Acclimation of photosynthesis to elevated CO_2 in five C_3 species. Plant Physiology 89: 590–596.
Salisbury, F. B. and C. W. Ross. 1992. Plant Physiology. Wadsworth Publishing Company, Belmont, California, USA. Fourth Edition.
Smolander, H., P. Oker-Blom and S. Kellomäki. 1990. Typpipitoisuuden vaikutus männyn neulasten fotosynteesiin ja verson itsevarjostukseen. Abstract: The effect of nitrogen concentration on needle photosynthesis and within shoot shading in Scots pine. Silva Fennica 24(1): 123–128.

Stitt, M. 1992. Enhanced CO_2, photosynthesis and growth; what should we measure to gain a better understanding of the plant's response. pp. 3–28. *In*: D. E. Schulze and H. A. Mooney (eds.). Design and Execution of Experiments on CO_2 Enrichment. Ecosystem Research Report 6. Commission of the European Communities, Brussels. Belgium.

Troeng, E. and S. Linder. 1982. Gas exchange in a 20-year-old stand of Scots pine. I. Net photosynthesis of current and one-year-old shoots within and between seasons. Physiological Plantarum 54: 7–14.

Vapaavuori, E., T. Oksanen, J. K. Holopainen, T. Holopainen, J. Heiskanen, R. Julkunen-Tiitto et al. 2002. Technical report. Open-top chamber fumigation of cloned silver birch (*Betula pendula* Roth) trees to elevated CO_2 and ozone: description of the fumigation system and the experimental site. Metsäntutkimuslaitoksen Tiedonantoja 838: 1–128.

von Caemmerer, S. and G. D. Farquhar. 1981. Some relationships between the biochemistry of photosynthesis and the gas exchange of leaves. Planta 153: 376–387.

Wallin, G., M. Hall, M. Slaney, M. Räntfors, J. Medhurst and S. Linder. 2013. Spring photosynthetic recovery of boreal Norway spruce under conditions of elevated [CO_2] and temperature. Tree Physiology 33: 1177–1191.

Wang, K. -Y., S. Kellomäki and K. Laitinen. 1995. Effects of needles age, long-term temperature and CO_2 treatments on the photosynthesis of Scots pine. Tree Physiology 15: 211–218.

Wang, K. -Y. 1996a. Apparent quantum yield in Scots pine after four years of exposure to elevated temperature and CO_2. Photosynthetica 32: 339–353.

Wang, K. -Y. 1996b. Effects of long-term CO_2 and temperature elevation on gas exchange of Scots pine. Ph. D. Thesis, University of Joensuu, Joensuu, Finland.

Wang, K. -Y., S. Kellomäki and K. Laitinen. 1996. Acclimation of photosynthetic parameters in Scots pine after three years of exposure to elevated temperature and CO_2. Agricultural and Forest Meteorology 82: 195–217.

Wang, K. -Y. and S. Kellomäki. 2003. CLIMFOR: A Climate-Forest Model and Its Application. Sichuan Publishing House of Science and Technology, Chengdu, People's Republic of China.

Wang, K. -Y., S. Kellomäki, L. Chunyang and T. Zha. 2003. Light and water-use efficiencies of pine shoots exposed to elevated carbon dioxide and temperature. Annals of Botany 92: 53–64.

7

Response of Respiration to Climate Change

ABSTRACT

Autotrophic respiration, releasing of CO_2, indicates the synthesis of new mass and use of energy to maintain the living functions of existing mass. The short-term variability in maintenance respiration is related to temperature in an exponential way. However, the sensitivity of maintenance respiration rate to temperature may reduce when trees are growing under elevated CO_2. Conversely, elevated temperatures may increase the sensitivity of the respiration rate to temperature. Any increase of nitrogen content in living tissues seems to increase the respiration rate, regardless of the CO_2 and temperature conditions in the growing environment.

Keywords: respiration, growth respiration, maintenance respiration, basal respiration, Q_{10}, sensitivity of respiration, impact of elevated CO_2 on respiration losses

7.1 Respiration Losses of Carbon in Trees

Carbon in the growth and maintenance of tree functions and structure

Autotrophic respiration or simply to respiration is related to the growth and survival of trees. The release of CO_2 from tree organs indicates the use of energy in the synthesis of new mass (growth respiration), and to the use of energy to maintain living functions (maintenance respiration) in foliage, branches, stems, coarse roots and fine roots. In respiration, energy is released in the oxidation of carbohydrates:

$$C_6H_{12}O_6 + 6O_2 + 6H_2O = 6CO_2 + 12H_2O \qquad (7.1)$$

In Equation (7.1), the carbohydrate is glucose, which is an idealized case of an oxidizing carbohydrate for energy releasing CO_2 and H_2O. In this reaction, 2880 kJ of energy is released per molecule of glucose. Energy is mostly combined in ATP (adenosine triphosphate, see Chapter 6). This is usable in biosynthesis, where intermediates produced in the oxidation of glucose are used in the synthesis of carbohydrates, lipids and proteins for the growth of new tissues in different organs (growth respiration, R_g). Energy is also used, for example, in protein synthesis for preparing cell membranes and maintaining ion gradients in the uptake and transfer of nutrients (maintenance respiration, R_m). In modeling, maintenance respiration is generally first subtracted from photosynthesis and the remaining carbon is used to biomass growth (Ryan 1990, 1991, Ryan et al. 1994, 1997):

$$Ph = R_m + R_g + \Delta W = R + \Delta W \tag{7.2}$$

where Ph is the carbon in photosynthates, ΔW is the mass growth, and R is the sum of maintenance and growth respirations (autotrophic respiration), respectively. Ryan et al. (1997) estimate that annual autotrophic respiration may consume 50–70% of the net annual photosynthesis (NPP) or 60–80% of the annual gross photosynthesis in boreal forest ecosystems characterized by selected deciduous and coniferous species.

Growth respiration

Growth respiration is mainly related to the growth rate of organs, whereas maintenance respiration is related to the mass of the organ and temperatures affecting the organ. According to Mohren (1987), growth respiration (R_g) is: R_g = CPF x ΔW, where CPF is the CO_2 production factor in kg CO_2 per kg ΔW of mass growth. He estimated the CPF values for the needles, branches, stems and roots of Douglas fir [*Pseudotsuga menziesii* (Mirb.) Franco] as falling in the range of 0.3469–0.3866 [kg CO_2 (kg ΔW)$^{-1}$], the values for stems and branches being smaller than those for needles and roots. For example, the CO_2 produced [kg CO_2] in needle growth is: R_g = 0.3866 x ΔW. Rearranging Equation (7.2): ΔW = Ph – R_m – R_g = Ph – R_m – 0.3866 x ΔW. This yields: ΔW = 0.721 x (Ph – R_m). Consequently, the share of growth respiration is 27.9% and the share of mass growth 72.1% of the usable photosynthesis. The share of growth respiration is close to the value (25%) suggested by Ryan (1991) for estimating its share in annual total respiration losses (see further Ryan et al. 1997).

Maintenance respiration

The maintenance respiration rate (R_m) is sensitive to temperature, for example, the synthesis rate of proteins and the subsequent emission of CO_2 increase exponentially as a function of temperature (T, °C):

$$R_m(T) = R_o \times Q_{10}^{\left[(T-T_o)/10\right]} \tag{7.3}$$

where R_o is the basal (or reference) respiration rate at a given temperature (e.g., at 0°C or at some other reference temperature) and Q_{10} is the parameter indicating the change in respiration rate per increase of 10°C in temperature. The basal rate and subsequent respiration rate may be given per unit of mass, volume or area of the organs or tissues, for example, [µmol m^{-2} s^{-1}] per area. The rate of respiration is closely related to the nitrogen content of organs or tissues through the basal respiration rate: the synthesis and replacement of proteins with nitrogen cover more than 60% of maintenance respiration (Penning de Vries et al. 1974).

The values of Q_{10} commonly fall in the range 1.6–3.0 shown by Ryan et al. (1994) for several pines (*Pinus* sp.) common in boreal, temperate and subtropical climates. However, the value of Q_{10} is dependent on the plant tissue; i.e., in foliage, the value of Q_{10} varies in the range 1.39–2.6, whereas the values in stems are 1.8–2.9. The value 2 is widely used in calculating the rate of maintenance respiration as a function of temperature (Waring and Running 2007). However, the values of Q_{10} may acclimate to short-term increase in temperature across diverse plant taxa including several boreal tree species. The consequent value of Q_{10} is: $Q_{10} = 3.22-0.047 \times T$, where T is the midpoint between the recorded temperature and the reference temperature (Tjoelkel et al. 2001, Atkin and Tjolker 2003, Wythers et al. 2005, Wythers et al. 2013). The temperature correction of Q_{10} is needed to improve assessment of how the carbon emissions in maintenance respiration are responding to climatic warming.

Seasonality of respiration

Under boreal conditions, the seasonal course of maintenance and growth respiration follows the annual pattern of temperature and the timing of growth. Using young boreal Norway spruce in a fertilizing and irrigation experiment, Stockfors and Linder (1998) found that the rate of growth respiration of stems (per the area basis) was the main source of respiratory CO_2 from mid-June through mid-September, with the largest values in July (Fig. 7.1). Most CO_2 emissions originated from phloem, with the highest rate in mid-May due to high air temperatures. Over the growing season, growth respiration comprised 59% (13.7 mol m^{-2}) of the total respiration, while the maintenance respiration in phloem was 30% (7.0 mol m^{-2}) and in xylem 11% (2.5 mol m^{-2}). Similarly, Maier (2001) found that in the fertilized loblolly pines (*Pinus taeda* L.) the annual growth respiration was 40% of the total respiration. Maier (2001) found further that nitrogen fertilization increased stem respiration substantially as a function of stem nitrogen content. Stem growth rate and nitrogen content in stems explained 75% of the seasonal variation in temperature-normalized respiration given on the stem surface area. Similarly, nitrogen fertilization increased the stem growth and the consequent respiration in boreal Norway spruce (Stockfors and Linder 1998).

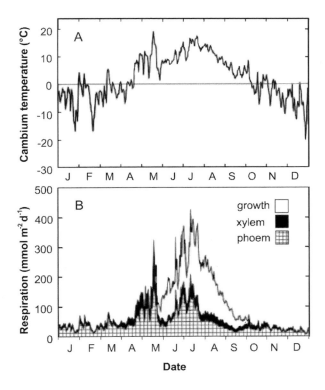

Fig. 7.1 Above: Annual variation of daily stem temperature in boreal Norway spruce, and Below: Daily stem respiration split into growth respiration in stem and maintenance respiration in xylem and phloem (Stockfors and Linder 1998). Permission of Oxford University Press.

Similarly, Zha et al. (2004) measured the intra- and inter-annual stem respiration in 50-year-old boreal Scots pines through April to November in 2001–2003. They found that diurnal, seasonal and inter-annual changes peaked at 1,600 hours during the day, with the highest values in July. The annual Q_{10} remained relatively constant at about 2 over the three years, while respiration at the 15°C (R_{15}) was higher in the growing than in the non-growing season (1.09 compared with 0.78 μmol C m^{-2}s^{-1} on stem surface), regardless of years. Maintenance respiration ranged from 76 to 80% of the total respiration (17.46–19.35 C mol m^{-2} on stem surface), respectively. The annual total stem respiration of the stand per unit ground area was 76.0–74.3 g C m^{-2}. Respiration of stem wood contributed 9% to total carbon loss from the whole ecosystem, comprising 8% of the carbon loss in gross primary production.

Balance of carbon between respiration and net primary production

In a whole-tree context, respiration is distributed unevenly between organs depending on their physiological activity and mass. This suggests that the maintenance respiration of foliage per biomass unit is greater than in other organs, as found by Janssens et al. (2005) for Scots pine seedlings grown in ambient CO_2 and elevated CO_2 (ambient + 400 ppm). Figure 7.2 shows that most carbon was allocated to foliage, and least to fine roots. In the third and fourth year since launching the experiment, the share of maintenance respiration was 58–64% of the total carbon (i.e., maintenance respiration, growth respiration and net primary production) used in ambient CO_2.

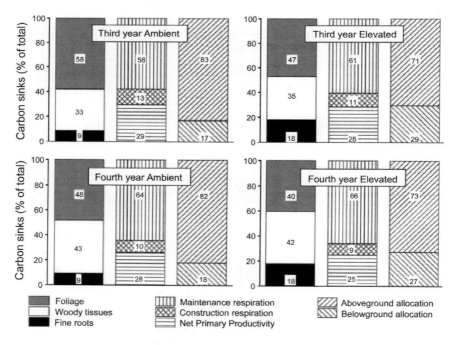

Fig. 7.2 Relative allocation of carbon (percentage of total accountable sink) to different compartments of young Scots pines and different processes during the third (upper panels) and fourth (lower panels) years since the launch of treatment, representing the ambient (left panels) or elevated (right panels) CO_2 (Janssens et al. 2005). Permission of Oxford University Press. Legend: in the upper and lower panels: (i) the bars on the left indicate the proportional carbon used in the different tissues; (ii) the bars in the middle indicate the relative carbon allocation to biomass production, maintenance respiration and construction (growth) respiration in different biomass compartments; and (iii) the bars on the right indicate the proportional carbon use in the aboveground tissue vs. belowground tissues.

Janssens et al. (2005) found further that in elevated CO_2 the share of maintenance respiration was slightly higher, even though the respiration rate was slightly lower than under ambient CO_2; i.e., the increased biomass outweighed the reduction in the respiration rate. The elevated CO_2 substantially enhanced the fine root growth, whose share was doubled compared to that under ambient CO_2. Regardless of the treatment and year, the share of maintenance respiration per year was 70–80%, indicating that the woody parts in stems and branches were mainly sapwood, with physiologically active cells. Similarly, Carey et al. (1996) found that in Ponderosa pine (*Pinus ponderosa* Douglas & C. Lawson) maintenance respiration was 79% of the total respiration under ambient conditions (CO_2, 350 ppm) and 83% under elevated CO_2 (ambient + 350 ppm).

7.2 Response of Respiration in Foliage under Elevated CO_2 and Temperature

Respiration responses over growing season in growing needles

Climate change affects plants through the temperature-controlled respiration rate, which may be changed (acclimated) in a growing condition with elevated CO_2 or temperature (Janssens et al. 2005). Furthermore, changes in the accumulation of mass under climate change affect the total maintenance respiration. At the same time, tree architecture may be modified with alterations in the amount and variation in the properties of leaves/needles and their position in the canopy. For example, Wang et al. (2002) found that the long-term exposure of Scots pines to elevated CO_2 and temperature affected the distribution of needles, their nitrogen content and other physiological properties controlling the rate of respiration and total respiration of foliage in response to prevailing temperature.

Under changing climate, the seasonal course of respiration follows the annual pattern of temperature and the timing of growth. Zha et al. (2001) found that the growth patterns of needles in boreal Scots pine were fairly similar, regardless of climatic treatment (Fig. 7.3). However, the elevation of CO_2 and temperature, either alone or combined, increased the area and dry weight of needles. At the end of the growing season, the needle area in trees grown under elevated CO_2 was 12% larger than under ambient conditions. Similarly, the increase was 35% under elevated temperature and 33% under the combined elevation of CO_2 and temperature. At the same time, the biomass of mature needles increased 20, 53 and 36% in these treatments. The changes in mass and area suggest that the specific needle area (SLA, cm^2 g^{-1}) decreased gradually during needle expansion. At the end of growing season, the value of SLA under ambient conditions was 45 cm^2 g^{-1} while 39, 52 and 51 cm^2 g^{-1} under elevated CO_2 and elevated temperature alone or combined. This implied that the value of SLA reduced under elevated CO_2 alone by 2–15% compared to that under ambient conditions, but increased by 3–16% under elevated temperature (alone or combined with CO_2).

Zha et al. (2001) showed that respiration rate in boreal Scots pine needles increased similarly in response to temperature, regardless of climatic treatment (Fig. 7.4, left). However, the level (intercept) was higher under elevated temperature alone or combined with elevated CO_2 than under other treatments. Furthermore, the sensitivity of respiration rate to temperature was slightly affected by climatic treatment. The mean Q_{10} value in June and July (midsummer) measurements was 2.09 for ambient conditions, 2.09 for elevated CO_2, 1.78 for elevated temperature and 1.85 for the elevation of both climatic factors. The values were slightly different in June than in July, but in both cases the impact of climatic treatment was similar (Zha et al. 2001). The reduction of the Q_{10} value under temperature elevation alone or combined with elevated CO_2 may imply the downward acclimation of the response of maintenance respiration to temperature (Tjoelker et al. 1999, 2001, Zha et al. 2003).

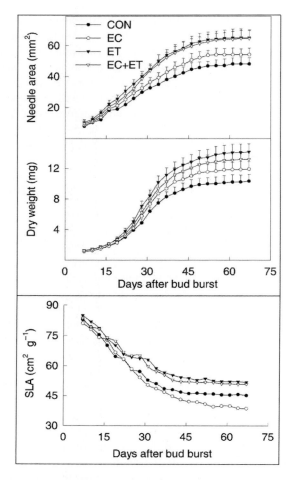

Fig. 7.3 Accumulation of area (upper), mass (middle) and specific needle area (lower) of boreal Scots pine needles grown under elevated CO_2 (EC) and temperature (T) and the combined elevation of CO_2 and T (CT + ET) (Zha et al. 2001). Permission of Oxford University Press.

The total respiration rate (combined growth and maintenance respiration) in the needles increased through the early expansion period, with the maximum values occurring in four weeks since the buds burst (Fig. 7.4, right). Thereafter, the respiration rates declined rapidly, and the values stabilized six weeks after the buds burst. Throughout the monitoring period, the values in the needles grown under ambient conditions were smaller than in those grown under elevated CO_2 or temperature. In the latter case, respiration was the greatest throughout the needle expansion. Similarly, the elevation of CO_2 increased respiration compared to ambient conditions, as was the case when the temperature alone or combined with CO_2 were elevated. In both cases, the respiration rate was, however, greater than in ambient conditions. Over the whole monitoring period, elevated temperature alone or combined with elevated CO_2 increased respiration rate by 5–30% on a dry weight and area basis, compared to that in ambient conditions. The elevation of CO_2 alone reduced respiration rate by 5–20% (Zha et al. 2001).

Response of Respiration to Climate Change 95

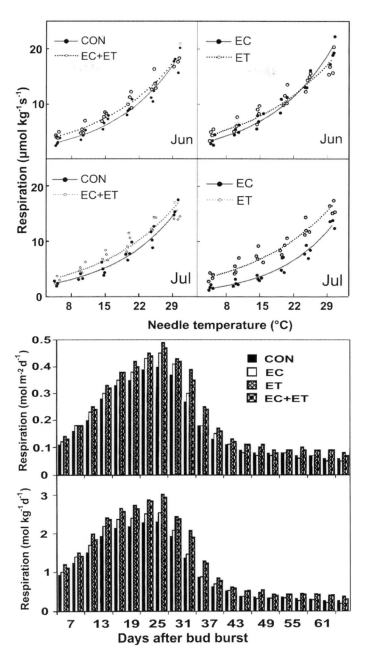

Fig. 7.4 Above: Respiration rate of Scots pine needles grown under ambient conditions (CON), elevated CO_2 (EC), elevated temperature (ET) and the combined elevation of CO_2 and temperature (EC + ET) as a function of measuring temperature. Measurements were done in the early (June 29–30) and late (July 30–31). Below: Daily values of combined growth and maintenance respiration rates on an area and dry weight basis through the expansion period of needles growing under elevated CO_2 and temperature (Zha et al. 2001). Permission of Oxford University Press.

Relationship between growth and maintenance respiration in needles

Zha et al. (2001, 2003) further used the two-component respiration model to determine how growing under elevated CO_2 and temperature affect needle respiration in boreal Scots pine, assuming that growth respiration is proportional to growth, and maintenance respiration to the mass of needles (Amthor 1989, Sprugel 1990, Wullschleger et al. 1992). The total respiration rate (R) over the growing season (the combined growth and maintenance ones) in the model is (Fig. 7.5; Table 7.1):

$$R = R_g \times \frac{dW}{dt} + R_m \times W \qquad (7.4)$$

where R_g (mol CO_2 kg^{-1}) and R_m (mol CO_2 kg^{-1} d^{-1}) are the coefficients for growth respiration and maintenance respiration, and W [kg] is the mass of needles grown in a given year. Dividing both sides by the mass yields:

$$\frac{R}{W} = R_g \times \frac{dW}{dt} \times \frac{1}{W} + R_m \qquad (7.5)$$

where R/W is the specific respiration rate (SRR, mol CO_2 kg^{-1} d^{-1}) and dW/dt x 1/W is the specific growth rate (SGR, d^{-1}). Consequently:

$$SRR = R_g \times SGR + R_m \qquad (7.9)$$

where the Specific Growth Rate (SGR) is:

$$SGR = \frac{\ln W_2 - \ln W_1}{t_2 - t_1} \qquad (7.6)$$

where W_1 and W_2 are the mass of needles at the points of time t_1 and t_2.

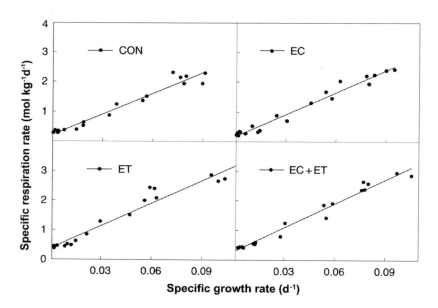

Fig. 7.5 Relationship between specific respiration rate (SRR) and specific growth rate (SGR) for boreal Scots pine needles grown under elevated temperature and elevated CO_2. The slope of the regression represents the growth coefficient (mol kg^{-1}) and the intercept represents the maintenance coefficient (mol kg^{-1} d^{-1}) (Zha et al. 2001). Permission of Oxford University Press. Legend: CON is ambient conditions, EC elevated CO_2, ET elevated temperature and EC + ET the combined elevation of both.

The two-component respiration model (Fig. 7.5; Table 7.1) showed that the specific respiration (SRR) in Scots pine was a linear function of the Specific Growth Rate (SGR) regardless of climatic treatment (Zha et al. 2001). The combined elevation of CO_2 and temperature increased the growth respiration by 16%, while elevated CO_2 and temperature alone increased the growth respiration by 6 and 11%, respectively. At the same time, maintenance respiration reduced significantly under elevated CO_2 (35%) alone opposite to elevated temperature alone and combined with elevated CO_2. Both growth and maintenance respiration correlated closely with the specific growth rate regardless of climatic treatment (Zha et al. 2001).

Table 7.1 Growth (R_g) and maintenance (R_m) coefficients of respiration for the growing needles of Scots pine exposed to elevated temperature (T) and CO_2 (Zha et al. 2001). The percentage change compared to the values for ambient conditions is in parenthesis.

Treatment	R_g, mol kg^{-1}	R_m mol kg^{-1} day^{-1}
Ambient	23.03	0.26
CO_2	24.35 (+6)	0.17 (−35)
T	25.51 (+11)	0.37 (+42)
CO_2 +T	26.77 (+16)	0.33 (+27)

Respiration vs. nitrogen content of foliage

There is a clear respiration/nitrogen scaling relationship regardless of higher land plants (Reich et al. 1998, Reich et al. 2008). However, the nitrogen concentration is organ-specific, e.g., lower in leaves than in stems and roots, but varying in organs. For example, the respiration in Scots pine needles is affected how the needles are distributed in the crown and how their nitrogen content varies in different parts of the crown. Wang et al. (2002) found that elevated CO_2 shifted the maximum needle area in Scots pine towards the crown base, whereas temperature elevation shifted the maximum needle area towards the crown top (Fig. 7.6). The total foliage area per tree increased by 11% under

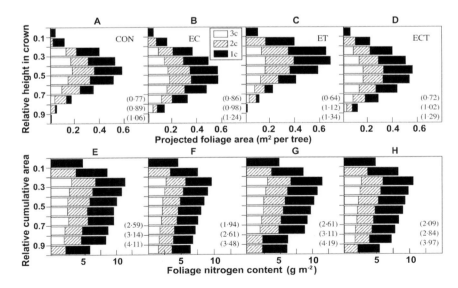

Fig. 7.6 Distribution of foliage area (upper) and foliage nitrogen (lower) in current-year (1c), one-year-old (2c), and older needles (3c) in the crowns of Scots pines grown under four climatic treatments: A and E are ambient CO_2 and temperature; B and F elevated CO_2 concentration (EC) but ambient temperature; C and G elevated temperature (ET) but ambient CO_2; and D and H elevated CO_2 and temperature (ECT). The values are the means of measurements from four trees per treatment. The numbers in parentheses refer to the needle area (m² in tree, upper Figure) per age class and nitrogen (g N m^{-2}, lower Figure) per needle age class (Wang et al. 2002). Permission of Oxford University Press.

elevated CO_2, 20% under elevated temperature and 14% under the combined elevation of CO_2 and temperature relative to ambient conditions. At the same time, both elevated CO_2 alone, or combined with elevated temperature, decreased the mean annual concentration of nitrogen in needles by 16 and 7%, whereas the reduction was only slight (1.2%) under elevated temperature alone. The total nitrogen content per tree clearly increased due to the increased needle area, regardless of treatment: 11% under elevated CO_2, 21% under elevated temperature and 14% under the combined elevation of CO_2 and temperature, compared to ambient conditions.

Regardless of climatic treatment, the respiration rate in Scots pine at a reference temperature of 20°C increased linearly along with the increasing nitrogen content (Fig. 7.7). Under elevated temperature alone or combined with elevated CO_2, the level (intercept) was clearly higher than under ambient conditions or elevated CO_2. In the latter cases the slope of regression was greater than in the former cases (Wang et al. 2002). The temperature elevation alone seemed to make the respiration more sensitive to the increase of nitrogen content, whereas the CO_2 elevation reduced the sensitivity. Zha et al. (2002a,b) found that elevated CO_2 increased the slope of respiration against the nitrogen content relative to ambient conditions. They also found that elevated temperature had no effect on the slope, but the combined elevation of temperature and CO_2 increased both the slope and the intercept.

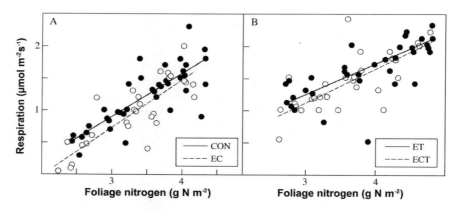

Fig. 7.7 Relationship between respiration and nitrogen content in Scots pine needles grown under ambient conditions (CON), elevated temperature (ET), elevated CO_2 (EC) and the combined elevation and temperature and CO_2 (ECT). The values of respiration rates are given at the reference temperature of 20°C (Wang et al. 2002). Permission of Oxford University Press. Legend: full circles are for ambient conditions and open circles for elevated CO_2 and temperature (T) and for the combination of both.

Respiration vs. needle age

Figure 7.8 shows the respiration response of boreal Scots pine needles of varying age to temperature after the growing season (August–September), when trees were grown under elevated CO_2 and temperature (Zha et al. 2002a,b). Regardless of the needle age class, the respiration rate increased exponentially in response to increasing temperature, but the response was much lower in older needles than in the current-year needles regardless of climatic treatment. This held for both the intercept (level) and the sensitivity (slope). Relative to respiration in older needles, the respiration in the current-year needles (at the middle and late parts of the growing season) was reduced slightly under elevated CO_2 but increased clearly in response to elevated temperature alone or combined with elevated CO_2. The age of needles had a greater effect on respiration than any climatic treatment, however. The difference in respiration between older and younger needles correlated with the difference in nitrogen content.

Response of Respiration to Climate Change 99

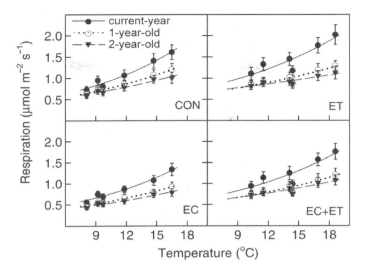

Fig. 7.8 Response of respiration to temperature in boreal Scots pine needles of varying age grown under current climate (CON), elevated temperature (T) and CO_2 (EC) or combined (EC + ET) (Zha et al. 2002a). Permission of Oxford University Press. The values are the means of four replicates. The curves are based on Equation $r = r_o \exp(kT)$, where r_o and k are coefficients specific to climatic treatment and age of needles and $\exp(kT)$ is the temperature coefficient (Q_{10}).

Table 7.2 Respiration (R_{20}) at a temperature of 20°C and the values of Q_{10}, as a function of needle age and climatic treatment (Zha et al. 2002a,b). The numbers in parenthesis indicate the percentage in relation to the values under ambient conditions.

Climatic treatment	R_{20} for needles of varying age, µmol m^{-2} s^{-1}			Q_{10} for needles of varying age		
	Current	One year	Two years	Current	One year	Two years
Ambient	2.22	1.57	1.29	2.44	2.08	1.83
CO_2	1.82 (82)	1.21 (77)	1.02 (79)	2.50 (102)	2.14 (103)	1.94 (106)
T	2.16 (97)	1.37 (87)	1.16 (90)	2.03 (83)	1.68 (81)	1.44 (79)
CO_2 + T	1.91 (86)	1.29 (82)	1.14 (88)	2.13 (87)	1.83 (88)	1.64 (99)

Table 7.2 shows further how climatic treatment affected the values of Q_{10} and respiration at a temperature of 20°C (R_{20}) as a function of needle age. CO_2 elevation increased the Q_{10} value, whereas elevated temperature alone or combined with elevated CO_2 reduced the value regardless of needle age. At the same time, the value of respiration at a temperature of 20°C reduced under elevated CO_2 alone or combined with elevated temperature. The reduction pattern was similar for all the needle age classes included in the measurements. Several studies show that the Q_{10} values tend to acclimate to elevated temperature, thus reducing carbon losses under warming climate if no acclimation occurs (Teskey and Will 1999, Tjoelker et al. 1999, Tjoelker et al. 2001, Nedlo et al. 2009).

Respiration vs. needle position in crown

Zha et al. (2002b) also found that the respiration rate of current-year needles of boreal Scots pine reduced from the crown surface towards the stem and from the crown top towards the crown bottom regardless of climatic treatment (Fig. 7.9). The respiration rate reduced in elevated CO_2 relative to the needles grown in ambient conditions, and increased under elevated temperature alone or combined with elevated CO_2. Zha et al. (2002b) found that the respiration rate increased along

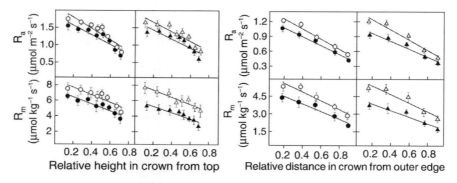

Fig. 7.9 Response of respiration to temperature in boreal Scots pine needles grown in ambient conditions (closed circles), elevated temperature (T, open circles), CO_2 (EC, closed triangles) or combined (EC + ET, open triangles) (Zha et al. 2002b) as a function of the position in crown. Left: from stem top downwards, and Right: from crown surface towards stem. The values are the means of four replicates. The curves are based on Equation $r = r_o \exp(kT)$, where r_o and k are coefficients specific to climatic treatment and age of needles and $\exp(kT)$ is the temperature coefficient (Q_{10}). Permission of Oxford University Press.

with the increasing nitrogen content regardless of position in the crown or climatic treatment. Furthermore, the reduction in the specific leaf area indicated the reduction in respiration rate. This may suggest that the acclimation of respiration rate is affected by the light conditions under which the expanding needles (current-year needles) are growing (Zha et al. 2002a,b).

7.3 Response of Respiration in Woody Parts to Elevated CO_2 and Temperature

Respiration in woody parts of trees involves the CO_2 efflux from sapwood, cambium and phloem in stems, branches and coarse roots. In boreal Scots pines, for example, branches survive 40–50 years: longer on poor sites than on fertile sites and longer in northern than in southern sites (Kellomäki and Väisänen 1988). In maturing branches, the share of phloem and sapwood tissues with a high respiration rate (Pruyn et al. 2002) will decrease and inactive heart wood increase even though the total mass of branches increases in maturing trees. This is true also for stems and probably for coarse roots (the supporting roots) of maturing Scots pines. The sapwood dominates the stem mass in young trees, whereas in mature trees the percentage of sapwood is less than 50% (Kärkkäinen 2007). The percentage of phloem and sapwood tissues in mature Scots pines reduces from the stem apex down to the stem butt.

The CO_2 efflux from the stem wood of coniferous and deciduous trees in varying climates seems to be linearly related to the volume and temperature of sapwood (Ryan 1990, Ryan and Waring 1992, Ryan et al. 1995, 1996, Lavigne and Ryan 1997, Bosc et al. 2003). This holds for boreal Scots pine as shown by Zha et al. (2005). They found that the total respiration in Scots pine stems was closely correlated to the temperature in sapwood just below the bark following the seasonal pattern of air temperature, regardless of climate treatment (Fig. 7.10). Respiration rates had the maximum values in early July, with values ranging from 1.0 to 1.4 µmol m^{-2} s^{-1}, with the highest values under elevated CO_2 and temperature. Respiration for the reference temperature (R_{15}) was greater in the growing season (May–August) (0.4–0.5 µmol m^{-2} s^{-1}) than in the non-growing season (September) (0.2–0.4 µmol m^{-2} s^{-1}), whereas the temperature coefficient (Q_{10}) was lower in the growing season (1.8–2.2) than in the non-growing season (2.2–2.6). The integration of the respiration rate of Scots pine stem wood over a year showed that elevated CO_2 and temperature alone increased respiration by 18 and 15%, and the combined elevation of both factors by 21% (Zha et al. 2005).

The total CO_2 efflux from stem is closely related to the increasing biomass, which may outweigh the climate-induced changes in the respiration rates. Zha et al. (2005) found that maintenance respiration comprised 65–69% of the annual total respiration in mature Scots pines, the lowest value representing trees grown under ambient conditions and the largest values the trees grown under elevated CO_2. The elevation of CO_2 alone or concurrently with temperature increased the maintenance respiration (22%) and the growth respiration (9%). Edwards et al. (2002) found that growth under elevated CO_2 mainly increased maintenance respiration in stem wood, probably as a result of new supporting tissues from the growth.

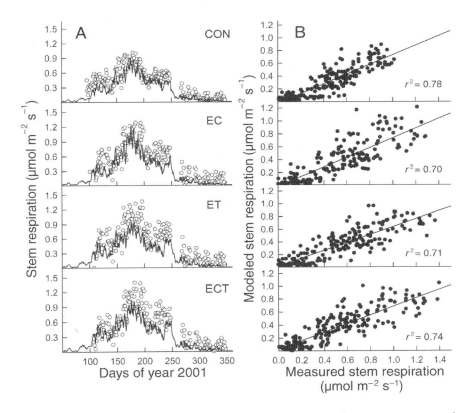

Fig. 7.10 A: Measured (dots) and calculated (line) values of respiration from Scots pine stems over a year under ambient conditions (CON), elevated CO_2 (EC), elevated temperature (ET) and combined both (ECT) climatic treatments. B: Relationship between the measured and calculated values for respiration in Scots pine stems (Zha et al. 2005). Permission of Oxford University Press.

Table 7.3 Annual growth (R_g) and maintenance (R_m) respiration and total respiration (Total) rate in Scots pine stem wood grown under varying climatic conditions in the Mekrijärvi chamber experiment (Zha et al. 2005). The numbers in parenthesis show the percentage contribution of maintenance and growth respiration to the annual respiration.

Treatment	R_g, mol m^{-2} yr^{-1}	R_m, mol m^{-2} yr^{-1}	Total, mol m^{-2} yr^{-1}
Ambient	2.32 (35)	4.22 (65)	6.55 (100)
CO_2	2.53 (33)	5.16 (67)	7.69 (100)
T	2.29 (31)	5.20 (69)	7.50 (100)
CO_2 + T	2.61 (33)	5.29 (67)	7.90 (100)

7.4 Response of Respiration in Roots to Elevated CO_2 and Temperature

Fine roots form a major sink of the total carbon taken up by trees, as demonstrated by Högberg et al. (2001) for Scots pine. They girdled trees by stripping the stem bark to the depth of xylem, thus curbing the supply of photosynthates to roots with mycorrhizal fungi. Consequently, the soil respiration declined about 50% compared to the stand with ungirdled trees, but was otherwise similar. Thus, respiration in fine roots contributed greatly to the soil respiration, and fine roots cycled a large part of carbon in the current canopy photosynthesis back to the atmosphere in a few days (Pregitzer et al. 2000, Högberg et al. 2001, George et al. 2003). Combining the findings about girdling with the carbon budget in the Swedish Coniferous Forests (SWECON) project, Högberg et al. (2002) suggested that 75% of the carbon allocated to the roots of Scots pines is respired and 25% used for growth. Janssens et al. (2002) estimated that in the temperate conditions the growth of fine roots of a 70-year-old Scots pine stand used 9–12% of the gross primary production and 26–31% of the net primary production. Root respiration rate, nitrogen uptake rate, and specific root length (i.e., root length per root mass) are closely correlated with each other, thus showing the close links between root respiration and primary production in canopy (Reich et al. 1998).

In general, elevated CO_2 increases the growth and mass of fine roots and the consequent fine root sink (Saxe et al. 1998, Janssens et al. 2005). This suggests an increase in soil respiration as found by Drake et al. (2008) in the Duke FACE experiment for loblolly pine (*Pinus taeda* L.). In this case, fine root respiration comprised 43% of the yearly soil respiration under ambient CO_2, whereas the same share was 50% under elevated CO_2. However, the specific respiration rate in fine roots is likely to be reduced as found by Hamilton et al. (2002) in the Duke FACE experiment. In this case, elevated CO_2 reduced mean respiration rate (measurements done at 25°C) by 23% relative the values in the ambient CO_2. The reduction was relative to the reduced nitrogen in roots under elevated CO_2. Similarly, Qi et al. (1994) found the fine root respiration in Douglas fir [*Pseudotsuga menziesii* (Mirb.) Franco] to reduce by 4–5 nmol CO_2 per gram of root dry weight for every doubling of CO_2 in the soil. Respiration rate at a CO_2 concentration of 5000 ppm in soil was only one tenth of that under ambient conditions. Similarly, root respiration in sugar maple (*Acer saccharum* Marshall) reduced rapidly along with increasing soil CO_2 (Burton et al. 1997). The respiration rate was the most sensitive below soil CO_2 concentrations < 1500 ppm.

In general, soil respiration rate including root respiration is related to temperature like above-ground organs but the response of respiration of roots to temperature seems to be more sensitive to climate warming than that of the organs above ground. Boone et al. (1998) found that the Q_{10} for root respiration and rhizosphere decay in a mixed temperate forest is 4.6. The value is clearly higher than that (2–3) commonly used for other organs of trees and higher than that found in some other studies (Zogg et al. 1996). High sensitivity of fine roots to warming is likely limiting carbon sequestration in soil, whereas elevated CO_2 and temperature are likely to increase carbon uptake and forest growth in boreal zone (Melillo et al. 2002).

7.5 Concluding Remarks

Autotrophic respiration involves CO_2 emissions due to the use of energy in the synthesis of new mass (growth respiration) and the use of energy to maintain the living functions of existing mass (maintenance respiration). Respiration occurs throughout living tissues in tree organs (foliage, branches, stems and roots), but its response to climate change (elevated CO_2 and temperature) is variable, related to the growth rate and the properties and amount of the accumulated tissue. In this respect, the response of non-woody (leaves/needles, fine roots) and woody (branches, stems, coarse roots) organs are most likely different. However, the respiration rate in different organs seems to be related to the nitrogen content of any living tissue in the organs (Ryan et al. 1996, Makita et al. 2009).

Elevated CO_2 seems to reduce the specific respiration rate in boreal Scots pine needles, mostly in older needles (Janssens et al. 2005). Such a downward acclimation is probably related to length of exposure with reduced nitrogen content. The impacts of climate change on respiration rate are further dependent on the position of leaves/needles in crown (e.g., upper/lower crown, surface/inner crown), making it unclear how the whole crown responds to elevated CO_2 and temperature. The elevation of CO_2 also seems to increase and the elevated temperature to decrease the life span of leaves/needles, e.g., in boreal Scots pine (Wang et al. 1996). The longevity of needles and the changes in longevity may thus play an important role in adapting conifers to a warming climate.

Even in seedlings, the stems form the bulk of the woody tissues in trees. Edwards et al. (2002) found that CO_2-induced increase in stem respiration is related to the enhanced supply of photosynthates and consequent stem growth. Similarly, Zha et al. (2005) found that the increase in stem respiration in Scots pine coincided with an increase in stem growth under elevated CO_2 and temperature. However, elevated CO_2 alone, or combined with elevated temperature, increased only maintenance more than growth respiration, whereas elevated temperature only increased maintenance respiration. This refers to the fact that the elevated temperature increases the biosynthesis of compounds in living tissues of cambium, phloem and wood on the stem surface just below bark.

References

Amthor, J. S. 1989. Respiration and Crop Productivity. Springer-Verlag, New York, USA.
Atkin, O. K. and M. G. Tjoelker. 2003. Thermal acclimation and the dynamic response of plant respiration to temperature. Trends in Plant Science 8(7): 343–351.
Boone, R. D., K. J. Nadelhoffer, J. D. Canary and J. P. Kaye. 1998. Roots exert a strong influence on the temperature sensitivity of soil respiration. Nature 396: 570–572.
Bosc, A., A. de Grandcourt and D. Loustau. 2003. Variability of stem and branch maintenance respiration in a *Pinus pinaster* tree. Tree Physiology 23: 227–236.
Burton, A. J., G. P. Zogg, K. S. Pregitzer and D. R. Zak. 1997. Effect of measurement CO_2 concentration on sugar maple root respiration. Tree Physiology 17: 421–427.
Carey, E. V., E. H. DeLucia and J. T. Ball. 1996. Stem maintenance and construction respiration in *Pinus ponderosa* grown in different concentrations of atmospheric CO_2. Tree Physiology 16: 125–130.
Drake, J. E., P. C. Stoy, R. B. Jackson and E. H. DeLucia. 2008. Fine-root respiration in a loblolly pine (*Pinus taeda* L.) forest exposed to elevated CO_2 and N fertilization. Plant, Cell and Environment 31: 1663–1672.
Edwards, N. T., T. J. Tschaplinski and R. J. Norby. 2002. Stem respiration increases in CO_2-enriched sweetgum trees. New Phytologist 155: 239–248.
George, K., R. J. Norby, J. G. Hamilton and E. H. DeLucia. 2003. Fine-root respiration in a loblolly and sweetgum forest growing in elevated CO_2. New Phytologist 160: 511–522.
Hamilton, J. G., E. H. DeLucia, K. George, S. L. Naida, A. C. Finzi and W. H. Schlesinger. 2002. Forest carbon balance under elevated CO_2. Oecologia 131: 250–260.
Högberg, P., A. Nordgren, N. Buchmann, A. F. Taylor, A. Ekblad, M. N. Högberg et al. 2001. Large-scale forest girdling shows that current photosynthesis drives soil respiration. Nature 411: 789–792.
Högberg, P., A. Nordgren and G. I. Ågren. 2002. Carbon allocation between tree growth and root respiration in boreal pine forests. Oecologia 132: 579–581.
Janssens, I. A., D. A. Sampson, J. Curiel-Yuste, A. Carrara and R. Ceulemans. 2002. The carbon cost of fine root turnover in a Scots pine forests. Forest Ecology and Management 168: 231–240.
Janssens, I. A., B. Medlyn, B. Gielen, I. Laureysens, M. E. Jach, D. Van Hove et al. 2005. Carbon budget of *Pinus sylvestris* sapling after four years' exposure to elevated atmospheric carbon dioxide concentration. Tree Physiology 25: 325–337.
Kärkkäinen, M. 2007. Puun rakenne ja ominaisuudet. Metsäkustannus. Karisto, Hämeenlinna, Finland.
Kellomäki, S. and H. Väisänen. 1988. Dynamics of branch population in the canopy of young Scots pine stands. Forest Ecology and Management 24: 67–83.
Lavigne, M. B. and M. G. Ryan. 1997. Growth and maintenance respiration of rates of aspen, back spruce and jack pine at northern and southern BOREAS sites. Tree Physiology 17: 543–551.
Maier, C. A. 2001. Stem growth and respiration in loblolly pine plantations differing soil resource availability. Tree Physiology 21: 1183–1193.

Makita, N., Y. Hirano, M. Dannourra, Y. Kominami, T. Mizoguchi, H. Ishii et al. 2009. Fine root morphology traits determine variation in root respiration of *Quercus serrata*. Tree Physiology 29: 579–585.

Melillo, J. M., P. A. Steudler, J. D. Aber, K. Newkirk, H. Lux, F. P. Bowles et al. 2002. Soil warming and carbon-cycle feedbacks to climate systems. Science 298: 2173–2175.

Mohren, G. M. J. 1987. Simulation of forest growth applied to Douglas fir stands in the Netherlands. Ph. D. Thesis, Agricultural University of Wageningen, The Netherlands.

Nedlo, J. E., T. A. Martin, J. M. Vose and R. O. Teskey. 2009. Growing season temperatures limit growth of loblolly pine (*Pinus taeda* L.) seedling across a wide geographic transect. Trees 23(4): 751–759.

Penning de Vries, F. W. T., A. Brunsting and H. H. Laar. 1974. Products, requirements and efficiency of biosynthesis: a quantitative approach. Journal of Theoretical Biology 45: 339–377.

Pregitzer, K. S., J. S. King, A. J. Burton and S. E. Brown. 2000. Responses of tree fine roots to temperature. New Phytologist 147: 105–115.

Pruyn, M. G., B. L. Gartner and M. E. Harmon. 2002. Respiratory potential in sapwood of old versus young ponderosa pine trees in the Pacific Northwest. Tree Physiology 22: 105–116.

Qi, J., J. D. Marshall and K. G. Mattson. 2004. High soil carbon dioxide concentrations inhibit root respiration of Douglas fir. New Phytologist 128: 435–442.

Reich, P. B., M. B. Walters, M. G. Tjoelker, D. Van Der Klein and C. Buschena. 1998. Photosynthesis and respiration rates depend on leaf and root morphology and nitrogen concentration in nine boreal species differing in relative growth rate. Functional Ecology 12: 395–405.

Reich, P. B., M. G. Tjoelker, K. S. Pregitzer, I. J. Wright, J. Oleksyn and J. -L. Machado. 2008. Scaling of respiration to nitrogen in leaves, stems and roots of higher land plants. Ecology Letters 11: 793–801.

Ryan, M. G. 1990. Growth and maintenance respiration in stems of *Pinus contorta* and *Picea engelmanii*. Canadian Journal of Forest Research 20: 48–57.

Ryan, M. G. 1991. Effects of climate change on plant respiration. Ecological Applications 1: 157–167.

Ryan, M. G. and R. H. Waring. 1992. Maintenance respiration and stand development in subalpine lodgepole pine forest. Ecology 73(6): 2100–2108.

Ryan, M. G., S. Linder, J. M. Vose and R. M. Hubbard. 1994. Dark respiration in pines. Ecological Bulletin 43: 50–63.

Ryan, M. G., S. T. Gower, R. M. Hubbard, R. H. Waring, H. L. Golz, W. P. Cropper, Jr. et al. 1995. Woody tissue maintenance respiration of four confers in contrasting climates. Oecologia 101: 133–140.

Ryan, M. G., R. M. Hubbard, S. Pongracic, R. J. Raison and R. E. McMurtie. 1996. Foliage, fine-root, woody-tissue and stand respiration in *Pinus radiata* in relation to nitrogen status. Tree Physiology 16: 333–343.

Ryan, M. G., M. B. Lavigne and S. T. Gower. 1997. Annual carbon cost of autotrophic respiration in boreal forest ecosystem in relation to species and climate. Journal of Geophysical Research 102(24): 28, 871–28, 883.

Saxe, H., D. S. Ellsworth and J. O. Heath. 1998. Tree and forest functioning in an enriched CO_2 atmosphere. New Phytologist 139: 395–436.

Sprugel, D. G. 1990. Components of woody-tissue respiration in young *Abies amabilis* (Dougl.) Forbes trees. Trees 4: 88–98.

Stockfors, J. and S. Linder. 1998. Effect of nitrogen on the seasonal course of growth and maintenance respiration in stems of Norway spruce trees. Tree Physiology 18: 155–166.

Teskey, R. O. and R. E. Will. 1999. Acclimation of loblolly pine (*Pinus taeda*) seedlings to high temperatures. Tree Physiology 19: 519–525.

Tjoelker, M. G., L. Oleksyn and P. B. Reich. 1999. Acclimation of respiration to temperature and CO_2 in seedlings of boreal tree species in relation to plant size and relative growth rate. Global Change Biology 49: 679–691.

Tjoelker, M. G., J. Oleksyn and P. B. Reich. 2001. Modelling respiration of vegetation: evidence for a general temperature-dependent Q_{10}. Global Change Biology 7: 223–230.

Wang, K. -Y., S. Kellomäki and K. Laitinen. 1996. Acclimation of photosynthetic parameters in Scots pine after three years of exposure to elevated temperature and CO_2. Agricultural and Forest Meteorology 82: 195–217.

Wang, K. -Y., T. Zha and S. Kellomäki. 2002. Measuring and simulating crown respiration of Scots pine with increased temperature and carbon dioxide enrichment. Annals of Botany 90: 325–335.

Waring, R. H. and S. W. Running. 2007. Forest Ecosystems. Analysis at Multiple Scales. 2007. Elsevier Academic Press, Amsterdam. The Netherlands.

Wullschleger, S. D., R. J. Norby and C. A. Gunderson. 1992. Growth and maintenance respiration in leaves of *Liriodendron tulipifera* L. exposed to long-term carbon dioxide enrichment ion field. New Phytologist 121: 515–523.

Wythers, K. R., P. B. Reich, M. G. Tjoelker and P. B. Bolstad. 2005. Foliar respiration acclimation to temperature and temperature variable Q_{10} alter ecosystem carbon balance. Global Change Biology 11: 435–449.

Wythers, K. R., P. B. Reich and J. B. Bradford. 2013. Incorporating temperature-sensitive Q_{10} and foliar respiration acclimation algorithms modifies modeled ecosystem response to global change. Journal of Geophysical Research: Biogeosciences 118: 1–14.

Zha, T., A. Ryyppö, K. -Y. Wang and S. Kellomäki. 2001. Effects of elevated carbon dioxide concentration and temperature on needle growth, respiration and carbohydrate status in field-grown Scots pines during the needle expansion period. Tree Physiology 21: 1279–1287.

Zha, T., K. -Y. Wang, A. Ryyppö and S. Kellomäki. 2002a. Impact of needle age on the response of respiration in Scots pine to long-term elevation of carbon dioxide concentration and temperature. Tree Physiology 22: 1241–1248.

Zha, T., K. -Y. Wang, A. Ryyppö and S. Kellomäki. 2002b. Needle dark respiration in relation to within-crown position in long-term elevation of CO_2 concentration and temperature. New Phytologist 156: 33–41.

Zha, T., S. Kellomäki and K. -Y. Wang. 2003. Seasonal variation in respiration of 1-year-old shoots of Scots pine exposed to elevated carbon dioxide and temperature for 4 years. Annals of Botany 92: 89–96.

Zha, T., S. Kellomäki, K. -Y. Wang, A. Ryyppö and S. Niinistö. 2004. Seasonal and annual stem respiration of Scots pine trees under boreal conditions. Annals of Botany 94: 889–896.

Zha, T., S. Kellomäki, K. -Y. Wang and A. Ryyppö. 2005. Respiratory responses of Scots pine stems to 5 years exposure to elevated CO_2 concentration and temperature. Tree Physiology 25: 49–56.

Zogg, G. P., D. R. Zak, A. J. Burton and K. S. Pregitzer. 1996. Fine root respiration in northern hardwood forests in relation to temperature and nitrogen availability. Tree Physiology 16: 719–725.

Response of Transpiration to Climate Change

ABSTRACT

Transpiration is controlled by climate conditions and stomatal conductance, and only little water vapor is lost through leaf/needle surfaces outside the stomata. In boreal conditions, stomata respond to several environmental factors, including photon flux density, air temperature, leaf-to-air vapor pressure difference, CO_2 concentration and soil water potential. The long-term exposure of boreal Scots pine to elevated CO_2 has reduced the instant conductance of stomata, whereas exposure to elevated temperatures has increased stomata conductance. Climate change is likely to increase water-use efficiency, thus reducing the effects of possible drought episodes on carbon uptake and the growth of trees.

Keywords: stomatal conductance, transpiration, impacts of climate change, water use efficiency

8.1 Transpiration and Climate Change

Biophysical background and stomatal conductance

Transpiration is the loss of water through stomata in leaves/needles. It is a part of water flow through plants, maintaining the turgor of cells and providing water and nutrients to foliage and other organs for use in physiological and growth processes. Water flow is driven by the difference in water potential between soil and atmosphere through roots, stems and foliage. In transpiration, water vapor in the stomatal cavity is evaporated through the stomata to the atmosphere depending on the energy balance on surfaces of leaves/needles:

$$R - H - \text{Hv} \times E = 0 \tag{8.1}$$

where R is the net radiation on the surfaces of leaves/needles [W m^{-2}], H is the sensible heat [W m^{-2}] related to temperature differences between air and the surfaces of leaves/needles, and Hv × E is the latent heat (i.e., energy used for evaporation, W m^{-2}), with Hv being the heat of evaporation

[2,453,000 J kg⁻¹] and E the evaporation rate [g H₂O m⁻² s⁻¹]. Based on the energy balance, the daily potential evaporation (transpiration) rate through stomata is (Monteith and Unsworth 1990):

$$E = \frac{s \times R + \rho \times C_p \times D_a \times g_b}{s + \gamma \times (1 + g_b/g_s) \times Hv} \tag{8.2}$$

where s is the derivate of the slope of saturation vapor pressure curve [Pa °C⁻¹], ρ is the air density [1.220 kg m⁻³], C_p is the specific heat capacity of air at a constant pressure [1004.0 J kg⁻¹ °C⁻¹], Da is the current vapor pressure deficit [Pa], γ is the psychrometric constant [66.0 Pa °C⁻¹], g_b is boundary layer conductance [m s⁻¹] and g_s is stomatal conductance [m s⁻¹].

Transpiration is tightly controlled by stomatal conductance, and little water vapor is lost elsewhere through leaf/needle surfaces. Under boreal conditions, stomata respond to several environmental factors including photon flux density, temperature, leaf-to-air vapor pressure difference, CO_2 concentration and leaf/needle (or soil) water potential. According to Jarvis (1976), the stomatal conductance [g_s, mmol m⁻² s⁻¹] is:

$$g_s = f_1(Q) \times f_2(T) \times f_3(D_l) \times f_4(C_a) \times f_5(\Psi) \tag{8.3}$$

where the variables $f_1 \ldots f_5$ are the functions of flux density [Q, μmol m⁻² s⁻¹], leaf temperature [T, °C], leaf-to-air water vapor pressure deficit [D, kPa], ambient atmospheric carbon dioxide [C_a, μmol mol⁻¹] and leaf water potential [ψ, MPa] affecting stomatal conductance (Fig. 8.1).

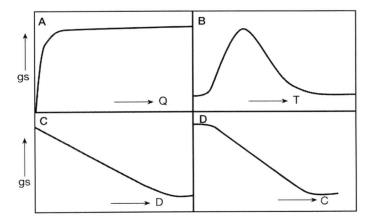

Fig. 8.1 Schematic presentation of how stomatal conductance (g_s) responds to selected environmental factors, based on Jarvis (1976). A: flux density (Q), B: leaf temperature (T), C: leaf-to-air vapor pressure deficit (D) and D: ambient atmospheric CO_2 concentration.

Impact of climate change on stomatal functions

In general, stomatal conduction is sensitive to light that is less than half that of full sun light while saturating in light above (Morison 1985). Stomatal conductance may increase along with air temperature up to 25–30°C, beyond which high temperatures may reduce. Under boreal conditions, the frequency and duration of such high temperatures are likely to increase as a result of climate change (Jylhä et al. 2009), but their direct effects are probably less than those of the changes in atmospheric water vapor pressure deficit caused by higher temperatures during the main growing season.

In crop plants grown under current CO_2 levels, stomatal conductance responds directly to the elevated CO_2 as shown by Morison (1985). In such experiments, the transfer of plants from ambient CO_2 concentration to elevated CO_2 reduced stomatal conductance by 40% (the mean over 50 plants), with a 30% reduction in transpiration. Similarly, Field et al. (1995) showed that in over 23 conifer tree species (including conifers from both boreal and tropical conditions) stomatal conductance declined, on average by 23%, when the concentration of CO_2 in the growing environment was doubled. Climate change also indirectly affects stomatal conductance through leaf/needle water potential as a function of soil moisture controlling leaf/needle water potential (Granier et al. 2000, Kellomäki and Wang 2000).

8.2 Response of Stomatal Conductance to the Elevation of CO_2 and Temperature

Responses to photon flux density and vapor pressure deficit

Using the submodels of the Jarvis-type stomatal conductance model (Equation 8.3), Wang and Kellomäki (1997) found that the elevation of CO_2 reduced the response of stomatal conductance to high radiation (the maximum conductance at the flux density > 1,000 µmol m^{-2} s^{-1}), whereas elevated temperature increased the response (Fig. 8.2). In the former case, the reduction was 22% and in the latter cases the increase was 10%. The concurrent elevation of both factors increased the response of stomatal conductance (6%), but the increase remained smaller than under the elevated temperature alone. The elevation of CO_2 and temperature increased the sensitivity of stomatal conductance to the radiation, especially under low radiation (flux density < 400 µmol m^{-2} s^{-1}).

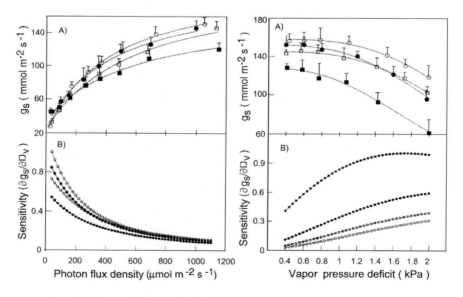

Fig. 8.2 Stomatal conductance (A Figures) and relative sensitivity (B Figures) of stomatal conductance as a function of (left) photon flux density and water vapor pressure deficit (right) in Scots pines grown under ambient conditions and under the long-term elevation of CO_2 and temperature (Wang and Kellomäki 1997). Permission of Canadian Science Publishing. Measurements were done in a laboratory at a leaf-air water vapor pressure difference of 0.4–06 kPa, leaf temperature 20 ± 0.5°C and ambient partial pressure of CO_2 (35 Pa for control and elevated temperature and 70 Pa for the elevated CO_2 and combined elevation of temperature and CO_2). Legend: filled circle is ambient conditions; open circle elevated temperature; open square elevated temperature and CO_2; and filled square elevated CO_2.

Stomatal conductance was fairly insensitive to water vapor deficit at low values of deficit regardless of the climate treatment (Wang and Kellomäki 1997) (Fig. 8.2). However, the values of stomatal conductance clearly declined along with increasing sensitivity, when water vapor deficit increased. In a range from 0.6 to 2.0 kPa, the increasing vapor deficit reduced stomatal conductance by 36% in the trees grown under ambient conditions, but the reduction was 50% in trees grown under elevated CO_2. Under elevated temperature alone or combined with elevated CO_2, the conductance reduced by 25 and 30%, respectively. On the other hand, the maximum stomatal conductance reduced by 16% in the trees grown under elevated CO_2, but only 5% in the trees grown under the elevated temperature and CO_2. The sensitivity of stomatal conductance was greatest in trees grown under elevated CO_2 and the smallest in the trees grown under the elevated temperature.

Responses to CO_2 and temperature

Medlyn et al. (1999, 2001) found that the stomatal conductance of several major European tree species reduced by 21% in response to elevated CO_2. This was also the case for Scots pine, whose conductance declined substantially in trees grown under the current climate, when the partial intercellular pressure of CO_2 increased from 3 to 100 Pa in a measuring situation. Wang and Kellomäki (1997) found that the increase of the intercellular CO_2 pressure from 3 Pa to 18 Pa halved stomatal conductance (Fig. 8.3). Thereafter, stomatal conductance (minimum conductance) was 13% less in trees grown under elevated CO_2 than in trees grown under ambient conditions. The sensitivity of stomatal conductance to the intercellular CO_2 was similar in trees under different climatic treatments. However, only the response to increased CO_2 was significantly different from the responses to other climatic treatments. In response to the elevated CO_2, the reduction of stomatal conductance in Scots pine was substantially smaller than in white spruce [*Picea glauce* (Moench) Voss] seedlings, where there was a great increase in the water use efficiency (Dang et al. 2008).

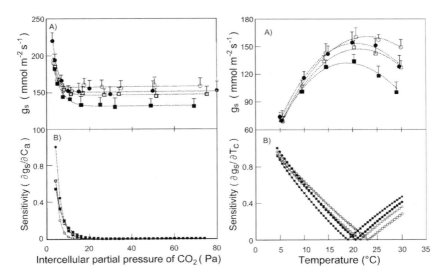

Fig. 8.3 Stomatal conductance (A Figures) and relative sensitivity (B Figures) of stomatal conductance as a function of (left) intercellular partial pressure of CO_2 and temperature (right) in Scots pines grown under ambient conditions and under the long-term elevation of CO_2 and temperature (Wang and Kellomäki 1997). Permission of Canadian Science Publishing. Measurements of response to the increased CO_2 were done in a laboratory at a leaf-air water vapor pressure difference of 0.4–06 kPa, leaf temperature 20 ± 0.5°C and photon flux density 1,400 µmol m^{-2} s^{-1}. Measurements of the response to the increased temperature were taken under similar conditions but the ambient partial pressure of CO_2 was 35 Pa for the ambient conditions. For the elevated CO_2 and the combined elevation of CO_2 and temperature, the partial CO_2 pressure was 75 Pa (Wang and Kellomäki 1997). Legend: filled circle is ambient conditions; open circle elevated temperature; open square elevated temperature and CO_2; and filled square elevated CO_2.

Figure 8.3 shows further that stomatal conductance in Scots pine increased with increasing needle temperature under suboptimal temperatures (5–20°C) but declined under temperatures over the optimal (Wang and Kellomäki 1997). In contrast, stomatal sensitivity in relation to temperature performed in an opposite way, with high sensitivity under temperatures below and over optimal for stomatal conductance. In trees grown under elevated CO_2 alone or combined with elevated temperature, the maximum conductance reduced by 14 and 4% compared to the ambient conditions. At the same time, the stomatal conductance under elevated temperature alone increased by 5%.

Responses to soil water potential

Low soil water potential results in water stress (drought) and reduces the leaf water potential. This reduces both stomatal conductance and the CO_2 fixation as found by Teskey et al. (1986). Figure 8.4 shows how stomatal conductance responses to the soil water potential in boreal Scots pines grown under elevated CO_2 and temperature based on the Mekrijärvi chamber experiment. Stomatal conductance reduced rapidly when the soil water potential dropped below –0.5 MPa. The general pattern was similar regardless of climatic treatment, but stomatal conductance in the trees exposed to elevated CO_2 was clearly smaller than if the trees were not exposed. Stomatal conductance was systematically greater when Scots pines were grown under elevated temperature. This was also the case when the growing conditions represented the combined elevation of CO_2 and temperature (Kellomäki and Wang 2000).

Stomatal conductance is closely coupled to changes in photosynthesis, but the CO_2 fixation may further be limited by biochemical processes controlling the carbon fixation. Kellomäki and Wang (1996) found that the reduced needle water potential during drought depressed the carbon uptake through both stomatal and non-stomatal limitations. The decline in photosynthesis with decreasing needle water potential was a function of stomatal limitation during the early stages of drought, and that of non-stomatal limitation during severe water stress. Thus, conductance adjusted quickly to decreasing water potential, while non-stomatal components responded slowly (Kellomäki and Wang 2000).

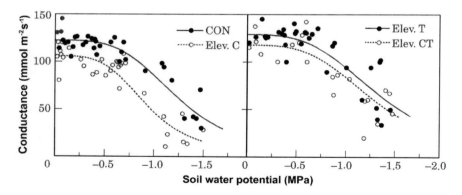

Fig. 8.4 Response of stomatal conductance to the soil water potential in Scots pines grown under ambient conditions (CON), elevated CO_2 (Elev. C), elevated temperature (Elev. T) and combined both (Elev. CT) in the Mekrijärvi chamber experiment (Kellomäki and Wang 2000). Permission of Oxford University Press. Measurements were taken at a photon flux density of 1,000 µmol m^{-2} s^{-1}, at temperature 20°C, and at the vapor pressure deficit 0.6 kPa. Each point is the mean of four separate measurements. The lines are the best fit of the Equation: $f(\psi_s) = [1 + (\psi_s/k_1)^{k_2}]^{-1}$, where ψ_s is the soil water potential and k_1 and k_2 parameters.

8.3 Interaction between Transpiration and Photosynthesis

Responses of transpiration

The effects of long-term elevation of CO_2 and temperature on stomatal conductance in boreal Scots pine and subsequent transpiration were evident when the transpiration is regressed against the water vapor deficit. Figure 8.5 shows that the transpiration rate was nearly a linear function of vapor deficit with low values. However, the transpiration rate culminated when the measurements were taken under dryer and dryer air (Kellomäki and Wang 2000). The maximum transpiration rate was clearly lowest in trees grown under elevated CO_2 and highest in trees grown under elevated temperature. At the same time, the culmination occurred earlier in trees grown under elevated CO_2 than under other climate treatments. These changes in transpiration response are probably related to the changes in stomatal functions, such as the increase of maximal stomatal conductance and its optimal temperature, under elevated temperatures and their reduction under elevated CO_2 (Wang and Kellomäki 1997).

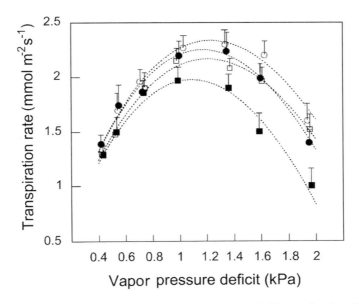

Fig. 8.5 Transpiration rate in Scots pines grown under elevated temperature and CO_2 as a function of water vapor deficit of the air. The measurements were done at 20°C under saturating photon flux density (1,400 μmol m^{-2} s^{-1}) (Wang and Kellomäki 1997). Permission of Canadian Science Publishing. Legend: filled circle is ambient conditions; open circle elevated temperature; filled square elevated CO_2; and open square combined elevation of temperature and CO_2.

Water use efficiency

The Mekrijärvi chamber experiment showed that long-term exposure to elevated CO_2 and temperature changed the rates of photosynthesis and transpiration in Scots pine in response to CO_2, photon flux density, temperature, water vapor pressure deficit and soil and needle water potential. The effects of variable climatic treatments were found in water use efficiency, which indicates the photosynthetic production per unit of water lost in transpiration. Figure 8.6 shows that the photosynthetic rate of Scots pine was dependent on the transpiration rate regardless of the growing conditions (Wang and Kellomäki 1997). This curvilinear relationship demonstrates that water use efficiency was clearly greater in trees grown under elevated CO_2 than in trees grown under other climatic treatments. This

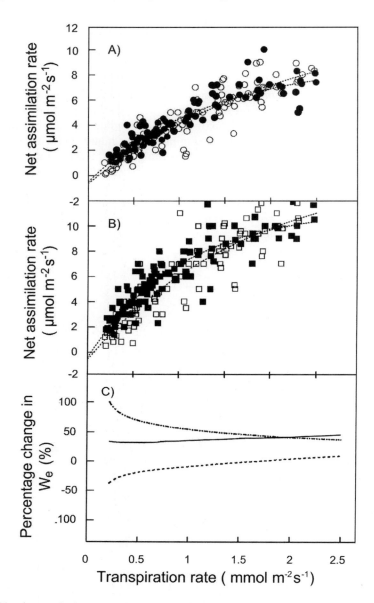

Fig. 8.6 A and B: Net photosynthetic rate, and C: Water use efficiency (W_e, % of that under the ambient conditions) of Scots pine grown under elevated temperature and CO_2 as a function of transpiration rate (Wang and Kellomäki 1997). Permission of Canadian Science Publishing. Measurements were done under water vapor deficit of 0.3–2.0 kPa, temperature 5–20°C, irradiation 0–1,400 µmol m^{-2} s^{-1} and the partial pressure CO_2 380 Pa. Legend for A and B: filled circle is ambient conditions, open circle elevated temperature; filled square elevated CO_2; and open square elevated temperature and CO_2. Legend for C: dotted line with two dots is elevated CO_2; solid line is elevated temperature and CO_2; dotted line is elevated temperature.

was especially clear when the transpiration rate was low, but even at higher transpiration rates water use efficiency increased up to 40–50% compared to that under the ambient conditions. The situation was the opposite when only the temperature was elevated.

Wang et al. (2003) further analyzed water use efficiency in Scots pine shoots grown in closed chambers with elevated CO_2 (the ambient CO_2 plus 350 µmol mol^{-1}) and temperature (the ambient temperature plus 2–6°C, more in winter than in summer), and the combination of both. They found that elevated CO_2 increased the daily total carbon flux per unit projected area by 17 to 21%. This implied a 16–24% increase in light-use efficiency and a reduction of 1–12% in the water flux, with an increase of 13–35% in water use efficiency. Similarly, the elevated temperature increased the carbon flux per unit needle area by 10 to 18%, which increased light-use efficiency by 8–19%, and reduced water use efficiency by 19–34%. Wang et al. (2003) found no interaction between elevated CO_2 and temperature; i.e., carbon and water fluxes were similar to those under the elevation of single factors. Light-use efficiency was mainly related to the increasing temperature, while water use efficiency was mainly related to high water vapor deficit.

8.4 Concluding Remarks

In general, elevated CO_2 is likely to reduce stomatal conductance, as found by Scarascia-Mugnozza and De Angelis (1998) for selected European deciduous and coniferous tree species. The reduction varied substantially (10–70%) depending on tree species and modifications of leaf/needle anatomy, and morphology (e.g., specific leaf/needle area). For example, the density of stomata in birch (*Betula pendula* Roth) and beech (*Fagus sylvatica* L.) reduced by 8–16% under elevated CO_2. Paoletti et al. (1998) found further that stomatal density in white oak (*Quercus alba* L.) reduced by a factor of 1.5 when atmospheric CO_2 increased from 350 ppm to 700 ppm as a function of the distance around a natural CO_2 well. Unlike angiosperms, stomatal density in gymnosperms seems to be only slightly sensitive to elevated CO_2. For example, Luomala et al. (2005) found that stomatal density in Scots pine was only slightly reduced under elevated CO_2 (700 ppm) but under elevated temperature (the mean increase over a year was 4°C) the reduction was clear. Under a combined elevation of CO_2 and temperature, the reduction was less than under elevated temperature only. In the latter case, the reduction of stomatal density was associated with the change in mesophyll, which was thinner on abaxial surfaces, as was the case for the vascular cylinder.

However, Lin et al. (2001) found that stomatal conductance reduced 7.4% in seven-year-old Scots pine seedlings when grown for four years under elevated CO_2 (ambient vs. 700 ppm). The reduction took place on both adaxial and abaxial needle surfaces. Coincidently with this reduction, the cross-sectional area of needles increased by 10% due to an increase in the thickness and width of needles. The increase in needle thickness was due to a large increase in mesophyll tissue. Lin et al. (2001) also found that "the relative area (i.e., proportion of the total area) of epidermis plus hypodermis, of resin canal, of xylem and of central cylinder decreased, whereas the relative area of needle phloem significantly increased." Kouwenberg et al. (2003) also showed that stomatal density (i.e., the number of stomata per millimeter of needle length) in western hemlock [*Tsuga heterophylla* (Raf.) Sarg.] decreased clearly as a function of increasing CO_2 in the growing environment. Reduced transpiration in conifers under elevated CO_2 may thus be affected by several other changes in the structural and functional properties of needles, rather than being due only to a reduction in stomatal density.

References

Dang, Q. -L., J. M. Maepea and W. H. Parker. 2008. Genetic variation of ecophysiological responses to CO_2 in *Picea glauca* seedlings. The Open Forest Science Journal 1: 68–79.
Field, C. B., R. B. Jackson and H. A. Mooney. 1995. Stomatal response to increased CO_2: implications from the plant to the global scale. Plant, Cell and Environment 18: 1241–1225.
Granier, A., D. Loustau and N. Bréda. 2000. A generic model of forest canopy conductance dependent on climate, soil water availability and leaf area index. Annals of Forest Science 57(8): 755–765.
Jarvis, P. G. 1976. The interpretation of the variations in leaf water potential and stomatal conductance found in canopies in the field. Philosophical Transactions of the Royal Society B 273: 593–610.

Jylhä, K., K. Ruosteenoja, A. Venäläinen, H. Tuomenvirta, L. Ruokolainen, S. Saikku et al. 2009. Arvioita Suomen muuttuvasta ilmastosta sopeutumistutkimuksia varten. ACCLIM—hankkeen raportti 2009. Finnish Meteorological Institute Reports 2009: 4.

Kellomäki, S. and K. -Y. Wang. 1996. Photosynthetic responses to needle water potentials in Scots pine after a four-your exposure to elevated CO_2 and temperature. Tree Physiology 16: 765–772.

Kellomäki, S. and K. -Y. Wang. 2000. Modelling and measuring transpiration from Scots pine with increased temperature and carbon dioxide enrichment. Annals of Botany 85: 263–278.

Kouwenberg, L. L. R., J. C. McElwan, W. M. Kuschner, F. Wagner, J. David, D. Beerling et al. 2003. Stomatal frequency adjustment of four conifer species to historical changes in atmospheric CO_2. American Journal of Botany 90(4): 610–619.

Lin, J., M. E. Jach and R. Ceulemans. 2001. Stomatal density and needle anatomy of Scots pine (*Pinus sylvestris*) are affected by elevated CO_2. New Phytologist 150: 665–674.

Luomala, E. -M., K. Laitinen, S. Sutinen, S. Kellomäki and E. Vapaavuori. 2005. Stomatal density, anatomy and nutrient concentrations of Scots pine needles are affected by elevated CO_2 and temperature. Plant, Cell and Environment 28: 733–749.

Medlyn, B., P. Jarvis, F. Badeck, D. Pury, C. Barton, M. Broadmeadow et al. 1999. Effects of elevated CO_2 on photosynthesis in European forest species: a meta-analysis of model parameters. Plant, Cell and Environment 22: 1475–1495.

Medlyn, B. E., C. V. M. Barton, M. S. J. Broadmeadow, R. Ceulemans, P. De Angelis and M. Forstreuter. 2001. Stomatal conductance of forest species after long-term exposure to elevated CO_2 concentration: a synthesis. New Phytologist 149(2): 247–264.

Monteith, J. L. and M. H. Unsworth. 1990. Principles of Environmental Physics. Edward Arnold, London, UK.

Morison, J. I. L. 1985. Sensitivity of stomatal and water use efficiency to high CO_2. Plant, Cell and Environment 8: 467–474.

Paoletti, E., G. Nourrisson, J. P. Garrec and A. Raschi. 1998. Modifications of the leaf surface structures of *Quercus ilex* L. in open, naturally CO_2-enriched environments. Plant, Cell and Environment 21: 1071–1075.

Scarascia-Mugnozza, G. and P. De Angelis. 1998. Is water used more efficiently? pp. 192–214. *In*: P. G. Jarvis (ed.). European Forests and Global Change. The Likely Impacts of Rising CO_2 and Temperature. Cambridge University Press, Cambridge, UK.

Teskey, R. O., J. A. Fites, L. J. Samuelson and B. C. Bongarten. 1986. Stomatal and nonstomatal limitation of net photosynthesis in *Pinus taeda* L. under different environmental conditions. Tree Physiology 2: 131–142.

Wang, K. -Y. and S. Kellomäki. 1997. Stomatal conductance and transpiration in shoots of Scots pine after 4-year exposure to elevated CO_2 and temperature. Canadian Journal of Botany 75: 552–561.

Wang, K. -Y., S. Kellomäki, C. Li and T. Zha. 2003. Light and water-use efficiencies in pine shoots exposed to elevated carbon dioxide and temperature. Annals Botany 92: 53–64.

9

Response of Whole Tree Physiology to Climate Change

ABSTRACT

The interaction between entire trees and their immediate environment is a process, in which the structure of trees modifies the environment, providing feedback to the growth and structural development of trees. Under elevated CO_2, the total foliage of boreal Scots pine increases, as does the photosynthetic rate. The elevation of temperature lengthens the growing season, but the total photosynthesis may remain clearly smaller than that under elevated CO_2. The growth of Scots pine under elevated CO_2 may decrease total diurnal and cumulative sap flow, whereas elevated temperature increases both.

Keywords: whole tree physiology, canopy photosynthesis and respiration, sap flow and crown transpiration, modeling physiological responses at canopy level

9.1 Response of Whole Tree to Climate Change

The growth and development of trees are affected by single physiological processes working in the whole tree context. The interaction between entire trees and their immediate environment is a complex process, in which the structure of trees modifies the properties of the environment with feedback to the growth and structural development of trees. Whole tree physiology integrates physiological impacts on the growth and development of the crown and other tree organs over longer periods than those represented by single physiological processes at the organ level (Leuzinger and Bader 2012). Figure 9.1 demonstrates the main interactions which a tree may have with the climate and the weather controlling physiological processes and affecting the growth and development of trees.

The crown of a single tree or the canopy formed by several trees growing in a stand (crown/canopy level) provide the interface for the interaction between climatic and edaphic factors driving physiological processes behind the growth and development of trees (Fig. 9.1). The amount of leaf/needle area and its spatial distribution modify the microclimate in the crown/canopy, thus affecting the distribution of radiation, temperature, CO_2 concentration, air humidity and wind velocity in the

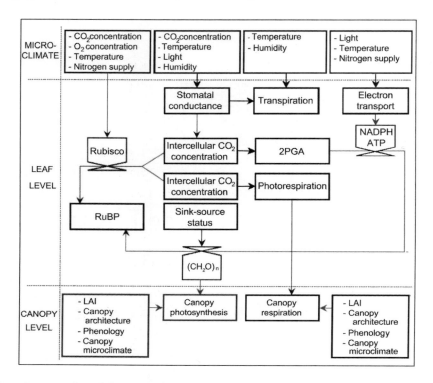

Fig. 9.1 Schematic presentations of photosynthesis, respiration and transpiration in single leaves/needles and the subsequent whole-tree physiology with the focus on photosynthetic production in the context of the tree and canopy structure controlling the microclimatic conditions in a tree stand (Wang 1996a,b). Legend: PGA refers to three-carbon intermediates that further reduce to carbohydrates using photochemically generated ATP and NADPH$_2$. RuBP (ribulose bisphosphate) involves the five-carbon acceptor molecules in generating PGA. The reaction between carbon and RuBP is catalyzed by the Rubisco enzyme (ribulose bisphosphate carboxylase/oxygenase). Further details, see Box 6.1, Chapter 6.

crown/canopy. Furthermore, the distribution of nitrogen in foliage affects the rates of photosynthesis and respiration at the leaf/needle level in the canopy. Stomatal conductance and subsequent photosynthesis, respiration and transpiration may further acclimate to the prevailing microclimate, as in biochemical processes of photosynthesis. In scaling from leaf to crown of trees and canopy of stand, processes can be modeled by applying multi-layer crown/canopy models, the parameters of which represent leaf/needle-level measurements. The foliage area and leaf area index integrate the output of leaf level processes to the crown and canopy-level output.

9.2 Canopy Photosynthesis—Model Calculations

Wang (1996a) applied a tree-layer canopy model to calculate the distribution of photosynthesis and respiration over a year for single Scots pines grown under elevated CO$_2$ and temperature. In the calculation of instantaneous photosynthesis and respiration, the crown area and radiation are distributed in the canopy layers and the radiation is divided into direct and diffuse radiation in the layers for sunlit and shaded areas. The hourly canopy photosynthesis (A_c (t)) and further over a day (A_c (d)) is:

$$A_c(t) = \sum_{i=1}^{i}\left[\left(A_{n/suni} \times L_{suni}\right) + \left(A_{n/shade} \times L_{shadei}\right)\right] \tag{9.1}$$

$$A_c(d) = \int_0^t A_c(t)dt \tag{9.2}$$

where $A_{n/suni}$ is the photosynthetic rate on a sunlit leaf area L_{suni} in layer i, and $A_{n/shadei}$ is the photosynthetic rate on a shaded leaf area L_{shadei} in the layer i. For further details of the calculations, see Box 9.1.

Figure 9.2 shows how elevated temperature and CO_2 affected the total photosynthesis and respiration of Scots pine over a year (Wang 1996b,c). In the calculations, the crown was divided into three layers (upper, middle and lower), which were used in calculations (e.g., dynamics of stomatal functions and photosynthesis in different layers). The same layers were used in defining the distribution of needle mass in the crown and the properties of needles in different layers. For example, the elevation of CO_2 alone, or concurrently with temperature, increased the foliage area by 35 and 28%, especially in the lower crown. The share of older needles clearly increased, and needles were substantially larger than under ambient conditions, thus indicating the lengthened life span under elevated CO_2. In the tree grown under elevated temperature, needles were also large, but in this case the needle area increased only by 5%, and only in the upper crown. The increased temperature clearly reduced the life span of needles compared to that under elevated CO_2 alone or combined with temperature elevation.

Changes in the photosynthesis were due to three factors (Fig. 9.2; Wang 1996a,b,c): (i) changes in total needle area; (ii) changes in the length of growing season; and (iii) changes in the rates of

---

Box 9.1 Calculation of photon flux density on sunlit leaves and on shaded leaves in the canopy, applying the layered crown structure

Let Q_o refer to the total radiation, Q_{odir} to the direct radiation and Q_{odiff} to the diffuse radiation above the crown (Wang 1996a):

$$Q_o = Q_{odir}(t) + Q_{odiff}(t) \tag{9.3}$$

Furthermore, let $Q_{dir(i)}(t)$ refer to the direct radiation and $Q_{diff(i)}(t)$ to the diffuse radiation on the crown layer i at the time t. The values of both factors are the function of Q_{odir} and Q_{odiff}, the cumulative leaf area index (L_c), extinction coefficient (1/4 sin β), solar declination (θ), solar elevation (β), latitude (Φ) and date (D_n). The crown layers are horizontally homogenous and needles have a bisected form and randomly distributed and oriented in crown (Oker-Blom and Kellomäki 1981). The photosynthetic flux density on the sunlit needles ($Q_{sun(i)}$) is the sum of direct ($Q_{dir(i)}$) and diffuse ($Q_{diff(i)}$) radiation, and the photon flux density on shaded needles ($Q_{shade(i)}$) is the same as diffuse radiation (Wang 1996a,b):

$$Q_{sun(i)}(t) = Q_{dir(i)}(t) + Q_{diff(i)}(t) = Q_{odir}(t) \times \exp\left[-\frac{L_{c(i)}}{4\sin\beta(t)}\right] + Q_{diff(i)}(t) \tag{9.4}$$

$$Q_{shade(i)}(t) = Q_{diff(i)}(t) = Q_{odiff}(t) \times \frac{6}{7} \times \int_0^{\pi/2} (1 + 2 \times \sin\theta) \times \exp\left[-\frac{L_{c(i)}}{4 \times \sin\theta}\right] \sin\theta \cos\theta d\theta \tag{9.5}$$

The projected needle area of sunlit ($L_{sun(i)}$) and shaded ($L_{shade(i)}$) foliage in a layer i is a function of the cumulative leaf area index (L_c), the needle surface area (L_s) and the extinction coefficient. Following Oker-Blom and Kellomäki (1981):

$$L_{sun(i)} = 4 \times \sin\beta \times \left[e^{-\frac{L_{c(i)}}{4 \times \sin\beta}} - e^{-\frac{L_{c(i-1)}}{4 \times \sin\beta}}\right] \tag{9.6}$$

$$L_{shade(i)} = L_{c(i)} - L_{sun(i)} \tag{9.7}$$

photosynthesis and respiration per unit needle area. Under elevated CO_2, the total foliage increased substantially but the photosynthetic rate also increased. The increased temperature lengthened the growing season, but photosynthesis remained only half that under elevated CO_2 as summarized in Table 9.1. The reduction was mainly due to the enhanced rate of respiration and the reduction of needle area. The effect of respiration was also clear when the temperature and CO_2 were elevated at the same time; i.e., the photosynthesis was 60–70% of that when CO_2 was elevated alone. When the values of the photosynthesis were integrated over the year, it appeared that elevated CO_2 alone, or together with temperature, increased the annual photosynthesis by 42 and 51% compared to that under ambient conditions. Under elevated temperature alone, the increase was only 5%. At the same time, the total respiration increased by 4% under elevated CO_2, by 12% under elevated temperature alone and by 8% when both were elevated. Elevated temperature also increased respiration in the winter.

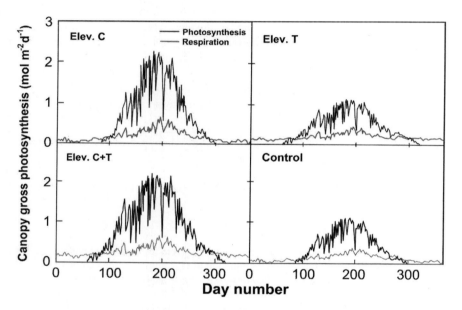

Fig. 9.2 Annual courses of daily photosynthesis and respiration in Scots pine grown under current climate (Control), elevated temperature (Elev. T) and CO_2 (Elev. C) and combined both (Elev. C + T) in the Mekrijärvi chamber experiment. The calculations were made with the model presented in Box 9.1 in such a way, that during the growing season (April 15–September 15) the CO_2 concentration in the air was 550–600 μmol mol^{-1}, and outside the growing season it was the same as under ambient conditions (350 μmol mol^{-1}). Under elevated temperature, the air temperature was 2°C higher than outside the chambers. In other cases, the temperature was equal to that outside the chambers (Wang 1996c).

Table 9.1 Annual photosynthesis and respiration of Scots pine grown in the Mekrijärvi chamber experiment (Wang 1996a,b). The growing season was the period between days D121–D243 from the beginning of the year.

Season	Photosynthesis, mol m^{-2} yr^{-1}			
	CO_2	CO_2 + T	T	Ambient
Growing season (D121–D243)	106.6	113.7	77.9	75.4
• Outside growing season	12.7	20.2	15.7	9.6
• Over the year	119.3	133.9	93.6	85.0
	Respiration, mol m^{-2} yr^{-1}			
Growing season (D121–D243)	37.2	38.7	40.1	35.7
• Outside growing season, photosynthesis	11.8	25.6	25.9	12.7
• Outside growing season, no photosynthesis	10.5	18.0	22.8	15.3
• Over the year	59.5	82.3	88.8	63.7

9.3 Canopy Respiration—Measurements and Model Calculations

Wang et al. (2002) developed further a multi-layer model to extrapolate the response of respiration of individual needles to the whole crown, and used to estimate daily and annual crown respiration. In the calculations, the crown structure and microclimate data were used as input (Box 9.2).

Figure 9.3 shows that the total projected foliage area varied substantially over the year. The foliage areas were 10–15% higher in the early autumn than in spring, thus showing the reduction of older needles in autumn and winter, especially in the lower crown. On the other hand, the elevation of temperature and CO_2 alone or combined increased the total projected foliage area. Respiration measurements showed that elevated CO_2 alone led to higher Q_{10} values (4.6%) relative to the trees grown under ambient conditions, but lower respiration rates at 20°C ($R_{l.d}$ (20)) (−7.1%) during the

---

Box 9.2 Model used to calculate respiration in the foliage of Scots pine

Wang et al. (2002) calculated the daily crown respiration for Scots pines grown under elevated temperature and CO_2 over a year, considering the within-crown distribution of the foliage area and nitrogen content in needles per needle age classes. Daily crown respiration ($R_l(d)$) included light respiration ($R_{l.l}$) and dark respiration ($R_{l.d}$), when five layers were separated in the crown:

$$R_l(d) = \sum_{i=1}^{5}\left[\int_0^{td} R_{i.d.i}(t)dt + \int_0^{24-td} R_{i.l.i}(t)dt\right] \qquad (9.8)$$

Dark respiration in layer i of crown was assumed to be a function of the nitrogen content of needles (N_l), air temperature (T_a), ambient CO_2 concentration (C_a) and foliage area in the layer ($L_{a.i}$):

$$R_{id}(t) = f(L_{a.i}, T_a, C_a, N_l) \qquad (9.9)$$

The response of $R_{l.d}$ to temperature was:

$$f(T_a) = R_{i.d}(20) \times Q_{10}^{(T_a(t)-20/10)} \qquad (9.10)$$

where $R_{i.d}(20)$ is the values of $R_{l.d}$ at the reference temperature 20°C, and Q_{10} the change in respiration for a 10°C change in temperature. The dependence of $R_{l.d}$ was related to N_l mainly through the basal respiration rate, which was a linear function of N_l:

$$R_{l.d}(20) = c_{nl} \times N_l + c_{n2} \qquad (9.11)$$

where c_{nl} and c_{n2} are parameters. Nitrogen content reduced downwards in the crown following the relative height in the crown (H_r):

$$N_l(H_r) = N_{l.o} \times \exp(-k_n \times L_c / L_t) \qquad (9.12)$$

where $N_{l.o}$ is the nitrogen per unit of the projected foliage area in the uppermost whorl in the crown (g N m^{-2}), L_c (m^2 m^{-2} ground) and L_t (m^2 m^{-2} ground) are the cumulative foliage area index from the top of crown and the total foliage area index, and k_n is the allocation coefficient of nitrogen (Kellomäki and Wang 1997). Consequently, the total nitrogen per unit ground area is (N_t, g N m^{-2}):

$$N_t = \int_o^{Lc} N_l(H_r)dL = \frac{1}{k_n} \times N_{l.o} \times L_t\left[1-\exp(-k_n)\right] \qquad (9.13)$$

The distribution of N_l within the crown is:

$$N_l(H_r) = \frac{k_n \times N_t \times \exp(k_n \times L_c/L_t)}{L_t \times \left[1-\exp(-k_n)\right]} \qquad (9.14)$$

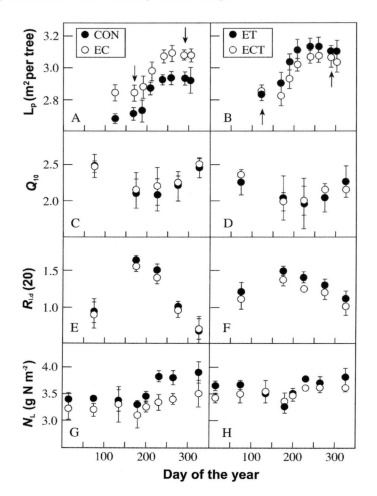

Fig. 9.3 Annual variation of total projected foliage area (L_p) (A and B), Q_{10} (C and D) and $R_{l,d}(20)$ (respiration rate at 20°C) (E and F), and foliage nitrogen content of one-year-old needles in the middle of the crown (N_L, g N m^{-2}) (G and H) in 20-year-old Scots pines grown four years under ambient (CON) conditions, elevated CO_2 (EC) and temperature (ET) alone and both combined (ECT) (Wang et al. 2002). Permission of Oxford University Press. The values of L_p were estimated based on hemispherical photographs (indicated by arrows in Figures A and B) from the crown base and combined with measurements on the sample branches.

main growth season (D121–D243). Elevated temperature alone and combined with elevated CO_2 reduced the values of Q_{10} (−12.0 and −9.8%) throughout the year, and increased $R_{l,d}(20)$ (27.2 and 21.6%) during the period of no growth, and slightly reduced $R_{l,d}(20)$ (−1.7 and −2.8%) during the main growth season (Wang et al. 2002).

Daily total respiration was highest from mid-June to mid-August, accounting for over 40% of annual total respiration 210 g C per tree in the ambient conditions (Fig. 9.4). The climatic treatment increased the annual sum of respiration; i.e., under elevated CO_2 and temperature alone and the combined elevation of both factors the increase was 16, 35 and 27%, respectively. The increase was mainly due to respiration during the main growing season (D121–D243), which accounted for 73, 68, and 65% of respiration under elevated CO_2 and temperature alone and the combined elevation of both factors. Regardless of the climatic treatment, one-year-old needles contributed to respiration more than needles in other age classes. The contribution of this age class was 47% under ambient conditions, 52% under elevated CO_2 alone, 64% under elevated temperature alone and 58% under

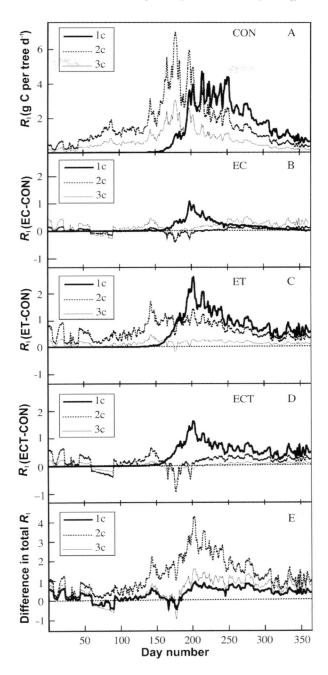

Fig. 9.4 Calculated respiration over crown per day over a year for young Scots pines as a function of needle age class (1c, 2c and 3c) grown under ambient conditions (A, CON), and the differences in respiration between the trees grown under elevated CO_2 (B, EC), elevated temperature (C, ET) and the combined elevation of both factors (D, EC + ET) and the trees grown under ambient conditions (e.g., R_l(ECT − CON)). Figure E summarizes the differences in respiration over crown per needle age classes regardless of climatic treatment (Wang et al. 2002). Permission of Oxford University Press.

the combined elevation of both. The contribution from current-year needles was large only after mid-June, when the needle growth was completed.

The model computations showed further that the total annual respiration over crown increased by 16% under elevated CO_2, of which 92% was attributable to the increase in foliage area (Fig. 9.4). Under elevated temperature alone, the increase was 35%, of which 66% was related to the increase in the foliage area and 17% to the rise in temperature. Under the combined elevation of CO_2 and temperature, the increase was 27%, of which 43% was due to the increase in foliage area and 29% due to the rise under ambient temperature. The changes in the respiration parameters for individual needles that were caused by treatments contributed little to the increase of annual respiration over crown compared to the increased foliage area.

9.4 Canopy Transpiration—Measurements and Model Calculations

Kellomäki and Wang (1998) used sap flow measurements in studying how growth under elevated CO_2 and temperature affected the total transpiration from boreal Scots pines. The sap flow measurements involved monitoring the velocity of the mass flow of water through the stem cross-section due to transpiration over the crown. The constant-power heat balance method was used in the measurements to monitor the sap flow. The method allows the variability in sap flow and its response to variability in environmental conditions to be monitored continuously without disturbing the natural functions of the experimental trees. In the trees being measured, the crown needle area varied from 1.62 m² to 1.85 m² per tree. Under elevated temperature alone and combined with elevated CO_2, the areas were 5.5 and 5.3% larger than under ambient conditions, whereas under elevated CO_2 alone the area was 1.4% larger than under ambient conditions.

Figure 9.5 shows a large diurnal variation in sap flow throughout the measuring period (from July 25 to August 25, 1997) (Kellomäki and Wang 1998). The flow ranged from 0.15 to 2.82 kg tree^{-1} d^{-1} in such a way that the CO_2 elevation reduced the flow 4–14% on most days, but on some days elevated CO_2 even slightly increased the flow, depending on weather conditions. Over the whole measuring period (32 days), elevated CO_2 reduced the total transpiration 14%, suggesting that elevated CO_2 strongly influenced the seasonal water use efficiency of experimental trees. Based on a longer monitoring period, Wang et al. (2005) found similarly that over the main growing season (D150–D240), the elevated CO_2 increased the total transpiration by 14% and reduced by 13 and 16% in the years 1999, 2000 and 2001, respectively, the increase being related to the increase of foliage area, and the reduction to the reduction of crown conductance.

Kellomäki and Wang (1998) found further that under a short monitoring elevated temperature increased the daily sap flow by 11–36%. Consequently, the cumulative transpiration over the period (from July 25 to August 25, 1997) increased by 33%. Under the combined elevation of CO_2 and temperature, daily variability was slightly less than under elevated temperature alone, showing that the temperature elevation eliminated the effect of elevated CO_2 to a large extent. In a longer monitoring period (1997–2001), elevated temperature alone or combined with CO_2 increased the sap flow per main growing season by 45–57% (Wang et al. 2005). The increase was related to the increase of foliage area, high canopy conductance, low stomatal sensitivity to high water vapor deficit and increased transpiration demand due to elevated temperature.

The impact of elevated CO_2 and temperature was similar if the sap flow was calculated per needle area or ground area over the measuring period (Table 9.2). Under ambient conditions, the cumulated flow rate per unit needle area was 20 kg H_2O m^{-2} and per unit ground area 19 kg H_2O m^{-2}. Under elevated CO_2, the cumulated sap flow reduced 15–25%, more per needle area than per ground area. The elevation of temperature alone or concurrently with CO_2 increased the cumulated sap flow 11–23%, and was lowest per needle area under elevated CO_2 alone (Kellomäki and Wang 1998).

Kellomäki and Wang (1998) further found that the sap flow was linearly related to vapor pressure deficit regardless of the climatic treatment, if radiation (R_s) in the chambers was less than

300 Wm^{-2} (Fig. 9.6). Under high radiation (R$_s$ > 300 Wm^{-2}), elevated temperature alone (or with CO$_2$) increased further the flow rate under high water vapor deficit, whereas under elevated CO$_2$ alone the flow rates were reduced. Thus, the growth of Scots pine under increased CO$_2$ led to a decrease in total diurnal and cumulative sap flow, whereas elevated temperature increased both diurnal and cumulative sap flow (Table 9.2). There was no significant interaction between elevated CO$_2$ and temperature, but elevated temperature played the dominant role in the combination of CO$_2$ and temperature effecting in the present experimental setup (Wang and Kellomäki 1997, Kellomäki and Wang 1998).

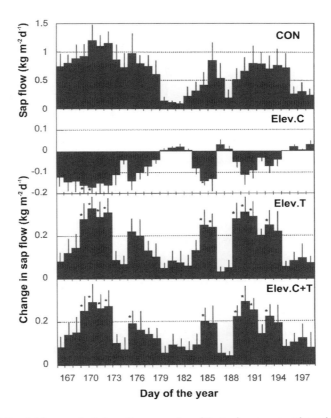

Fig. 9.5 Sap flow (kg H$_2$O m^{-2} d^{-1}) per projected needle area per day of Scots pines grown under ambient (CON) conditions, elevated temperature (Elev. T), elevated CO$_2$ (Elev. C) and both combined (Elev. C + T) in the Mekrijärvi chamber experiment (Kellomäki and Wang 1998). Permission of Wiley & Sons. The change indicates the difference between the values for ambient conditions and elevated temperature and CO$_2$.

Table 9.2 Cumulative sap flow per needle area and ground area from July 25 to August 25, 1997 under elevated CO$_2$ and temperature in the Mekrijärvi chamber experiment (Kellomäki and Wang 1998). The numbers in parenthesis indicate the percent value under elevated CO$_2$ and temperature of the mean value under ambient conditions.

Climatic treatment	Cumulative sap flow of H$_2$O		
	kg tree^{-1}	kg m^{-2} of needle area	kg m^{-2} of ground area
Ambient	34.9 ± 3.2	20.1 ± 1.9	18.5 ± 1.6
CO$_2$	30.9 ± 4.3 (89)	17.1 ± 1.6 (85)	15.8 ± 1.2 (75)
T	46.2 ± 5.0 (132)	25.2 ± 3.0 (125)	20.9 ± 2.7 (113)
CO$_2$ + T	45.3 ± 4.8 (130)	24.8 ± 2.3 (123)	21.2 ± 2.2 (115)

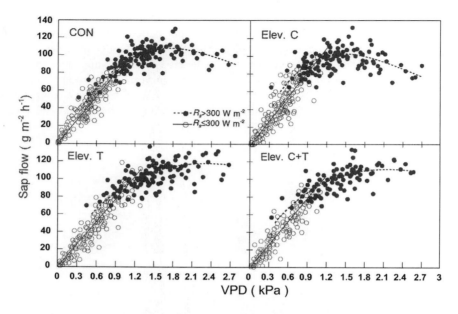

Fig. 9.6 Sap flow (g H_2O m^{-2} h^{-1}) per projected needle area per hour in Scots pines grown under ambient conditions (CON), elevated CO_2 (Elev. C), elevated temperature (Elev. T) and combined elevation of CO_2 and temperature (Elev. C + T) in relation to the daytime hourly mean of vapor pressure (VPD) deficit over the previous 1.5 hours (Kellomäki and Wang 1998). Permission of Wiley & Sons. The data includes all measurements made on days D167–D198, 1997. The bold line is fitting a linear equation: Sap flow = A1 + B1 VPD, to the data (open circles) with R_s < 300 Wm^{-2}. The dashed line (full circles) is fitting a polynomial equation: Sap flow = A2 VPD + B2 VPD^2 + C VPD^3, to the data with R_s > 300 Wm^{-2}.

9.5 Concluding Remarks

The response of whole trees to climate change integrates the physiological impacts of single physiological processes on the growth and development of trees. In this context, climate warming and elevating CO_2 are likely to increase carbon uptake due to: (i) lengthening of growing season; (ii) reducing the dormant period in spring and autumn; and (iii) increasing foliage area and light intercepting. The increase of the foliage area under elevated CO_2 and temperature also increases the total transpiration, but acclimation to climatic treatment affects the rate of water flow through foliage. Wang et al. (2005) found that the year when the Scots pines were enclosed in the chambers increased the total transpiration in linear relation to the increase of foliage, but in subsequent years elevated CO_2 reduced the total transpiration due to decreased crown conductance, thus off-setting the increase in foliage area. Conversely, the elevated temperature increased total transpiration due to the increased foliage area and lengthened growing season. This was also the case for trees grown under the combined elevation of CO_2 and temperature, which offset the reducing effect of the CO_2 elevation.

References

Kellomäki, S. and K. -Y. Wang. 1997. Effects of long-term CO_2 and temperature elevation on crown nitrogen distribution and daily photosynthetic performance of Scots pine. Forest Ecology and Management 450: 309–327.

Kellomäki, S. and K. -Y. Wang. 1998. Sap flow in Scots pine growing under conditions of year-round carbon dioxide enrichment and temperature elevation. Plant, Cell and Environment 21: 969–981.

Leuzinger, S. and M. K. -F. Bader. 2012. Experimental vs. modeled water use in mature Norway spruce (*Picea abies*) exposed to elevated CO_2. Frontiers in Plant Science 3(229): 1–11.

Oker-Blom, P. and S. Kellomäki. 1981. Light regime and photosynthetic production in the canopy of a Scots pine during a prolonged period. Agricultural and Forest Meteorology 24: 185–199.

Wang, K. -Y. 1996a. Apparent quantum yield in Scots pine after four years of exposure to elevated temperature and CO_2. Photosynthetica 32: 339–353.

Wang, K. -Y. 1996b. Canopy CO_2 exchange of Scots pine and its seasonal variation after four-year exposure to elevated CO_2 and temperature. Agricultural and Forest Meteorology 82: 1–27.

Wang, K. -Y. 1996c. Effects of long-term CO_2 and temperature elevation on gas exchange of Scots pine. Ph. D. Thesis, University of Eastern Finland, Finland.

Wang, K. -Y. and S. Kellomäki. 1997. Stomatal conductance and transpiration in shoots of Scots pine after 4-year exposure to elevated CO_2 and temperature. Canadian Journal of Botany 75: 552–561.

Wang, K. -Y., T. Zha and S. Kellomäki. 2002. Measuring and simulating crown respiration of Scots pine with increased temperature and carbon dioxide enrichment. Annals of Botany 90: 325–335.

Wang, K. -Y., S. Kellomäki, T. Zha and H. Peltola. 2005. Annual and seasonal variation of sap flow and conductance of pine trees in elevated carbon dioxide and temperature. Journal of Experimental Botany 56(409): 155–165.

Growth and Structure of Trees under Climate Change

ABSTRACT

Elevated CO_2 and temperature enhance photosynthesis under boreal conditions, thus increasing biomass growth of foliage, branches, foliage, stems and roots. This implies changes in the dynamics of tree structure, e.g., causing more branches to be born in the whorls of boreal Scots pines. At the same time, the elongation and radial growth of branches and the forking of branches increase. Elevated temperature reduces and elevated CO_2 increases the life span of needles. Under elevated CO_2, light interception is likely to be greater than under elevated temperature due to the differences in shoot structure induced by climatic treatment.

Keywords: climate change, biomass growth, growth of branches and shoots, forking of branches, properties of needles, Scots pine, light interception in shoots

10.1 Responses of Growth to Climate Change

Photosynthesis, respiration and growth

Biomass growth is the mass of new tissue synthesized over whole tree in a given time step. Biomass growth is a function of the photosynthates remaining after autotrophic respiration, driving the maintenance of existing tissues and the growth of new tissues: $\Delta W = P - (R_m + R_g)$, where ΔW is the biomass growth, P is the gross amount of carbon available for maintenance and growth in the whole tree or in a given organ (allocated to organ), R_m is the maintenance and R_g the growth respiration. The impacts of climate change on the biomass growth are evident, but the mechanisms are complex as related to the biochemical/physiological processes behind the growth. Biomass growth is further manifested in the growth of dimensions such as height and diameter, the former representing the primary, and the latter secondary growth.

Biomass growth under elevated CO$_2$

In general, the elevation of CO$_2$ increases the biomass growth of C$_3$ plants due to enhanced photosynthesis, but the response is variable. Based on the available literature, Kimball (1983) showed that in crop plants the growth increase ranged from 0 to 200%, with the mean 40%, under an elevation of CO$_2$ up to 1200 µmol mol^{-1} compared to the growth under the ambient CO$_2$ of 300 µmol mol^{-1}. Similarly, Ceulemans and Mousseau (1994) demonstrated in their review, that the growth of seedlings in the genera of *Quercus, Populus, Eucalyptus, Acer, Betula, Pinus* and *Picea* increased under elevated CO$_2$. The mean enhancement of growth was 38% for coniferous and 63% for deciduous species under the doubling of CO$_2$ (700 µmol mol^{-1}) compared to that under ambient CO$_2$ (350 µmol mol^{-1}). However, the increase was variable from 0 up to 300% depending on species, seedling age, experimental infrastructure (e.g., greenhouse, growth chamber, open top chamber, Free-Air CO$_2$ Enrichment (FACE)), light and nutrient supply and duration of treatment. Photosynthesis per unit leaf area increased by 40% in coniferous, and by 61% in deciduous species, but the variability in photosynthesis was less than that of biomass growth.

Most studies of growth under elevated CO$_2$ deal with seedlings or young trees. For example, the total biomass of seedlings of Sitka spruce [*Picea sitchensis* (Bong.) Carr.] (Centritto et al. 1999), Scots pine (*Pinus sylvestris* L.) (Utriainen et al. 2000), loblolly pine [*Pinus taeda* L.] (Tissue et al. 1997), Norway spruce [*Picea abies* (L.) Karst.] (Zheng et al. 2002) and white fir [*Abies alba* Miller] (Hättenschwiler and Körner 2000) and European beech [*Fagus sylvatica* L.] (Overdieck et al. 2007) increased under elevated CO$_2$, with a concurrent increase of height and diameter growth. The FACE experiments with trees passed their juvenility, however, showed a smaller growth increase in response to elevated CO$_2$ than those obtained with the seedlings and/or trees in the juvenile phase (Norby et al. 2005). The same was found by Hättenschwiler et al. (1997), who studied the tree ring chronologies of Mediterranean oaks (*Quercus ilex* L.) grown close to a natural CO$_2$ well. In this study, trees had been grown under an atmospheric CO$_2$ of 650 µmol mol^{-1} over 25–30 years since their establishment. The final stem diameter was 12% greater than that in the trees grown in the similar conditions but under ambient CO$_2$ (< 400 µmol mol^{-1}). Elevated CO$_2$ increased the radial growth mainly in the seedling phase, and later the difference between these two CO$_2$ treatments gradually disappeared. At any given tree age, the effect was greater in the years with a dry spring (April and May) than in a spring with enough precipitation. Similarly, Körner et al. (2005) found small or no extra carbon accumulation in common Central-European deciduous trees (*Fagus, Quercus, Carpinus*) in mature phase when exposed to free-air elevation of CO$_2$.

Biomass growth under elevated temperature

Based on a meta-analysis, Lin et al. (2010) found that climatic warming increased the growth of plants 8–16% depending on the plant species and its functional type. The growth increase in deciduous trees was 56% and 28% in coniferous species. In general, elevated temperature enhances biomass growth, wherever the current growth is very limited by low summer temperatures and a low supply of nitrogen, as in boreal forests (Bergh et al. 1999, Jarvis and Linder 2000). The increasing growth of boreal trees under climatic warming was also noted by Sigurdsson et al. (2013), if the nitrogen supply is high enough. They showed that elevated temperature alone increased the biomass growth of boreal Norway spruce if nitrogen fertilization was used in the experiment. When excluding the fertilization, the response to elevated temperature was marginal or non-existent, as was the case for elevated CO$_2$. These findings are contrary to those of Peltola et al. (2002) and Kilpeläinen et al. (2003, 2005), who found elevated CO$_2$ and temperature to increase the biomass growth of Scots pine in a chamber experiment.

In general, the findings from the boreal chamber experiments agree with those of Way and Oren (2010), showing a clear growth increase in relation to warming (Fig. 10.1). They studied the growth response of several functional plant groups and biomes to changes in temperature. They found that elevated temperature increased growth in deciduous species more than in coniferous ones. Furthermore, the growth of boreal trees (cool-adapted trees) is likely to increase more than tropical trees (warm-adapted trees) if water or nitrogen does not limit growth. On the other hand, tropical trees were more likely to reduce growth under higher temperatures than boreal trees (Ghannoum and Way 2011). Based on a meta-analysis of experiments with elevated CO_2 and temperature, Stinziano and Way (2014) concluded that the biomass growth of boreal trees may increase along with increased temperature, if the temperature increase remains less than 5°C. Higher temperature increase in the growing season seemed to reduce the growth under both ambient and elevated CO_2.

Similarly, Reich and Oleksyn (2008) claimed that climatic warming is likely to affect the growth and survival of Scots pine depending on the geographical location of the site on which trees were originally grown. Based on the data representing Scots pine grown in common-garden experiments, they analyzed how growth and survival responded to the difference in Mean Annual Temperature (MAT) between the growing site and the original growing site (climatic transfer). They found that "climate transfers equivalent to warming by 1–4°C markedly increased the survival of populations in northern Europe (≥ 62° N, < 2°C MAT) and modestly increased height growth ≥ 57° N but decreased survival at < 62° N and modestly decreased height growth at < 54° N latitude in Europe. Thus, even modest climate warming will likely influence Scots pine survival and growth, but in distinct ways in different parts of the species range." Reich and Oleksyn (2008) concluded further that "modest warming will likely enhance the growth and survival of Scots pines in northern Europe (> 62° N), but reduce its growth and survival at lower latitudes."

Wertin et al. (2011) also studied experimentally the effects of elevated temperature on the net assimilation and biomass production of southern read oak (*Quercus rubra* L.) seedlings grown near the southern limit of the species distribution. They found that elevated growing temperature reduced photosynthesis, increased respiration and reduced the height and diameter growth and biomass production. Reduced growth correlated with reduced carbon assimilation. The reduced total biomass at the end of the growing season was directly related to the temperature elevation. The total biomass growth reduced 6% per 1°C increase of the mean temperature over the growing season. Wertin et al. (2011) concluded that the temperature increase above the current conditions in the southern limit of southern red oak seedling is harmful.

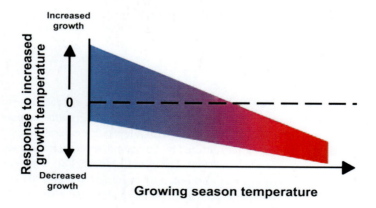

Fig. 10.1 Schematic presentations of Way and Oren (2010) showing how the growth of trees in different thermal conditions (thermal niche indicated by growing season temperature) responds to an increase of temperature under the changing climate. Species from colder environments (in blue) would generally have a positive growth response to warming but could show reduced growth if water or nutrients were limiting. Species from warmer environments (in red) would experience decreased growth, with less variation between individuals or species. Permission of Oxford University Press.

Biomass growth under elevated CO_2 and temperature

The Mekrijärvi chamber experiment provided measurements to estimate how the long-term elevation of CO_2 and temperature may affect the total biomass growth and its allocation to needles, shoots (branches) and stems (Seppo Kellomäki, University of Eastern Finland, unpublished). The data included the annual count of shoots and the annual measurements of the diameter of permanent sample shoots selected for monitoring in the period 1997–2002. In the year 2003, the trees grown in the chambers were harvested and measured in detail (Box 10.1).

Box 10.1 Characterizing the crown structure of experimental Scots pines

During monitoring, the crown of experimental trees was treated as a population of shoots (Fig. 10.2); i.e., the shoots making stems were of zero order, those attached to stem were of first order (branches), shoots attached to branch were of second order, etc. (Kellomäki and Kurttio 1991).

The mass (M) and growth (Growth) of shoot axis in trees were based on the number of shoots (n(k,j,t)) and the mean mass (m(k,j,t)) of shoots in different orders:

$$M(k,t) = \sum n(k,j,t) \times m(k,j,t) \qquad (10.1)$$

$$Growth(k,t) = \sum n(k,j,t) \times m(k,j,t) - \sum n(k,j,t-1) \times m(k,j,t-1) \qquad (10.2)$$

where k is the tree, j is the order of shoot and t is the year. Mass (NM) and growth of needles ($Growth_N$) over trees were calculated based on the length of shoot, the density of needles on shoot, the unit mass of needles, the survival of needles and the number of shoots:

$$NM(k,t) = \sum ND_o(k,j,t) \times L_o(k,j,t) \times p(k,j,t) \qquad (10.3)$$

$$Growth_N(k,t) = \sum ND_o(k,j,t) \times L_o(k,j,t) \times NM_o(k,j,t) \qquad (10.4)$$

where ND_o is the mean needle density of shoot when born, NM_o is the mean mass of single needles when born, L_o is the total length of shoot in different order when born and p the probability of needles to survive to the next year.

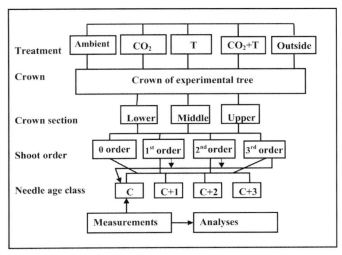

Fig. 10.2 System of sampling shoots for characterizing the crown structure of the trees. First, the total number of shoots with their position in the crown hierarchy was determined. Second, sample shoots in each hierarchical position were selected, and they were marked for annual measurements. A section of length 3 cm was marked permanently on the sample shoots, and the needle density on the shoot was measured annually (Seppo Kellomäki, University of Eastern Finland, unpublished).

Figure 10.3 shows that the mean total growth per tree (four replicates for each climatic treatment) was most increased by the combined elevation of temperature and CO_2, with 300 g more mass growth than under ambient conditions. Under elevated CO_2 alone, the growth was smaller (200 g) but still greater than that under elevated temperature (100 g). The response pattern of different mass components was similar to the total growth, except needle growth, which was similar to that under ambient conditions. Climatic treatment shifted growth from needles to shoots and stems. The growth shift to shoots was 17% under elevated temperature and CO_2 alone and 19%, when the both factors were combined. In the case of stems, the shift was largest under elevated temperature (12%), whereas the shift under the elevation of CO_2 alone or combined with temperature elevation was basically the same (3%) (Table 10.1).

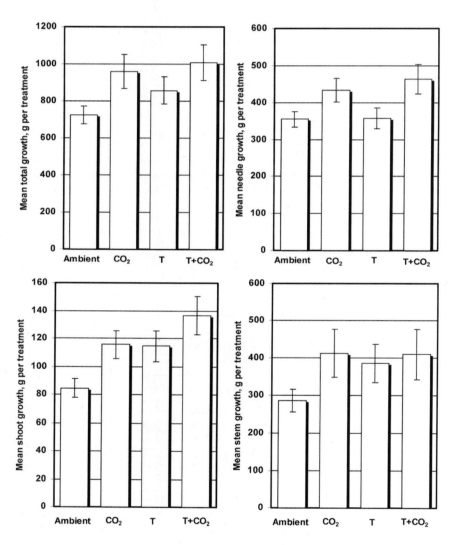

Fig. 10.3 Mean total growth and the growth of different mass components of experimental Scots pines over the period (1997–2002) grown under the current conditions (Ambient), elevated CO_2 and temperature (T) and the combination of both (T + CO_2). The bars indicate the standard error of the mean values (Seppo Kellomäki, University of Eastern Finland, unpublished).

Table 10.1 Mean percentages of the growth of needles, shoots and stems from the total growth and the shift of growth allocation in percentages to that under ambient conditions as a function of climatic treatments (Seppo Kellomäki, University of Eastern Finland, unpublished).

Treatment	Needles		Shoots		Stem	
	Share, %	Change in share, %	Share, %	Change in share, %	Share, %	Change in share, %
Ambient	50.26		11.69		38.05	
CO_2	48.17	−4	12.62	+8	39.22	+3
T	43.78	−13	13.67	+17	42.55	+12
T + CO_2	48.08	−4	13.89	+19	39.04	+3

10.2 Response of Stem Growth to Climate Change

Figure 10.4 shows the cumulative height and radial growth of Scots pine in the Mekrijärvi chamber experiment when Scots pines were exposed to elevated CO_2 and temperature. After three years, the stem diameter was 26 and 57% under elevated temperature and CO_2 alone and 67% greater under the combined treatment (Peltola et al. 2002). The relative growth response to climatic treatments remained fairly similar over a six-year monitoring period (Kilpeläinen et al. 2005): the elevation of CO_2 alone increased the total diameter by 66%, which was clearly more than that under the combined elevation of temperature and CO_2 (47%) compared to the ambient conditions. Obviously, the temperature elevation with enhanced respiration reduced the amount of photosynthates for growth, just as under the elevation temperature alone. In this case, the diameter growth increased by 19% compared to ambient trees.

Figure 10.4 shows further that the diameter increased fairly linearly as a function of time in the early phase of monitoring but later tended to level off. At the same time, the height increased exponentially in the early phase, peaking in the middle of the monitoring period. In both cases, climatic treatment affected the level, but the patterns of height and diameter were similar regardless of climatic treatment. This indicates that the distribution of diameter growth along the stem is affected by climatic treatment, as shown in Fig. 10.5. In the calculations, the elongation of stem (L (0, j), cm) in the year j was related to the total length of stem in the previous year (TL (j−1), cm).

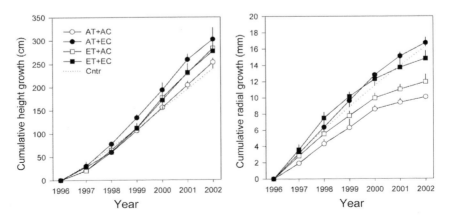

Fig. 10.4 Development of height (left) and diameter at 1.3 m above soil level (right) in Scots pines growing under elevated CO_2 and temperature in the period 1997–2002 after closed into chambers in 1996 (Kilpeläinen et al. 2005). Permission of Oxford University Press. Legend: Cntr is ambient conditions; AT is ambient temperature; AC is ambient CO_2; ET is elevated temperature; and EC is elevated CO_2.

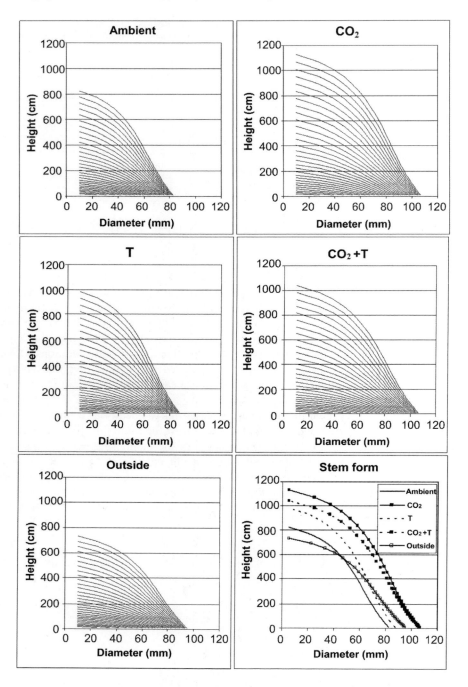

Fig. 10.5 Distribution of diameter growth along stem of young Scots pines over the calculations of 25 years with Equations (10.5) and (10.6) as a function of climatic treatment, representing the current conditions (Ambient), elevated atmospheric carbon (CO_2) and temperature (T) and the combination of both (CO_2 + T) including the conditions outside the chambers in the Mekrijärvi experiment. The calculations are based on the assumption that the height and diameter growth of a daughter shoot are related to the length and diameter of the mother shoot (Box 10.1). The initial length and butt diameter of seedlings were 10 cm and 10 mm (Seppo Kellomäki, University of Eastern Finland, unpublished).

Similarly, the diameter (D (0, j), mm) of each was related to the diameter (D (0, j–1), mm) in the previous year. Consequently:

$$L(0, j) = \frac{a}{1+b \times e^{-c \times TL(0, j-1)}} \qquad (10.5)$$

$$D(0, j) = k(j) \times D((0, j-1)) \qquad (10.6)$$

where a, c and b are parameters and k (j) = 1 – exp(-b) x D ((0, j–1)).

The calculations launched with small seedlings showed, that the combined elevation of CO_2 and temperature or elevated CO_2 alone increased height by 20–25% by compared to the ambient conditions (Fig. 10.5). At the same time, the diameter at the stem butt was 25–30% greater than under ambient conditions. Elevated temperature alone increased clearly less the height and diameter than other climatic treatments, excluding the ambient conditions as found by Peltola et al. (2002) and Kilpeläinen (2005). Throughout the calculations, the diameter growth was greater in the upper stem than in the lower stem and the stem form at the end of the simulation was similar regardless of climatic treatment. The stem form in trees enclosed in the chambers was, however, more slender than those grown outside the chambers, where trees were exposed to wind sway and snow load.

10.3 Response of Dynamics of Branches in Crown to Elevated CO_2 and Temperature

In conifers, height growth is predetermined, and a new whorl with a varying number of branches is formed every year. In boreal Scots pine, for example, a new whorl has three to six branches, higher numbers representing more fertile sites and wider spacing than lower ones (Flower-Ellis et al. 1976). The number and diameter of branches in a whorl are inversely related; i.e., the allocation of total branch growth among a larger number of branches will give a small initial mean diameter and wide diameter differences between branches (Kellomäki and Kurttio 1991). The diameter differences between initial branches remain throughout the life span of the whorl, and they affect the mortality of branches. The risk of mortality increases more rapidly for small branches than for larger ones when the whorl is deeper in the canopy (Kellomäki and Väisänen 1988).

Volanen et al. (2006) found that under elevated CO_2 the birth rate of branches was 28% higher than that before treatment (Fig. 10.6). The elevation of temperature also increased the average birth rate, which was 16% greater than that before exposure. Under the concurrent elevation of CO_2 and temperature, the birth rate reduced slightly (5%). Exclusion of the effects of differences in tree size (height, diameter) and growth on the birth rate before treatment, the elevation of CO_2 alone increased the birth rate by 19% and the elevation of temperature alone by 28%, whereas their combination increased the birth rate only marginally (< 1%). At the same time, the average mortality of branches was 0.3 ± 0.10 branches per year under ambient conditions and 0.6 ± 0.13 branches per year for elevated CO_2. Similarly, 0.5 ± 0.26 branches died per year under elevated temperature, mainly during the later phase of exposure. Under the combined elevation of CO_2 and temperature, the mortality rate was highest; i.e., 1.8 ± 0.48 branches per year (Volanen et al. 2006).

Diameter growth of branches in Scots pine is linearly related to the stem growth (Kellomäki and Väisänen 1988). For example, Jach and Ceulemans (1999) found that a two-year exposure to elevated CO_2 increased the branch diameter up to 29% along with the enhanced total growth of Scots pine seedlings. Volanen et al. (2006) also found that exposure to elevated CO_2 increased the initial branch diameter 20% compared to the branches born under ambient conditions. This pattern held for elevated temperature alone and concurrently with elevated CO_2 during the first two years. Volanen et al. (2006) also found that the mean diameter of the oldest branches (born 1993) was similar regardless of climatic treatment (Fig. 10.6); i.e., no differences were found in the three branch cohorts born before climatic treatment (1994–1996). However, the initial branch diameter had a strong effect on the final branch diameter, which was related directly to the stem diameter and inversely to the number of branches in the whorl (Volanen et al. 2006).

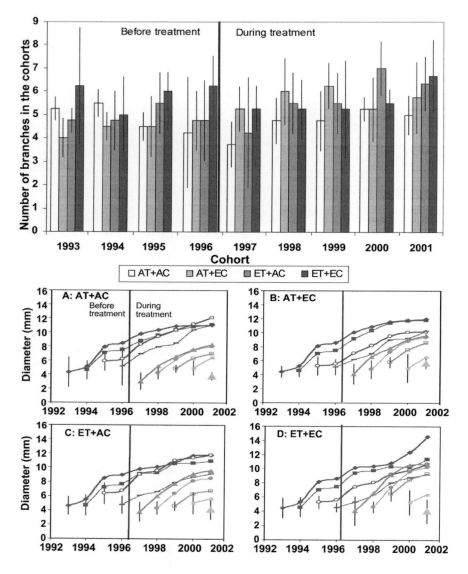

Fig. 10.6 Above: Birth rate of branches, and Below: Development of mean diameter of branches per cohorts in Scots pines grown under ambient conditions (A, AT + AC), elevated CO_2 (B, AT + EC), elevated temperature (C, ET, AC) and combined both (D, ET + EC) (Volanen et al. 2006). Permission of Taylor & Francis. In Figures, the lines upwards indicate the year (1996), when climatic treatment began.

10.4 Response of Shoot Properties to Climate Change

Impacts of elevated CO_2 and temperature on shoot structure

Light interception in the coniferous crown is affected by the structure of shoots and their distribution in the crown space. Over time, the shoot structure is a function of the needle area of a shoot (TNA (i, j), cm²) as affected by the survival rate of needles (p (i, j)), the density of needles on shoot (ND (i, j), number of needles cm⁻¹), length of shoot (L (i, j), cm), and the projection area of single needles (SNA (i, j), cm² needle⁻¹):

$$TNA(i, j) = p(i, j) \times ND(i, j) \times L(i, j) \times SNA(i, j)/100 \tag{10.7}$$

where i refers to the order of shoots and j to the year. The needle area per shoot reduces to zero when the age of the shoot exceeds the length of life span of needles on the shoot.

Figure 10.7 shows that in Scots pine the survival of needles from year zero to year one (from birth to the age of one year) was fairly similar regardless of climatic treatment (Seppo Kellomäki, University of Eastern Finland, unpublished data). However, elevated CO_2 alone increased survival from the year one to the year two (Fig. 10.7, upper part), whereas elevated temperature alone or concurrently with elevated CO_2 reduced survival (Wang 1996). The effect was especially large in the third-order shoots, where elevated CO_2 alone increased survival, as was the case for the fourth-order shoots. In general, the closure of trees in the chambers slightly reduced survival, but the survival pattern under ambient conditions was fairly similar to that outside the chambers (outside control).

The mean needle density per unit length of shoot was slightly reduced under elevated temperature alone or combined with elevated CO_2 (Fig. 10.7, middle part). Under elevated CO_2 alone, the density

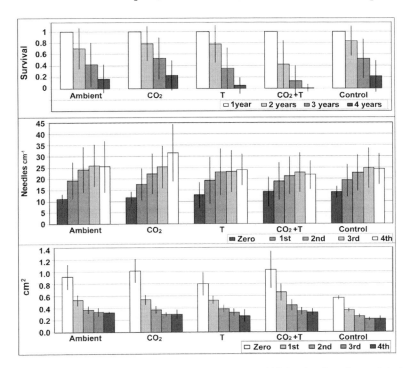

Fig. 10.7 Mean needle survival of varying age (upper), and needle density (middle) and projected area of single needle (lower) in varying shoot orders in Scots pines grown under ambient conditions (Ambient) and elevated of CO_2 and temperature (T) and combined both (CO_2 + T). The values of needle density and area indicate the situation at the age of one year in the autumn after the summer when the needles were born (Seppo Kellomäki, University of Eastern Finland, unpublished).

Table 10.2 Mean length and diameter of the shoots with percentage of that under ambient conditions at birth as a function of treatment and shoot order. Standard deviations are in parenthesis (unpublished, Seppo Kellomäki, University of Eastern Finland).

Treatment and shoot order	Length of shoot		Diameter of shoot axis	
	cm	% of ambient	cm	% of ambient
Ambient				
0	23.8 (12.0)		0.9 (0.3)	
1	11.0 (6.6)		0.2 (0.1)	
2	4.3 (2.7)		0.1 (0.1)	
3	2.1 (1.3)		0.1 (0.1)	
Elevated CO_2				
0	26.3 (14.5)	110	1.1 (0.2)	123
1	12.0 (8.1)	109	0.3 (0.2)	114
2	4.7 (3.1)	108	0.2 (0.1)	102
3	2.4 (1.4)	115	0.1 (0.1)	93
Elevated temperature (T)				
0	24.6 (13.6)	103	0.9 (0.3)	99
1	12.6 (8.2)	115	0.2 (0.1)	105
2	5.2 (3.5)	121	0.1 (0.1)	100
3	2.7 (1.7)	130	0.1 (0.1)	96
Elevated CO_2 + T				
0	25.1 (12.6)	105	1.1 (0.3)	116
1	12.6 (8.4)	115	0.3 (0.2)	113
2	5.1 (3.6)	118	0.1 (0.1)	99
3	2.5 (1.6)	120	0.1 (0.1)	91

was similar to that under ambient conditions or outside the chambers, except for the shoots of higher order, where the needle retention tended to increase. At the same time, the projected needle area in the stem main leader (the order 0) was slightly reduced under the elevated temperature (Fig. 10.7, lower part), whereas elevated CO_2 alone or combined with elevated temperature only slightly affected the projected needle area regardless of shoot order.

Climatic treatment also affected the length and diameter of shoot axis. As may be expected, the mean length of the zero-order shoot (the stem main leader) was the largest, regardless of climatic treatment, whereas the mean length reduced rapidly following the increase of shoot order (Table 10.2). Elevated CO_2 alone increased the length and diameter of stem leader more than did temperature alone or combined with elevated CO_2. The growth increase was the greatest in the third-order shoots, in which diameter growth remained the smallest. The length of higher order shoots increased substantially in response to elevated CO_2 and temperature, whereas the response of diameters remained smaller.

Light interception in shoots under elevated CO_2 and temperature

Light interception in a shoot is the function of incoming radiation and the effective needle area, which is the part of total needle area intercepting light. The effective needle area is a function of the projection area of the shoot and light transmission through the shoot (Box 10.2). Regardless of treatment, potential interception is the highest in second-order shoots (shoots attached to the main axis of branches) (Fig. 10.8). Elevated CO_2 slightly increased light interception compared to shoots grown under other treatments. Four years later, the interception was substantially less, regardless of treatment, but still greatest under elevated CO_2. Interception reduced most under elevated temperature and CO_2, due to the shorter survival of needles under this treatment. In the same way, interception was reduced under elevated temperature alone, but the reduction was not as pronounced as under elevated temperature and CO_2.

Growth and Structure of Trees under Climate Change 137

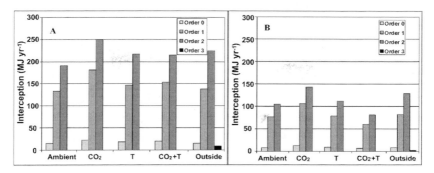

Fig. 10.8 Left: Potential light interception in the shoots of different orders in Scots pine with a full cover of needles around the shoot axis grown under the ambient conditions (Ambient) and elevated atmospheric carbon dioxide (CO_2), temperature (T), the combination of both (CO_2 + T) and outside of chambers. Right: Potential light interception by shoots of different orders in the fourth years, when needle survival was included in the calculations. In the calculations, the angle of the needles relative to the shoot axis was 60° regardless of the treatment. The calculations were made for latitude 63° N over the period from April 15 to September 15, when the amount of direct radiation was 1,962 MJ m^{-2} and the diffuse radiation 845 MJ m^{-2}. The calculations apply to single representative shoots, which are reached by radiation all over the hemisphere, excluding any shading from neighboring shoots (Seppo Kellomäki, University of Eastern Finland, unpublished).

Box 10.2 Potential light interception in a Scots pine shoot

Light interception (Intercept (i, j)) is a function of the incoming radiation and the effective needle area (Fig. 10.9) (Kellomäki and Strandman 1995):

$$Intercept(i,j) = \sum_{l=1}^{9}\sum_{r=1}^{24} SAS0(i,j)(\Phi) \times TOT0(i,j) \qquad (10.8)$$

where SAS0(I, j) (Φ) is the effective needle area of a shoot [m²], TOT0(i, j) is the total incoming radiation to the shoot, r is the azimuth of a sector (1...24) of hemisphere, and l is the sector (1...9) of the hemisphere (i.e., 15° wide and 10° high). The effective needle area is a function of the projection area of a shoot SAC (i, j) (Φ) [cm²] in relation to the angle (Φ) between the shoot and the projection surface and light transmission through a shoot (STRM (i, j)):

$$SAS(i,j)(\Phi) = (1 - STRM(i,j)) \times SAC(i,j) \qquad (10.9)$$

Light transmission is further:

$$STRM(i,j) = \exp(-K(i,j)(\Phi) \times SL(i,j)(\Phi)) \qquad (10.10)$$

where SL (I, j) is the needle area index of a shoot, which is the function of the total needle area of a shoot (TNA (i, j)) and the projection area of a shoot:

$$SL(i,j) = TNA(i,j)/(SAC(i,j)(\Phi)) \qquad (10.11)$$

The total needle area is a function of the needle survival rate (p (i, j)), density of needles on the shoot (ND (i, j), needles per cm of shoot), length of the shoot (L (i, j)) and the projection area of a single needle (SNA (i, j)). The projection area of a shoot (SAC (i, j) (Φ)) is:

$$\begin{aligned}SAC(i,j)(\Phi) = &\ 2 \times \left[L(i,j) + NL(i,j) \times \cos(NK)\right] \times \cos(\Phi) \\ &\times \left[NL(i,j) \times \sin(NK) + 0.5 \times D(i,j)\right]^2 \\ &+ \pi \times \left[NL(i,j) \times \sin(NK) + 0.5 \times D(i,j)\right]^2 \times \sin(\Phi)\end{aligned} \qquad (10.12)$$

where NK is the angle of needles in relation to shoot axis [degrees], NL (i, j) is the length of needles [cm] and D is the diameter of the shoot in the middle of the axis [cm].

Box 10.3 contd....

Box 10.3 contd....

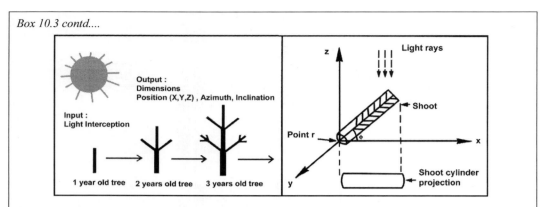

Fig. 10.9 Left: Schematic growth of crown in Scots pine, and Right: Interception of light in a shoot with a given structure (Oker-Blom and Kellomäki 1982, Ikonen et al. 2009). Courtesy of the Finnish Society of Forest Science and the Finnish Forest Research Institute.

10.5 Concluding Remarks

In general, climatic warming and elevating CO_2 are likely to increase the biomass growth of trees under boreal conditions wherever there are no water and/or nitrogen limitations (Bergh et al. 1999). Enhanced stem growth also implies the enhanced growth of branches, but the elevation of CO_2 and temperature, either alone or concurrently, had little effect on the number of branches in whorls, or the growth of branches in boreal Scots pine. Changes in the dimensions of needles correlated with those in the dimensions of the shoot axis. However, light interception in current shoots grown under elevated temperature or CO_2 was similar once the differences in size between the shoots were taken into account. Model calculations showed that after four years, light interception in shoots of Scots pine grown under elevated temperature alone or concurrent with elevated CO_2 was substantially smaller than in the other cases. The temperature elevation substantially reduced the life expectancy of the needles, thus reducing the total needle mass and foliage area in the crown.

Climate change may increase both height and diameter growth of stem, but the allometric structure of trees (e.g., the relationship between stem and foliage mass) is likely to remain as current. Findings based on small trees and shoots indicate that climate change may enhance the maturation of trees, and trees have their inherent maximum dimensions (e.g., maximum height and diameter) earlier than under the current climate but experimental findings are still scarce. However, the increasing maturity likely reduces the response of stem growth, including diameter and height growth, to the elevating CO_2 and temperature as found by Hättenschwiler et al. (1997). A similar response was found by Kellomäki et al. (2005, 2008), who used a process-based model (FinnFor) for estimating how the combined elevation CO_2 and temperature affected the diameter growth of Scots pine in different phases of maturity indicated by stem diameter (see Chapter 16, Fig. 16.3). The simulations showed that the growth culminated at the diameter 10–14 cm, earlier under the current than warming climate. In this developmental phase, the maximum increase of diameter growth was 25–30% compared to the growth under the current climate. Growth increase reduced as a function of increasing stem diameter, faster under warming than current climate. In large trees (diameter > 35 cm), the growth increase in response to climate change virtually disappeared.

References

Bergh, J., S. Linder, T. Lundmark and B. Elfving. 1999. The effect of nutrient availability on the productivity of Norway spruce in northern and southern Sweden. Forest Ecology and Management 119: 51–62.

Centritto, M., H. S. J. Lee and P. G. Jarvis. 1999. Long-term effects of elevated carbon dioxide concentration and provenance on four clones of Sitka spruce (*Picea sitchensis*). I. Plant growth, allocation and ontogeny. Tree Physiology 19: 799–806.

Ceulemans, R. and M. Mousseau. 1994. Effects of elevated atmospheric CO_2 on woody plants. New Phytologist 127(3): 425–446.

Flower-Ellis, J., A. Albrektsson and L. Olsson. 1976. Structure and growth of some young Scots pine stands: (1) dimensional and numerical relationships. Technical Report 3, Swedish Coniferous Forest Project, Uppsala, Sweden.

Ghannoum, O. and D. A. Way. 2011. On the role of ecological adaptation and geographical distribution in the response of trees to climate change. Tree Physiology 31: 1273–1276.

Hättenschwiler, S., F. Miglietta, A. Raschi and C. Körner. 1997. Thirty years of *in situ* tree growth under elevated CO_2: a model for future forest responses? Global Change Biology 3: 463–471.

Hättenschwiler, S. and C. Körner. 2000. Tree seedling responses to *in situ* CO_2-enrichment differ among species and depend on understorey light availability. Global Change Biology 6: 213–226.

Ikonen, V. -P., S. Kellomäki and H. Peltola. 2009. Sawn timber properties of Scots pine as affected by initial stand density, thinning and pruning: a simulation based approach. Silva Fennica 43(3): 411–431.

Jach, M. E. and R. Ceulemans. 1999. Effects of elevated atmospheric CO_2 on phenology, growth and crown structure of Scots pine (*Pinus sylvestris*) seedlings after two years of exposure in the field. Tree Physiology 19: 289–300.

Jarvis, P. G. and S. Linder. 2000. Constraints to growth of boreal forests. Nature 405: 904–905.

Kellomäki, S. and H. Väisänen. 1988. Dynamics of branch population in the canopy of young Scots pine stand. Forest Ecology and Management 24: 67–83.

Kellomäki, S. and O. Kurttio. 1991. A model for the structural development of a Scots pine crown based on modular growth. Forest Ecology and Management 43: 103–123.

Kellomäki, S. and H. Strandman. 1995. A model for the structural growth of young Scots pine crowns based on light interception of shoots. Ecological Modelling 80: 237–250.

Kellomäki, S., H. Strandman, T. Nuutinen, H. Peltola, K. T. Korhonen and H. Väisänen. 2005. Adaptation of forest ecosystems, forests and forestry to climate change. Finnish Environment Institute, FinAdapt Working Paper 4: 1–50.

Kellomäki, S., H. Peltola, T. Nuutinen, K. T. Korhonen, H. Strandman. 2008. Sensitivity of managed boreal forests in Finland to climate change, with implications for adaptive management. Philosophical Transactions of the Royal Society B363: 2341–2351.

Kilpeläinen, A., H. Peltola, A. Ryyppö, K. Sauvala, K. Laitinen and S. Kellomäki. 2003. Wood properties of Scots pine (*Pinus sylvestris*) grown at elevated temperature and carbon dioxide concentration. Tree Physiology 23: 889–897.

Kilpeläinen, A., H. Peltola, A. Ryyppö and S. Kellomäki. 2005. Scots pines (*Pinus sylvestris*) responses to elevated temperature and on wood properties. Tree Physiology 25: 75–83.

Kimball, B. A. 1983. Carbon dioxide and agricultural yield: an assemblage and analysis of 430 prior observations. Agronomy Journal 75: 779–788.

Körner, C., R. Asshoff, O. Bignucolo, S. Hättenschwiler, S. G. Keel, S. Peláez-Riedl et al. 2005. Carbon flux and growth in mature deciduous forest trees exposed to elevated CO_2. Science 309: 1360–1362.

Lin, D., J. Xia and S. Wan. 2010. Climatic warming and biomass accumulation of terrestrial plants: a meta-analysis. New Phytologist 188: 187–198.

Norby, R. J., E. H. DeLucia, B. Gielen, C. Calfapietra, C. P. Giardina, J. S. King et al. 2005. Forest response to elevated CO_2 is conserved across a broad range of productivity. PNAS 102(50): 18052–18056.

Oker-Blom, P. and S. Kellomäki. 1982. Metsikön tiheyden vaikutus puun latvuksen sisäiseen valoilmastoon ja oksien kuolemiseen. Teoreettinen tutkimus. Abstract: Effect of stand density on the within-crown light regime and dying-off of branches. Folia Forestalia 509: 1–14.

Overdieck, D., D. Ziche and K. Böttcher-Jungclaus. 2007. Temperature response of growth and wood anatomy in European beech saplings grown in different carbon dioxide concentrations. Tree Physiology 27: 261–268.

Peltola, H., A. Kilpeläinen and S. Kellomäki. 2002. Diameter growth of Scots pine (*Pinus sylvestris*) trees at elevated temperature and carbon dioxide under boreal conditions. Tree Physiology 22: 963–972.

Reich, P. B. and J. Oleksyn. 2008. Climate warming will reduce growth and survival of Scots pine except in the far north. Ecology Letters 11: 588–597.

Sigurdsson, B. D., J. L. Medhurst, G. Wallin, O. Eggertsson and S. Linder. 2013. Growth of mature boreal Norway spruce was not affected by elevated [CO_2] and/or air temperature unless nutrient availability was improved. Tree Physiology 33: 1192–1205.

Stinziano, J. R. and S. A. Way. 2014. Combined effects of rising [CO_2] and temperature on boreal forests: growth, physiology and limitations. Botany 92: 425–436.

Tissue, D. T., R. B. Thomas and B. R. Strain. 1997. Atmospheric CO_2 enrichment increases growth and photosynthesis of *Pinus taeda*: a 4-year experiment in the field. Plant, Cell and Environment 20: 1123–1134.

Utriainen, J., S. Janhunen, H. -S. Helmisaari and T. Holopainen. 2000. Biomass allocation, needle structural characteristics and nutrient composition in Scots pine seedlings exposed to elevated CO_2 and O_3 concentrations. Trees 14: 475–484.

Volanen, V., H. Peltola, I. Rouvinen and S. Kellomäki. 2006. Impact of long-term elevation of atmospheric CO_2 concentration and temperature on the establishment, growth and mortality of boreal Scots pine branches. Scandinavian Journal of Forest Research 21: 115–123.

Wang, K. -Y. 1996. Canopy CO_2 exchange of Scots pine and its seasonal variation after four-year exposure to elevated CO_2 and temperature. Agricultural and Forest Meteorology 82: 1–27.

Way, D. A. and R. Oren. 2010. Differential responses to changes in growth temperature between trees from different functional groups and biomass: a review and synthesis of data. Tree Physiology 30: 669–688.

Wertin, T. M., M. A. McGuire and R. O. Teskey. 2011. Higher growth temperatures decreased net carbon assimilation and biomass accumulation of northern red oak seedlings near the southern limit of the species range. Tree Physiology 31: 1–12.

Zheng, D., M. Freeman, J. Bergh, I. Røsberg and P. Nilsen. 2002. Production of *Picea abies* in South-east Norway in response to climate change: a case study using process-based model simulation with field validation. Scandinavian Journal of Forest Research 17: 35–46.

Properties of Plant Material under Climate Change

ABSTRACT

The physical and chemical properties of wood and other plant materials are closely linked with the growth and thus with the changes in growth caused by climate change. Impacts on the properties of wood are however, small compared to changes in growth. The temperature elevation alone may increase fiber length and the density of wood, whereas elevated CO_2 alone may reduce fiber length. Under elevated CO_2, the cellulose content may reduce but the lignin content increase. The concentration of monoterpenes in Scots pine foliage is likely reduced under elevated CO_2, whereas the effects of elevated temperature is not clear. Climate change may substantially alter the emissions of volatile secondary compounds due to changes in tree species composition and an increase of foliage mass and emission rate of volatile compounds due to higher temperatures.

Keywords: climate change, physical and chemical properties of wood, secondary compounds, emissions of volatile secondary compounds

11.1 Response of Wood and Plant Material to Climate Change

Wood is the secondary xylem in the stems of trees. It provides a strong skeleton for the whole structure of the tree, supporting foliage and optimizing light interception for carbon uptake and growth. At the same time, the surface parts of xylem (sapwood) transfer water and nutrients to the foliage and other growing tissues. The properties of wood are characterized by structural tissues, which are widely used in manufacturing wooden items and materials or used in construction in the form of sawn timber, plywood, etc. Wood is further a composite of cellulose and lignin, which can be decomposed into cellulose, lignin and other compounds in chemical pulping for the manufacture of paper and related products. Wood is widely used throughout the world as solid fuel, even in developed and industrialized countries. Wood is also used as a raw material, producing chemicals for different purposes, including liquid fuels for heating and transportation.

In general, the properties of wood are divided into the physical and chemical ones. In mechanical wood processing, such as in the sawing and joinery industry, the amount and share of heart wood,

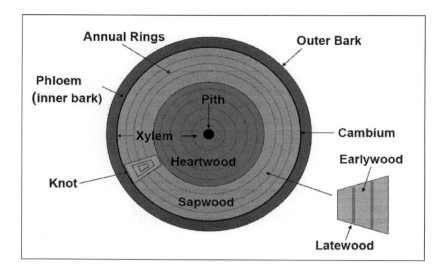

Fig. 11.1 Schematic presentation of the cross-section of a tree stem, with the main structural macroscopic components of wood (Gardiner 2011). Courtesy of Barry Gardiner.

knots in wood, wood density, width of annual growth rings and share of early and late wood are important properties affecting the strength, appearance and properties in processing (Fig. 11.1). In chemical wood processing, such as in the pulp and paper industry, the lignin and cellulose content and fiber length are important factors affecting manufacturing processes, yield and the properties of pulp, paper and related products. There are a variety of extractives in wood, including fatty acids, resin acids, waxes and terpenes, which are used in manufacturing, such as tall oil, turpentine and rosin. Extractives, such as rosin, reduce attacks from damaging insects and other organisms, thus protecting trees.

Both physical and chemical properties of wood are closely linked to growth and thus to changes in growth caused by climate change. The amount and size of sound and dead knots in wood are directly linked to changes in the dynamics of the crown: how climate change affects birth, growth, mortality and pruning-off of dead branches. On the other hand, climate change affects radial growth, modifying the amount and share of early and late wood and the density of wood. These factors further indicate the yield of pulp per unit amount of wood. Changes in growth further imply changes in cell walls, for example, changes in the amount of cellulose and lignin and fiber length affecting pulp and paper products.

11.2 Responses of Physical and Chemical Properties of Wood to Climate Change

Branches on stem and knots in wood under elevated CO_2 and temperature

The dynamics of the branch population in trees, like boreal Scots pine, with a predetermined growth pattern are controlled by the number and thickness of living and dead branches attached to the stem, and the consequent knots (living, dead) in wood (Kellomäki and Väisänen 1988, Volanen et al. 2006). Branchiness may be used to indicate likely impacts of branches on knots in wood; i.e., the share (%) of the total area of branch cross-sections related to the total area of stem surface from the apex to butt, as shown in Fig. 11.2 for Scots pine grown under elevated CO_2 and temperature.

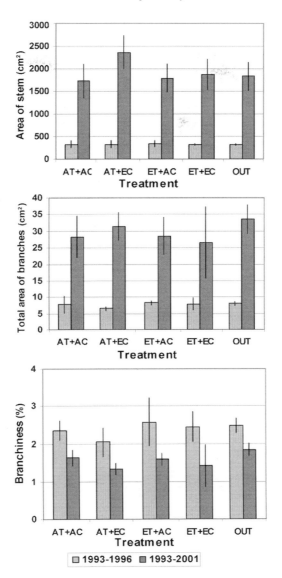

Fig. 11.2 Total stem surface (upper), total area of branch cross-sections (middle) and branchiness (lower) as indicated by the share (%) of branch cross-section area of the stem surface area before (1993–1996) and after exposure (1993–2001) under elevated CO_2 (EC) temperature (ET) and combined elevation of both factors compared to ambient conditions with no elevation of CO_2 (AC) and temperature (AT) (Seppo Kellomäki, University of Eastern Finland, unpublished).

Before exposure (1993–1996), the values of branchiness under ambient conditions were slightly less than in trees later exposed to elevated CO_2 and temperature. During exposure, these differences tended to reduce. The values of branchiness reduced substantially even before the treatment, due to stem growth. This tendency continued during climatic treatment until the values levelled off. The pattern was similar to that outside the chambers, demonstrating the reducing branchiness along with the maturing of trees and dying of branches in the lower crown, as occurred under the combined elevation of CO_2 and temperature. The enhanced turnover of the branch population under this treatment may indicate that the quality of timber in terms of branchiness and the quality of wood in terms of knots may be higher than can be expected based purely on the stem growth.

Early and late wood, and the density of wood under elevated CO_2 and temperature

Table 11.1 shows the response of selected properties of Scots pine wood to elevated CO_2 and temperature (Kilpeläinen et al. 2004). Height growth increased under elevated CO_2 and temperature, but the change was not statistically significant. This was opposite to the diameter growth, which increased substantially under elevated CO_2. At the same time, the thickness of early wood increased more than that of late wood. Under elevated temperature, the increase of early wood was, however, smaller than in late wood. The concurrent elevation of CO_2 and temperature increased the thickness in a similar way, in both early and late wood. The density of both early and late wood increased under elevated temperature, whereas they tended to reduce slightly under elevated CO_2. This pattern held for the minimum, maximum and mean value of wood density.

In general, the elevated CO_2 and temperature substantially increased diameter growth and the share of early and late wood but these changes had only small effects on wood density of boreal Scots pines past their juvenility (Table 11.1). This is in line with the findings of Ceulemans et al. (2002), who demonstrated a large increase in the biomass (49%) and volume (38%) of Scots pine seedlings grown three years under elevated CO_2 (ambient + 400 ppm). The strong effect on diameter growth was mainly due to the increase in the share of early wood in ring width. No changes in wood density were detectable, but the compression strength of wood was reduced due to large tracheids, with thinner walls and larger cavities.

The findings for Scots pine are partly different from those for mature Norway spruces. Kostiainen et al. (2004) found that under elevated CO_2 (700 ppm) the diameter of the radial lumen of wood cells decreased and annual ring width increased in the second year of exposure, while the CO_2 effect was later clear only in the unfertilized trees. At the same time, stem wood chemistry and structure were significantly affected by fertilization, which decreased C/N ratio, mean ring density, early wood density, latewood density, cell wall thickness and late wood percentage. Kostiainen et al. (2004, 2009) conclude that elevated CO_2 and temperature had only minor effects on wood properties of Norway spruce, while fertilization had more marked effects. However, the differences between the effects of elevated CO_2 and temperature are probably related to the effects of treatments on the seasonal course of xylogenesis. Kalliokoski et al. (2013) suggested that "the onset and rate of tracheid formation and differentiation during summer is primarily controlled by photoperiod, temperature and availability of nutrients, rather than supply of carbohydrates."

Table 11.1 Effect of elevated temperature and CO_2 on properties of wood in Scots pine grown in the Mekrijärvi chamber experiment (Kilpeläinen et al. 2003, 2004, 2007).

Variable	Effect		Change, % of that under ambient conditions		
	CO_2	T	CO_2	T	CO_2 + T
Height growth	ns	ns	19	12	9
Diameter growth	+	ns	66	19	47
Thickness of early wood	+	ns	70	10	46
Thickness of late wood	+	ns	56	39	50
Share of early wood, %	ns	ns	7	−8	−3
Share of late wood, %	ns	ns	−14	15	6
Mean density of wood	ns	+	−6	6	6
Density of early wood	ns	+	−6	2	6
Density of late wood	ns	ns	−1	2	2
Minimum density	ns	ns	−9	−3	6
Maximum density	ns	ns	−2	2	−1

ns = effect not statistically significant (p > 0.10), + or − = increase/decrease with statistically significant (p < 0.10).

Fiber length, and the content of cellulose and lignin under elevated CO_2 and temperature

Kilpeläinen et al. (2003, 2004, 2007) found that temperature elevation increased the length of fibers in Scots pine wood (Table 11.2), whereas elevated CO_2 slightly reduced fiber length. Similarly, the cellulose content was reduced but the lignin content increased under elevated CO_2. The effect of elevated temperature tended to be the opposite, but these changes were relative small compared to changes in growth. Nevertheless, temperature elevation alone or concurrently with CO_2 reduced extractives while increasing fiber length. This is line with the findings of Kostiainen et al. (2009), who showed temperature elevation to reduce the concentrations of acetone-soluble extractives and soluble sugars in Norway spruce, while the mean and early wood cell wall thickness and wood density were increased. They found no effects due to CO_2 elevation, as found by Kilpeläinen et al. (2003, 2004, 2007) for Scots pine. On the other hand, Ceulemans et al. (2002) found that the number of resin canals and the resin content of wood were reduced under elevated CO_2, thus affecting extractives in the wood of Scots pine seedlings.

Table 11.2 Effect of elevated temperature and CO_2 on properties of wood in Scots pine grown in the Mekrijärvi chamber experiment (Kilpeläinen 2003, 2004, 2007).

Variable	Effect CO_2	Effect T	Change, % CO_2	Change, % T	Change, % CO_2+T
Fiber length	ns	+	−2	5	6
Lignin	ns	ns	3	−1	3
Cellulose	−	ns	−3	2	−1
Hemicellulose	ns	ns	−3	0.4	−1
Extractives	ns	−	11	−15	−20

ns = effect not statistically significant (p > 0.10), + or − = increase/decrease with statistically significant (p < 0.10).

11.3 Secondary Compounds in Plant Material under Climate Change

Interrelations between primary production and secondary compounds

The response of ecosystems to climate change is closely linked to the dynamics of their structure and functions. A key question is how climate change affects the fixation and allocation of carbon to biomass and to secondary compounds (Fig. 11.3): how the production of primary and secondary compounds interacts in the dynamics of the ecosystem. Secondary compounds (e.g., terpenoids, phenolics, tannins) make the plant material bitter or toxic, thus providing a chemical defense against herbivory. Trees can further respond to herbivory by releasing a wide variety of volatile substances from resin ducts. Volatile substances emitted from insect-damaged plant organs may attract parasitic and predatory insects that are natural enemies of the feeding insects (Paré and Tumlinson 1999).

Secondary compounds refer to phenolics, alkaloids, terpenes, flavonoids and other compounds, which are produced in secondary metabolic pathways not producing lipids, carbohydrates, and amino acids important in primary production (Box 11.1) (Mazid et al. 2011). The secondary and primary productions are however, closely linked with each other, and with climate change. For example, the exposure of trees to elevated atmospheric CO_2 increases the rate of photosynthesis but decreases the rate of respiration compared to ambient conditions (Fig. 11.4). The higher rate of net photosynthesis under elevated CO_2 with increased carbon availability will therefore enhance the accumulation of non-structural carbohydrates, terpenoids, soluble phenolics and tannins over growth demands. This is especially the case if the availability of other resources (i.e., nitrogen) is not increased. The secretory organs of trees, such as the resin ducts of Scots pine needles and axial organs, contain large amounts of secondary compounds, as monoterpenes, sesquiterpenes and diterpenoid resin acids (Mazid et al. 2011).

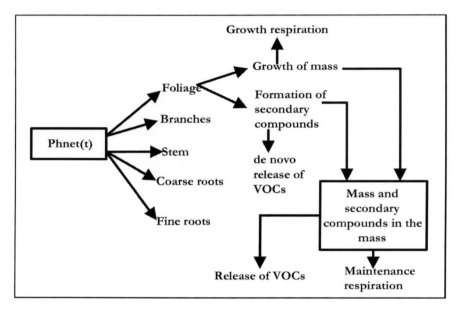

Fig. 11.3 Outline of the allocation and use of carbon (Phnet(t)) for growth in tree and the formation of volatile secondary compounds (VOC) in foliage.

Box 11.1 Links between primary production and the production of some main groups of secondary compounds important for the defense

Plants produce secondary compounds which have no direct functions in their growth and development. However, high concentrations of secondary compounds may make plants more resistant to herbivores and parasites. The production of defense chemicals needs a large amount of energy, thus reducing the growth and reproduction of plants.

Terpenes include a largest number of secondary compounds, with a common biosynthetic origin. The majority of terpenes are toxins and thus deterrents to a large number of plant feeding insects and mammals. In gymnosperms, monoterpenes accumulate in the resin ducts in needles, twigs and stems, mainly as α-pinene, β-pinene, limonene and myrecene. They are toxic to bark beetles, for example. Phenolics are a heterogenous group of compounds, which have a hydroxyl functional group on an aromatic ring. Phenolics provide a defense against pests and diseases including root parasitic nematodes. Alkaloids include compounds with nitrogen. Most of them are toxic, providing defense against microbial infections and herbivore attacks, including mammals (Mazid et al. 2011).

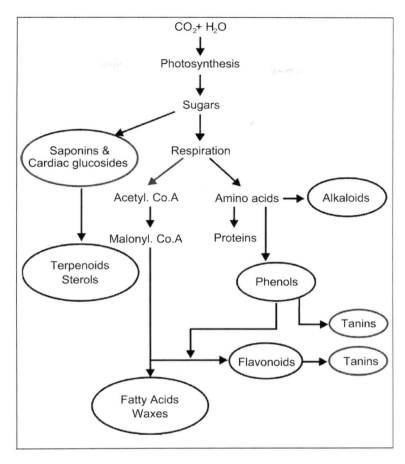

Fig. 11.4 Outline of the links between the primary production and production of the main groups of secondary compounds important for the defense of plants (Mazid et al. 2011). Courtesy of Journal and co-publisher OMICS Group. The ovals indicate selected groups of secondary compounds important for defense.

Secondary compounds in plant material under elevated CO_2 and temperature

Räisänen et al. (2008a) studied the way that elevated CO_2 and temperature affect the concentrations of monoterpenes, phenolics and condensed tannins in the needles of Scots pine in two consecutive years. Elevated CO_2 significantly reduced the concentration of total monoterpenes, whereas the effects of elevated temperature were not equally clear (Table 11.3). Under the concurrent elevation of CO_2 and temperature, the concentration of total monoterpenes was only half that under ambient conditions, but the effect was not statistically significant. Räisänen et al. (2008b) further found that the concentration of phenolics remained nearly unaffected under elevated CO_2 and temperature, but under the CO_2 elevation alone the concentration of condensed tannins increased, thus compensating for the reduced carbon allocation to monoterpenes. However, the responses of the selected secondary compounds to elevated CO_2 and temperature were variable as similarly found by Sallas et al. (2003).

Table 11.3 Concentration of total monoterpenes (mg per g of dry weight) in the fully grown needles of Scots pine grown under elevated CO_2 and temperature in late August in 2000 and 2001. The percentage change in concentration is related to the values under ambient conditions given in parenthesis (Räisänen et al. 2008b).

Climatic treatment	Year 2000, mg g^{-1}	Year 2001, mg g^{-1}	Significance
Ambient	4.47	3.67	
CO_2	3.76 (−16)	2.42 (−34)	< 0.10
T	4.89 (+9)	3.30 (−10)	> 0.10
CO_2 + T	6.19 (+38)	1.80 (−51)	> 0.10

11.4 Emissions of Secondary Compounds from Trees and Forests under Climate Change

Dynamics of forest ecosystem and emissions of secondary compounds

Forests and other vegetation emit a wide variety of volatile organic compounds (BVOCs) into the atmosphere, including terpenoids, carbonyls, etc. (Kesselmaier and Staudt 1999). Among others, isoprene and monoterpenes are the dominant BVOCs emitted from forest vegetation.

Tree species are roughly divided into isoprene (C_5H_8) emitters (many broad-leaved tree species, most *Quercus* species) and terpene ($C_{10}H_{16}$) emitters (e.g., silver birch, conifers, *Quercus ilex* L.). Some species, such as Norway spruce, emit both isoprene and monoterpenes, whereas Scots pine and silver birch emit only terpenes (Steinbrecher et al. 1999). Isoprene is formed in chloroplasts through the methyl-erythrito 4-phosphate pathway. Globally, 600 million tons of isoprene are emitted, which accounts for one third of all the carbohydrates emitted to the atmosphere. The emissions are sensitive to temperature and thus to climatic warming. For example, the emissions of monoterpene and sesquiterpene increase exponentially with increasing foliage temperature, thus making emissions susceptible to climate change. On the other hand, the emissions of BVOCs are related to the total dynamics of primary production in a forest ecosystem, and the changes as demonstrated in Fig. 11.5. In Finland, monoterpenes comprise 45% and isoprenes 7% of the total annual BVOC emissions (Lindfors and Laurila 2000).

In the forest ecosystem, the emissions of BVOCs are related to the dynamics of primary production as demonstrated by Räisänen et al. (2009) for a mature Scots pine forest under boreal conditions. They found that 74% of the total monoterpene flux originated from the canopy. In this case, the total ecosystem flux from June through September was 502 ± 76 mg m^{-2}, while the canopy emission was 373 ± 30 mg m^{-2} (Fig. 11.5). The differences between the ecosystem flux and canopy flux indicate the monoterpene emissions from the forest floor, such as litter, root system and ground vegetation (Hellén et al. 2006).

Natural BVOC emissions are sensitive to disturbances, such as herbivory or forest management and timber harvest (Schade and Goldstein 2003). Similarly, Räisänen et al. (2008c) found that in June through September the monoterpene concentration in the air was 2–3 times normal during the first days or month after a clear cut of Scots pine stand compared to the uncut area (Table 11.4). In this case, the logging residues (needles, branches, stumps, top part of stems) and pulpwood were left in the area. The increase was larger than in the thinned stands. The differences in monoterpene concentrations between clear cut and thinned areas were related to the differences in the amount of logging residues and pulpwood left in the areas. At the end of the monitoring period (four months after the cut), there were only small differences in monoterpene concentrations between the clear cut and uncut areas, as was also demonstrated by Haapanala et al. (2012).

Properties of Plant Material under Climate Change   149

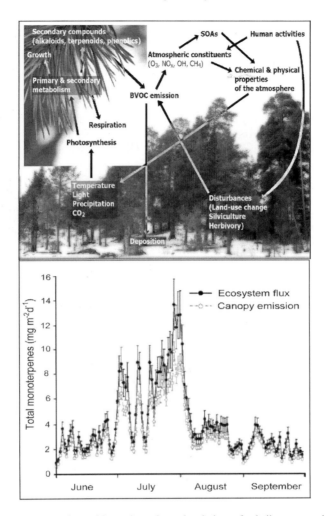

Fig. 11.5 Above: Outline of the interactions of forest dynamics and emissions of volatile compounds (BVOCs) with links to eco-physiology of trees and land use as management (Räisänen 2008). Below: Total monoterpene flux from the boreal forest ecosystem and the emission from the canopy of Scots pine dominating the ecosystem (Räisänen et al. 2009). Courtesy of the Finnish Society of Forest Science and the Finnish Forest Research Institute. Permission of Elsevier. SOAs in the above Figures refers to secondary organic aerosols.

Table 11.4 Effects of clear cut and thinning on monoterpene concentration ($\mu g\ m^{-3}$) in the air in 40-year-old Scots pine forests under boreal conditions (Räisänen et al. 2008c).

Period	Monoterpene concentration in air, $\mu g\ m^{-3}$						
	Clear cut		Thinning, 60%		Thinning, 30%		Control (uncut)
• June 22–23	9.2	(+197)	9.1	(+194)	10.0	(+223)	3.1
• June 29–July 3	15.7	(+196)	11.2	(+111)	13.0	(+145)	5.3
• July 22–24	17.1	(+134)	8.3	(+14)	8.5	(+16)	7.3
• July 27–29	16.3	(+94)	13.2	(+57)	15.5	(+84)	8.4
• August 16–20	5.3	(+47)	5.2	(+44)	5.3	(+47)	3.6
• September 22–24	2.1	(−16)	3.4	(+36)	2.8	(+12)	2.5
• Mean over periods	11.0	(+120)	8.4	(+68)	9.2	(+84)	5.0

Thinning intensity is indicated by the percentage of basal area reduced in thinning intervention. Numbers in parenthesis indicate the percentage increase/reduction in the monoterpene concentration due to clear cut or thinning compared to the emissions in the uncut area.

Emissions of secondary compounds under elevated CO_2 and temperature

In general, the monoterpene emissions in conifers are light-independent, mainly from monoterpenes stored in resin ducts (Monson et al. 1995), but emissions also take place as light-dependent biosynthesis. Emissions of monoterpenes from Scots pines foliage are thus physiologically controlled through photosynthesis and the rate of synthesis of intermediates availability for enzyme activity in forming monoterpenes. Consequently, the monoterpene emissions from Scots pine include both emissions from both stored and recently synthesized compounds, but the emission from storage is probably the major source of monoterpene emissions (Kesselmeier and Staudt 1999). Climate change may affect monoterpene emissions both through light-independent and light-dependent emissions based on the enhanced physiological activity under elevating CO_2 and temperature. Peñuelas and Estiarte (1998) have however, demonstrated that the effects of elevated CO_2 and temperature on secondary metabolites are variable, depending on, for example, the species and maturity of trees and the duration of exposure.

Räisänen et al. (2008a) measured the monoterpene emissions of Scots pine growing under elevated CO_2 and temperature in the Mekrijärvi chamber experiment. Monoterpenes were sampled from June through September in 2001 and from April through September in 2002. The chambers were used as open cuvettes for an automatic sampler, thus providing samples from emissions representing the whole tree. Räisänen et al. (2008a) found that the combined elevation of CO_2 and temperature significantly increased the normalized rate of monoterpene emissions over the whole growing period (+23%), whereas elevated CO_2 alone had no effect (–4%) and elevated temperature alone even decreased (–41%) the emission rate. The effect of the combined elevation of CO_2 and temperature was especially large during shoot elongation (+54%). Based on modeling (Box 11.2), Räisänen et al. (2008a) found that the total amount of monoterpenes emitted from May to September was 2.38 mg per gram of the dry mass of needles grown under ambient conditions. Growth under elevated CO_2 increased the total emissions by 5% and elevated temperature reduced it by 9% compared to that under ambient conditions. The combined elevation of CO_2 and temperature increased the total amount of monoterpenes emission over the growing period by 126% compared to that under ambient conditions.

Emissions over forest landscape under elevated CO_2 and temperature

At the scale of the forest landscape, the total emissions of BVOCs are linked to the tree species composition, stocking of tree biomass and canopy, and management and harvest of timber and biomass. Climate change affects the growth and development of forests, with consequent changes in tree species composition and foliage biomass emitting monoterpene and isoprene as shown by Kellomäki et al. (2001a,b) over the whole forest area of Finland. The simulations were based on the emission model of Guenther (1997) linked to a process-based forest ecosystem model (Kellomäki et al. 1992).

Seasonality between the active and dormant season and the transition from active to dormant and from dormant to active periods controlled the emissions from Scots pine and Norway spruce in year-round simulations. For birch, the calculations were made from the burst of foliage in spring to when they turned yellow in autumn. The calculations used the permanent sample plots of the Finnish National Forest Inventory, the database of which was used in calculating the foliage mass per tree species throughout the country in 1990 and in 2100, using the management rules recommended in 1990s in simulating the growth and development of forest until 2100. In the simulations, the current climate was a reference climate, and under climate change the temperature increased in spring (March–May) was 0.4°C, in summer (June–August) 0.3°C, in autumn (September–November) 0.4°C and in winter (December–February) 0.6°C per decade. At the same time, precipitation increased

Box 11.2 Monoterpene emissions from the needles of Scots pine

Räisänen et al. (2008a) compared the measured values of emission rates to the calculated values based on the assumption that the monoterpene emissions originated both from storage (E_{pool}, ng g^{-1} h^{-1}) and from synthesis ($E_{synthesis}$, ng g^{-1} h^{-1}):

$$E_{PS} = E_{pool} + E_{synthesis} \tag{11.1}$$

where E_{ps} is the total rate of monoterpene emission [ng g^{-1} h^{-1}]. Based on Guenther (1997), the emissions from storage were assumed to be a function of temperature:

$$Epool = EF_{mt} \times \exp\left[\beta \times (T - T_s)\right] \tag{11.2}$$

where EF_{mt} is the emission rate [ng g^{-1} h^{-1}] at temperature 25°C, T_s is the parameter [25°C] and β is the temperature dependency of emitted compound. Based on Guenther (1997), the emission from the synthesis ($E_{synthesis}$) was related to temperature and light:

$$E_{synthesis} = EF_{mt} \times C_L \times C_T \tag{11.3}$$

where EF_{mt} is the emission at temperature 25°C and flux density 1000 µmol m^{-2} s^{-1} and C_L and C_T are the correction factors dependent on flux density and temperature:

$$C_L = \frac{\alpha \times c_{LI} \times L}{\sqrt{1 + \alpha^2 \times L^2}} \tag{11.4}$$

$$C_T = \frac{\exp\left[c_{TI} \times (T - T_s)/R \times T_s \times T\right]}{c_{T3} + \exp\left[c_{T2} \times (T - T_M)/R \times T_s \times T\right]} \tag{11.5}$$

where L is the flux density [µmol m^{-2} s^{-1}], T leaf temperature [K], R is the gas constant, and α, c_{LI}, c_{TI}, c_{T2} and c_{T3} are parameters (Guenther 1997). Figure 11.6 shows the relationship between measured and calculated values of the emission rates of monoterpene from the needles of Scots pine in logarithmic scales (Räisänen et al. 2008a).

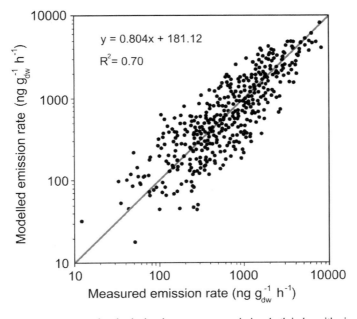

Fig. 11.6 Relationship between measured and calculated monoterpene emission, both in logarithmic scale, based on the assumption that the monoterpene emissions originated both from storage and from synthesis (Räisänen 2008, Räisänen et al. 2008a). Courtesy of the Finnish Society of Forest Science and the Finnish Forest Research Institute. Permission of Elsevier.

Table 11.5 Mean density of foliage mass per tree species over Finland in 1990 and 2100 and changes in mean mass density in response to climate change (Kellomäki et al. 2001b).

Region and species	Foliage in 1990, g m^{-2}	Foliage in 2100, g m^{-2}	Change, %
Southern Finland			
• Scots pine	228	47	−79
• Norway spruce	733	355	−52
• Birch	64	135	+109
Northern Finland			
• Scots pine	195	270	+38
• Norway spruce	380	882	+131
• Birch	40	68	+69
Whole country			
• Scots pine	217	105	−52
• Norway spruce	607	509	−16
• Birch	57	118	+108

0.5–2.0% per decade, least in spring and most in winter. Over the whole country, the annual mean temperature increased 4°C and precipitation 10% by the year 2100.

The simulations showed that the proportions of Scots pine and Norway spruce in southern Finland (60° < latitude < 64° N) may reduce from 40–50% in 1990 to less than 10–20%, with an increased dominance of birch (Table 11.5). In northern Finland (64° < latitude < 70° N), the proportions of Scots pine and Norway spruce may balance at a level of 40%; i.e., the proportion of Norway spruce increased from 21 to 37% and that of birch from 17 to 23%, whereas the proportion of Scots pine reduced from 63 to 40%. Consequently, the total mean emissions of monoterpene from Scots pine canopies reduced by 80% in the south but increased by 62% in the north. At the same time, the monoterpene emissions from Norway spruce canopies may increase by 4% in the south but by 428% in the north, while the emissions from birch canopies may increase by about 300 and 113%, respectively (Kellomäki et al. 2001a,b).

The overall emission of monoterpene throughout the whole country was 950 kg km^{-2} yr^{-1} under the current temperature. Under an elevating temperature and precipitation, the overall emission increased by 17% up to 1100 kg km^{-2} yr^{-1}, mainly due to the increase in the north. Under the current climate, the emissions of isoprene were related to the spatial distribution of spruce dominated forests. Emissions in the south were four times higher than in the north. Temperature elevation and the subsequent changes in the spatial distribution of Norway spruce increased isoprene emissions by 37% in the south and 435% in the north. Over the whole country, the mean annual isoprene emissions from Norway spruce canopies had increased 60% by 2100 (Kellomäki et al. 2001b).

11.5 Concluding Remarks

Under climate change, the productivity of boreal forests is likely to increase, but the effects on the physical and chemical properties of wood seem to be small compared to the effects on changes in growth. At the same time, the oxidative properties of the air may change due to the reduction of Scots pine and Norway spruce and the concurrent increase of birch, especially in southern Finland, where the share of birch may increase. These probable changes in tree species composition will alter the properties of forest canopies and the consequent emissions of volatile organic compounds. The decline in the dominance of Norway spruce alone may substantially reduce emissions of monoterpene and isoprene in southern Finland. However, emissions are greatly dependent on temperature, which may enhance the emission rate under climatic warming. A substantial increase in emissions may therefore be expected, especially in northern Finland, as a result of the increasing amount of coniferous foliage and the elevation in temperature.

References

Ceulemans, R., M. E. Jach, R. Van de Velde, J. X. Lin and M. Stevens. 2002. Elevated atmospheric CO_2 alters wood production, wood quality and wood strength of Scots pine (*Pinus sylvestris* L.) after three years of enrichment. Global Change Biology 8: 153–162.

Gardiner, B. 2011. Climate change, tree growth and timber properties. Presentation in Newcastle, 23rd February 2011. http://www.forestry.gov.uk. Visited 26.2.2014.

Guenther, A. 1997. Seasonal and spatial variations in natural volatile organic compounds emissions. Ecological Applications 7(1): 34–45.

Haapanala, S., H. Hakola, H. Hellén, M. Vestenius, J. Levula and J. Rinne. 2012. Is forest management a significant source of monoterpenes into the boreal atmosphere? Biogeosciences 9: 1291–1300.

Hellén, H., H. Hakola, K. -H. Pystynen, J. Rinne and S. Haapanala. 2006. C2-C10 hydrocarbon emissions from a boreal wetland and forests floor. Biogeosciences 3: 167–174.

Kalliokoski, T., H. Mäkinen, T. Jyske, P. Nöjd and S. Linder. 2013. Effects of nutrient optimization on intra-annual wood formation in Norway spruce. Tree Physiology 33: 1145–1155.

Kellomäki, S. and H. Väisänen. 1988. Dynamics of branch population in the canopy of young Scots pine stands. Forest Ecology and Management 24: 67–83.

Kellomäki, S., H. Väisänen, H. Hänninen, T. Kolström, R. Lauhanen, U. Mattila et al. 1992. A simulation model for the succession of the boreal forest ecosystem. Silva Fennica 26: 1–18.

Kellomäki, S., I. Rouvinen, H. Peltola and H. Strandman. 2001a. Density of foliage mass and area in the boreal forest cover in Finland, with applications to the estimation of monoterpene and isoprene emissions. Atmospheric Environment 35: 1491–1503.

Kellomäki, S., I. Rouvinen, H. Peltola, H. Strandman and R. Steinbrecher. 2001b. Impact of global warming on the tree species composition of boreal forests in Finland and effects on emissions of isoprenoids. Global Change Biology 7: 531–544.

Kesselmaier, J. and M. Staudt. 1999. Bionic volatile compounds (VOC): an overview on emissions, physiology and ecology. Journal of Atmospheric Chemistry 33: 23–88.

Kilpeläinen, A., H. Peltola, A. Ryyppö, K. Sauvala, K. Laitinen and S. Kellomäki. 2003. Wood properties of Scots pine (*Pinus sylvestris*) grown at elevated temperature and carbon dioxide concentration. Tree Physiology 23: 889–897.

Kilpeläinen, A., H. Peltola, A. Ryyppö and S. Kellomäki. 2004. Scots pines responses to elevated temperature: growth and on wood properties. Tree Physiology 25: 75–83.

Kilpeläinen, A., A. Zubizarreta-Gerendiain, K. Luostarinen, H. Peltola and S. Kellomäki. 2007. Elevated temperature and CO_2 concentration effects on xylem anatomy of Scots pine. Tree Physiology 27: 1329–1338.

Kostiainen, K., S. Kaakinen, P. Saranpää, R. D. Sigurdsson, S. Linder and E. Vapaavuori. 2004. Effect of [CO_2] on stem wood properties of mature Norway spruce grown at different soil nutrient availability. Global Change Biology 10: 1526–1538.

Kostiainen, K., S. Kaakinen, P. Saranpää, B. D. Sigurdsson, S. -O. Lundqvist, S. Linder et al. 2009. Stem wood properties of mature Norway spruce after 3 years of continuous exposure to elevated [CO_2] and temperature. Global Change Biology 15: 368–379.

Lindfors, V. and T. Laurila. 2000. Biogenic volatile organic compounds (VOC) emissions from forests in Finland. Boreal Environment Research 5: 95–113.

Mazid, M., T. A. Khan and F. Mohammad. 2011. Role of secondary metabolites in defence mechanisms of plants. Biology and Medicine 3(2): 232–249.

Monson, R. K., M. T. Lerdau, T. D. Sharkey, D. S. Schimel and R. Fall. 1995. Biological aspects of constructing volatile organic compound emission inventories. Atmospheric Environment 29: 2989–3002.

Paré, P. W. and J. H. Tumlinson. 1999. Plants volatiles as a defense against insect herbivores. Plant Physiology 121: 325–331.

Peñuelas, J. and M. Estiarte. 1998. Can elevated CO_2 affect secondary metabolism and ecosystem function? Trends in Ecology and Evolution 13(1): 20–24.

Räisänen, T. 2008. Impacts of climate change and forest management on monoterpene emission and needle secondary compounds of boreal Scots pine (*Pinus sylvestris* L.). Ph. D. Thesis, University of Eastern Finland, Joensuu, Finland.

Räisänen, T., A. Ryyppö and S. Kellomäki. 2008a. Effects of elevated CO_2 and temperature on monoterpene emission of Scots pine (*Pinus sylvestris* L.). Atmospheric Environment 42: 4160–4171.

Räisänen, T., A. Ryyppö, R. Julkunen-Tiitto and S. Kellomäki. 2008b. Effects of elevated CO_2 and temperature on secondary compounds in needles of Scots pine (*Pinus sylvestris* L.). Trees 22: 121–135.

Räisänen, T., A. Ryyppö and S. Kellomäki. 2008c. Impacts of timber felling on the ambient monoterpene concentration of A Scots pine (*Pinus sylvestris* L.) forests. Atmospheric Environment 42: 6759–6766.

Räisänen, T., A. Ryyppö and S. Kellomäki. 2009. Monoterpene emission of a boreal Scots pine (*Pinus sylvestris* L.) forest. Agricultural and Forest Meteorology 149(5): 808–819.

Sallas, L., E. -M. Luomala, J. Utriainen, P. Kainulainen and J. K. Holopainen. 2003. Contrasting effects of elevated carbon dioxide concentration and temperature on Rubisco activity, chlorophyll fluorescence, needle ultrastructure and secondary metabolites in conifer seedlings. Tree Physiology 23: 97–108.

Schade, G. W. and A. H. Goldstein. 2003. Increase of monoterpene emissions from a pine plantation as a result of mechanical disturbances. Geophysical Research Letters 30(7): 1380.

Steinbrecher, R., K. H. Hauff, H. Hakola and J. Rössler. 1999. A revised parametrization for emission modelling of isoprenes in boreal trees. pp. 29–43. *In*: T. Laurila and V. Lindfors (eds.). Biogenic VOC Emissions and Photochemistry in the Boreal Regions of Europe—Biphorep Scientific Final Report. EC, Brussels, Belgium.

Volanen, V., H. Peltola, I. Rouvinen and S. Kellomäki. 2006. Impact of long-term elevation of atmospheric CO_2 concentration and temperature on the establishment, growth and mortality of boreal Scots pine branches. Scandinavian Journal of Forest Research 21: 115–123.

PART IV
Eco-physiological Approaches to Modeling Responses of the Boreal Forest Ecosystem to Climate Change

Impact Mechanisms Linking the Dynamics of Forest Ecosystem to Climate Change

ABSTRACT

Globally, the distribution and structure of forests are closely related to the distribution of temperature and precipitation. Climate change modifies the environmental conditions and consequent long-term functioning and structure of forest ecosystems. New habitats may be more optimal for other species than for those currently dominating forest communities. Mechanistic (process-based) models, including physiological response mechanisms to changes in environmental conditions, help to understand how forests grow and develop under climate change and how management could be modified in order to avoid detrimental impacts and utilize the opportunities provided by climate change in boreal conditions.

Keywords: climate change, succession, models, modelling, forest ecosystem, functioning, structure

12.1 Factors Driving Processes in the Forest Ecosystem under Climate Change

Impact of climate change on the structure and functioning of forest ecosystem

Environmental factors affecting the success of organisms in the ecosystem may be divided into those that are energetic and material ones. Energetic factors involve the energy in radiation and heat, and the energy in mechanical forces, such as wind, snow load and gravity. Energy factors are driving forces which maintain the use of resources for the living functions of individuals, populations and communities in ecosystems. Material factors are resources used in the metabolism of organisms, and also for birth and growth. Material factors or resources include CO_2, O_2, H_2O and nutrients, the availability of which is affected by climate change. Except gravity, climate change may directly and indirectly modify energetic factors and the availability of resources affecting trees and other organisms occupying forest sites.

In primary production, the rates of metabolic functions (e.g., photosynthesis, respiration, transpiration, uptake of water and nutrients) in trees follow changes in energy and resource factors caused by climate change (Table 12.1). Responses are immediate, and they alter the growth and development of trees even in the short term (Kirschbaum 1999). Response rates themselves (specific response rate, sensitivity of response) may acclimate to changes in energy and resource factors (Medlyn et al. 2001, 2002), representing phenotypic plasticity in response to changes in the environment. For example, the response of photosynthesis to temperature is different in Scots pines grown under high CO_2 concentration than when grown under low CO_2 concentration (Wang et al. 1996). Primary production in the tree population may therefore be different from that expected on the basis of the immediate response (Oren et al. 2001). Changes in seasonality and the acclimation of the annual cycle of physiological activity and growth to the changing climate are among the key factors which control the success of trees under climate change. Metabolic responses are also linked to climate change through soil processes, e.g., soil moisture and soil temperature control the decay of litter and humus (soil organic matter) and the cycle of nutrients for reuse.

Metabolic responses to climate changes are mediated by the long-term dynamics of a forest ecosystem to regeneration (birth), growth and mortality, and thus to the management and harvest cycles in forestry (Table 12.1). These processes control the successional dynamics of the forest ecosystem, which are characterized by the rates, patterns and limits set by environmental conditions but modified by climate change. The current successional interactions may be disturbed by the natural invasion of new tree species in the time scale of tens to hundred years. Under boreal conditions, the invasion of new insects and pests is, however, much more probable than the natural invasion of new tree species (Parmesan et al. 2005). The expansion of existing insects and other pests northwards under climate change may increase the risk of uncontrolled damage. This may affect the trophic structure of ecosystems, with changes in the energy flow and nutrient cycle of the ecosystem.

Under boreal conditions, the dynamics of the forest ecosystem are further linked to climate change through mechanical forces as trees are broken and blown down by strong winds and excessive snow fall (Lexer et al. 2002). An increase in rainfall may reduce the carrying capacity of the soil and thus hamper harvest and management operations, and increase the risk of falling trees due to the reduction in anchorage. At the same time, soil warming will shorten the duration of soil frost, with a consequent reduction in the carrying capacity of soil and the anchorage of trees in winter. The risk of wildfire may also be increased under climate change, if evaporation in the summer increases more rapidly than does precipitation, with a consequent increase in the frequency of drought periods.

Table 12.1 Short- and long-term changes in the structure and dynamics of forest ecosystems under climate change.

Compartment of forest ecosystem	Immediate responses	Long-term responses
Impacts through populations and communities of organisms		
• Primary production in trees and other plants.	• Metabolism of trees and other plants, photosynthesis, respiration, uptake of water and nutrients, transpiration.	• Quantity and quality of matter produced in primary production, changes in geographical distribution of current species.
• Herbivory, secondary production in higher trophic levels.	• Quantity and quality of food used in higher trophic levels, effects on metabolism.	• Changes in species composition, invasion of new species, extinction of current species.
Impacts through environment		
• Atmosphere with material and energy factors.	• CO_2 concentration, temperature and temperature variability, precipitation and precipitation variability, air humidity, wind force and variability in wind force, quantity and quality of snow in precipitation.	• Fertility of site as affected by changes in climatic factors.
• Soil with material and energy factors.	• Soil moisture, soil temperature, nutrients, decay of soil organic matter, nutrient cycle, leaching.	• Fertility of site as affected by changes in soil factors.

Ecological succession under climate change

Globally, the distribution and structure of forests are closely related to the distribution of temperature and precipitation, which define the basic dimensions of niches occupied by different tree species and thus the functioning of forest ecosystems. According to Aber and Melillo (1991):

- "Temperature and the balance between precipitation and evaporation (including evaporation through plant surfaces, known as transpiration) are particularly important, since these largely determine the rate at which biological and chemical reactions occur. Both the production of new organic matter by plants and decomposition of dead organic matter by microbes are temperature- and moisture-dependent processes. The amount of water available for use by plants, relative to the potential for that water to evaporate, determines the amount of water stress encountered by plants and often limits the length of the growing season. The amount of water percolating down through the soil profile and the chemistry of that water play a large role in determining soil structure and nutrient content."

Climate change affects how the long-term functioning and the subsequent structure of forest ecosystems change over time in ecological succession. Succession is the long-term interaction between organisms occupying a site characterized by given climatic and edaphic conditions, which are changing as a result of changes in temperature and precipitation. The succession of an ecosystem is characterized by the following tendencies (Odum 1971):

- Directional and predictable changes in species structure and community process over time.
- Modifications of the physical environment by the community, controlled by the properties of the community, although it is the physical environment that determines the pattern, the rate of change and the limits to how far this development can proceed.
- Development culminates in a stabilized ecosystem in which the maximum mass and interactions between organisms are maintained per unit of available energy.

In succession, the properties of populations and communities of organisms change through regeneration, growth and mortality; i.e., the properties of the environment change, with the emergence of new habitats that are optimal for species other than those currently dominating the community (Odum 1971, van der Meer et al. 2002) (Fig. 12.1). Species with regeneration and growth properties more optimal for the new environment than those of other species may gain dominance. At the same time, species with a less optimal response are reduced in number or even excluded from the community, releasing space for the new species to occupy the site. Interaction

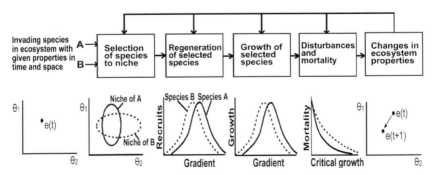

Fig. 12.1 Outline of genotype/environment interaction behind the ecological succession in a plant community, based on Reed and Clark (1976), and Kimmins (1987). Disturbances create variability in site conditions, providing varying niches for species to occupy the site. Only those most fit can survive and succeed, but these changes create conditions more optimal for other species than those currently occupy the site. This results in further changes in structure and functioning of the community, etc.

between the environment and the genotypes present in the community gives rise to an ecological succession in the course of time through niche differentiation controlled by disturbances which returns the developmental phase of the ecosystem to earlier phases. For example, a wildfire killing trees provides space for regeneration, with a later enhancement of growth and accumulation of mass. The structure of a boreal forest landscape widely represents a mosaic of tree populations and communities of varying age from seedlings to mature trees.

12.2 Modeling Climate Change Impacts on Forest Ecosystem

Models in investigating the dynamics of forest ecosystem

The responses of whole forest ecosystems to climate change are investigated only in Free Air Carbon Enrichment experiments (FACE experiments), where whole tree stands are exposed to elevated CO_2 (Ainsworth and Long 2005). Even in these cases, the exposure time is from only few years to a decade. The responses of whole ecosystems to climate change are therefore mainly explored through modeling, which utilizes the short-term physiological (e.g., rates of photosynthesis, respiration, transpiration, etc.) and ecological (e.g., timing of shoot flushing, elongation rate of shoots, diameter growth of stem, etc.) responses to climate change obtained in the FACE experiments or in chamber experiments (Jarvis 1998). In the latter case, exposure to elevated temperature, CO_2 or their combination, is again short compared to the time horizon in regard to the succession of a whole forest ecosystem or the production cycle in forestry, where the proper time horizon is decades or even centuries. Modeling can, however, provide a methodology to explore the long-term responses of forest ecosystems to climate change under a variety of site conditions, and to explain the complex interactions and feedback processes that give rise to the variability in ecosystem responses. Such models should include the main physiological and ecological processes driven by hydrological and nutrient cycles sensitive to climatic factors and thus to climate change (Mäkelä et al. 2000).

Modeling approaches

Ecosystem models describe or simulate the performance of an ecological system over time. A model or simulation model involves the performance of a system with the help of mathematic equations representing a set of factors and relationships between them; i.e., a model is a formal description of a system of interest. Model building is based on hypotheses, how major elements in the model vary over time and what processes control changes in system structure (state variables) (Swartzman and Kaluzny 1987). A model can be: (i) physical vs. abstract; (ii) dynamic vs. static; (iii) empirical (correlative) vs. mechanistic (explanatory); (iv) deterministic vs. stochastic; or (v) simulative vs. analytic. Models are often classified as empirical or mechanistic models, and the hybrids of these two:

- Empirical models (statistical) may use, for example, inventory data representing, the past growth and development of a forest. The applications of such models in simulating future growth and development assume that the future growing conditions are similar to those in the past. Any changes in the growing conditions may therefore bias the simulated growth and development of the forest.
- Mechanistic models (process-based) models include physiological response mechanisms to changes in environmental conditions. Process-based models can provide the same prediction capacity under practical management as empirical models. Moreover, process-based models may help to understand how forests grow and develop under climate change and how management could be modified in order to avoid detrimental impacts and utilize the opportunities probably

provided by climate change. Until now, the use of process-based models in forestry decision-making has been limited, because the application of these models may require, for example, data not provided by conventional forest inventories.

- Hybrid models include features of empirical and mechanistic and models; for example, sub-models for conventional growth and yield models include physiological mechanisms driving the growth of trees related to climatic and edaphic factors, making the growth and yield model applicable to climate change impact studies.

System boundaries and modeling the dynamics of forest ecosystem

A model concerns a system, which is defined by boundaries between the system and the surrounding environment. A model includes the relationships between the system and the surrounding environment, and the internal relationships in the system. Relationships include material and energy flows and the flow of information. The study problem dictates the system boundaries and the relations included in the model (Box 12.1; Fig. 12.2).

The properties and dynamics of a system are described by the concepts of state and change. System state indicates the properties of the system in terms of selected variables (state variables).

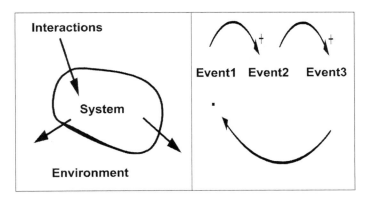

Fig. 12.2 Left: Schematic presentation of a system with external and internal relations, and Right: Sequence of events affecting system dynamics with the feeding in both directions from events.

Box 12.1 Main phases in building a forest ecosystem model

In general, the compilation of a model includes the following phases (Swartzman and Kaluzny 1987):

- Definition of the problem and purpose of modeling.
- Definition of system boundaries considering the purpose of modeling.
- Definition of the subsystems of the total system and the identification of subsystems to be modeled.
- Definition of the interaction between subsystems and between the total system and the environment in terms of material and information flow.
- Modeling the subsystems in terms of a set of equations which indicates the flow of material and information between subsystems.
- Solving the equations in a proper way and building algorithms.
- Evaluation of the performance of algorithms and validation of the model.
- Improvement of the model, and if the performance of the total model is not satisfactory, for example, the model structure should be improved.
- New round.

A change in system state indicates that the properties of the system have changed in some respect. In this context, a process refers to a series of changes one after another in the system state. In a system, the relationships may feed matter and information back or forward: feeding back (feedback) stabilizes the system processes and system performance, while feeding forward may destabilize the system performance (Fig. 12.2).

Eco-physiological approach to modeling the dynamics of the forest ecosystem

Forests are ecosystems where trees and other green organisms occupy sites and intercept solar energy under the control of climatic and edaphic factors. Solar energy flows from producers (green plants) to consumers (organisms other than green plants). In the ecosystem, different organisms form complex food webs, where the links between different organisms are dynamic. These links are the keys to the management of the forest ecosystem. Changing the structure and dynamics in management allows the ecosystem to produce and/or to enhance timber/biomass or other goods and services in forestry. Management includes measures to manipulate the properties of tree population and growing conditions. The ultimate goal is to sustain and enhance the biochemical processes behind the carbon uptake and the conversion of photosynthates for biomass utilized in forestry (Fig. 12.3).

In management, the choice of tree species (or genotype) and spacing has a major impact on the structure and growth of a tree population over time (rotation) (Fig. 12.3). Even in the early succession, spacing controls the growth of individual trees and thus the growth of whole populations. Later in the succession, thinning is used to control the stocking of the tree population and the allocation of resources among individual trees; i.e., affecting how fast remaining trees grow and achieve the necessary dimensions for timber (pulp wood, saw logs) and biomass harvest. Thinning effects are heavily based on changes in the canopy structure, which controls the within-population microclimate, the interception of radiation and water on the canopy, and the through fall of water on soil. Canopy structure and prevailing microclimate further control evaporation from the canopy

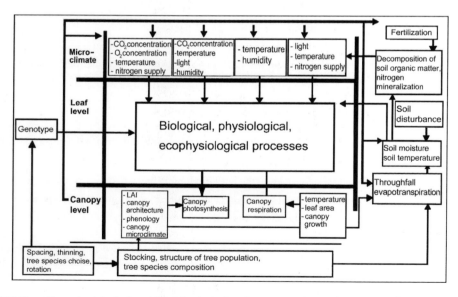

Fig. 12.3 Outline of eco-physiological modelling the dynamics of a managed forest ecosystem as an interacting system of tree populations and their environment subject to management, as modified from Wang (1996). Biological, physiological and eco-physiological processes affecting carbon uptake and use for growth in the whole tree context controlled in management, see Fig. 9.1, Chapter 9.

and soil surfaces, with impacts on the infiltration of water in the soil profile and soil moisture. The canopy structure thus indirectly affects the availability and uptake of water, and the transfer of nutrients from the soil to the physiological processes behind growth (Thornley and Cannell 2000). The within-population microclimate further controls the decay of organic matter in the soil, thus cycling nutrients for uptake. In management, the availability of nutrients may be added through fertilization and/or managing soil prior to regeneration through, e.g., describe burning or scarifying.

12.3 Concluding Remarks

Ecosystem models are problem-oriented, and they are able to answer only the questions facilitated by the structures and functions included in the model. This implies that the analysis behind model building has to identify the relevant factors and their interactions in controlling the dynamics of the forest ecosystem. Successful modeling also requires that it is possible to estimate the values of necessary parameters, and that the model is valid for simulating the dynamics of the ecosystem. In this context, model evaluation includes a set of tools for ascertaining: (i) how accurate the assumptions in building the model are (preciseness); (ii) how realistic the behavior resulting from those assumptions is (realism); and (iii) how sensitive model behavior is to changes in these assumptions (generality). Consequently, a model is precise if the values predicted by the model deviate from those measured as little as possible. This further implies that a precise model is necessarily a realistic or a general one. A model is realistic if the relationships included in the model are causal, and if they are of importance regarding the theory which is used to explain the performance of the phenomenon. Thus, a realistic model includes all the important factors required to explain the performance of the phenomenon as regards the selected approach to the study problem. A model is general if the model is applicable over a large range of spatial and temporal variability to describe the performance of a phenomenon. Thus, a general model is usually a realistic one, but it is not necessarily a precise one.

References

Aber, J. D. and J. M. Melillo. 1991. Terrestrial Ecosystems. Saunders College Publishing, Philadelphia, USA.
Ainsworth, E. A. and S. P. Long. 2005. What have we learned from 15 years of free-air CO_2 enrichment (FACE)? A meta-analytic review of the responses of photosynthesis, canopy properties and plant production to rising CO_2. New Phytologist 165: 351–372.
Jarvis, P. G. 1998. European Forests and Global Change. The Likely Impacts of Rising CO_2 and Temperature. Cambridge University Press, Cambridge, UK.
Kellomäki, S. (ed.). 2009. Forest Resources and Sustainable Management. Paper Engineer's Association/Paperi and Puu Oy, Helsinki Finland.
Kimmins, J. P. 1987. Forest Ecology. Macmillan Publishing Company, New York, USA.
Kirschbaum, M. U. F. 1999. CenW, a forest growth model with linked carbon, energy, nutrient and water cycles. Ecological Modelling 118: 17–59.
Lexer, M., K. Hönninger, H. Scheifinger, C. H. Matulla, H. Groll, H. Kromp-Kolb et al. 2002. The sensitivity of Austrian forests to scenarios of climate change: a large-scale risk assessment based on modified gap model and forest inventory data. Forest Ecology and Management 162: 53–72.
Mäkelä, A., J. Landsberg, A. R. Ek, T. E. Burk, M. Ter-Mikaelian, G. I. Ågren et al. 2000. Process-based models for ecosystem management: current state and the art and challenges for practical implementation. Tree Physiology 20: 289–298.
Medlyn, B., P. G. Jarvis, F. Badeck, D. Pury, C. Barton, M. Broadmeadow et al. 1999. Effects of elevated CO_2 on photosynthesis in European forest species: a meta-analysis of model parameters. Plant, Cell and Environment 22: 1475–1495.
Medlyn, B. E., C. V. M. Barton, M. S. J. Broadmedow, R. Ceulemans, P. Dangeles, M. Forstreuter et al. 2001. Stomatal conductance of forest species after long-term exposure to elevated CO_2 concentrations: a synthesis. New Phytologist 149(2): 247–264.
Odum, E. P. 1971. Fundamentals of Ecology. 3rd edition. W. B. Saunders Company, Philadelphia, USA.
Oren, R., D. S. Ellsworth, K. H. Johnsen, N. Phillips, B. S. Ewers, C. Maier et al. 2001. Soil fertility limits carbon sequestration by forest ecosystems in a CO_2-enriched atmosphere. Nature 411: 469–472.
Parmesan, C., S. Gaines, L. Gonzales, D. M. Kaufman, J. Kingsolver, J. T. Townsend et al. 2005. Empirical perspectives on species borders: from traditional biogeography to global change. Oikos 108: 58–75.

Reed, K. L. and S. G. Clark. 1976. SUCcession SIMUlator: a coniferous forests simulator. University of Washington. Biome Bulletin 11: 1–96.

Swartzman, G. L. and S. P. Kaluzny. 1987. Ecological Simulation Primer. MacMillan Publishing Company, New York, USA.

Thornley, J. H. M. and M. G. R. Cannell. 2000. Managing forests for wood yield and carbon storage: a theoretical study. Tree Physiology 20: 477–484.

van der Meer, P. J., I. T. M. Jorritsma and K. Kramer. 2002. Assessing climate change effects on long-term forest development: adjusting growth, phenology, and seed production in a gap model. Forest Ecology and Management 162: 39–52.

Wang, K. -Y. 1996. Effects of long-term CO_2 and temperature elevation on gas exchange of Scots pine. Ph. D. Thesis, University of Eastern Finland, Joensuu, Finland.

Wang, K. -Y., S. Kellomäki and K. Laitinen. 1996. Acclimation of photosynthetic parameters in Scots pine after three years of exposure to elevated temperature and CO_2. Agricultural and Forest Meteorology 82: 195–217.

13

Integrating Climate Change Impacts on Ecosystem Dynamics for Management Studies

ABSTRACT

A model for climate change impact studies is described, with a focus on the interaction between boreal tree populations and atmospheric and edaphic factors. Modeling involves the climate and weather, the soil moisture and temperature, the decay of litter and humus in circulating nitrogen and the growth and development of trees in response to changing climate and atmospheric CO_2. The model includes a set of management operations, which are used to control the growth and development of trees. Performance of the model in varying simulation conditions is demonstrated.

Keywords: modeling, boreal forest, climate change, weather generator, model validation, management, timber production

13.1 Main Structural and Functional Features of the Model

This chapter describes a process-based ecosystem model (FinnFor) for studying the impacts of climate change on the dynamics of a managed boreal forest ecosystem. The basic assumption was that climate change affects canopy photosynthesis and the consequent tree growth, including the growth of foliage, branches, stems, coarse roots and fine roots. Climate change implies an increase in atmospheric CO_2 concentration and temperature, with impacts on cloudiness, precipitation and air humidity and radiation falling on tree crowns (Fig. 13.1). Changes in atmospheric factors also affect soil moisture and temperature, which control decay of soil organic matter (litter, humus) and cycle of nitrogen (Strandman et al. 1993, Väisänen et al. 1994, Kellomäki and Väisänen 1997, Ge et al. 2010):

$$G_T = f(P_{nc}) = f(R_n) \times f(L) \times f(_{gcs}) \times f(W_a) \times f(N_{up}) \times f(T, C_a, P) \times f(M) \qquad (13.1)$$

where G_T is the total growth (foliage, branches, stems, coarse roots fine roots), P_{nc} is canopy photosynthesis, R_n is net radiation interception, L is leaf area, g_{cs} is canopy conductance, W_a is available soil water, N_{up} is uptake of nitrogen, T is air temperature, C_a is atmospheric CO_2, P is precipitation and M is management.

The two-dimensional model system represents a site (climatic and edaphic factors in a given place) and a tree stand, which may be a population of a single species or a mixture of several species (Table 13.1). The choice of tree species is, however, limited to boreal Scots pine, Norway spruce and birch. Management applications include regeneration (e.g., tree species choice and planting with given spacing, natural regeneration), tending of seedling stands (precommercial thinning with spacing and choice of tree species mixture for further growth). Maturing trees can be simply thinned (from below or above) or fertilized with nitrogen in several phases before the final cut (clear cut, cut for natural regeneration) at the end of the selected rotation length. Biomass and timber (saw logs, pulp wood) can be harvested in thinnings and final cut, with impacts on carbon sequestration. The model is the most applicable in stand-by-stand management.

The stand structure is comprised of one or several tree cohorts. Each cohort is represented by the object tree, which is characterized by species, diameter and height and is represented by a given number of similar trees per hectare (Fig. 13.1; Table 13.1). The structure of object tree also includes the mass of foliage, branches, stems, coarse roots and fine roots. The canopy surface is made up of crowns providing the boundary where climatic factors and trees interact by absorbing radiation, evaporating water from the canopy surfaces and through stomata (transpiration), taking up carbon in photosynthesis and emitting carbon in respiration. Similarly, the soil surface below the canopy provides a boundary for exchanges of heat and water between the atmosphere and soil below the canopy. The lowest layer in the selected soil profile provides the boundary for soil processes. The model applies an hourly time step for physiological (e.g., photosynthesis) and an annual time step for ecological (e.g., growth) and management processes.

Physiological and ecological processes include photosynthesis, autotrophic respiration, stomatal conductance with transpiration, growth and the mortality of trees and the litter fall from living trees (Fig. 13.1; Table 13.1). Physiological processes are driven directly and indirectly by climatic and

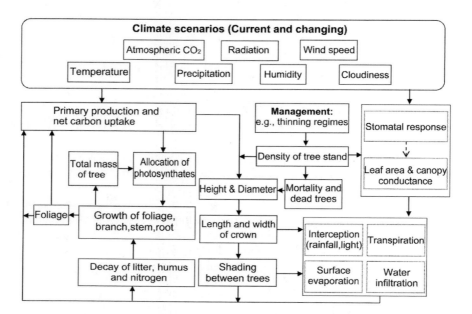

Fig. 13.1 Outline of the model showing how climate change may affect the dynamics of carbon uptake and hydrological and nutrient cycles with impacts on growth and yield in the forest ecosystem.

edaphic factors (e.g., radiation, CO_2, air temperature, air humidity, soil temperature, soil moisture). Photosynthates are allocated to different organs for growth and respiration. The mortality of whole trees is related to the crowding on the site, whereas organ litter of living trees indicates the organ-specific turnover of the mass. Soil processes and properties include water flow (vertical, horizontal) and availability of water in the rooting zone, and heat flow and consequent soil temperature.

Table 13.1 Outline of the structure and properties of a process-based model (FinnFor) built for simulating climate change impacts on the long-term dynamics of boreal forests, with management applications. The table was mainly compiled by Garcia-Gonzalo et al. (2008). For further details of the model, see Kellomäki and Väisänen (1997) and Matala et al. (2003).

Main modeling objectives and management options	
Modeling objectives	Dynamics of boreal forest ecosystems, controlled by stand structure, environmental conditions (climate, soil) and management.
Management options	Thinning (from below and above) and final cutting; nitrogen fertilization; natural regeneration, planting of Scots pine, Norway spruce and birch spp.
Ecosystem structure	
Stand structure	Cohorts of single tree species: number, age, height and diameter.
Tree structure	Foliage, branches, stems, course roots and fine roots.
Soil structure	Litter on soil, soil organic matter (humus), soil profile with different layers.
Model structure	
Time step	Hourly: physiological processes; annual: ecological and management processes.
Environmental control	Atmosphere: radiation, temperature, precipitation, air humidity, wind speed, CO_2 concentration; Soil: Soil moisture, soil temperature, available nitrogen.
Functioning of the model processes	
Tree and stand level processes	
Photosynthesis	Biochemical model for photosynthesis driven by atmospheric and soil factors.
Autotrophic respiration	Day respiration and maintenance respiration controlled by temperature, growth respiration as a fraction of photosynthesis allocated to growth.
Stomatal conductance	Controlled by radiation, temperature, air humidity, CO_2 concentration, soil temperature and moisture (the Jarvis type).
Transpiration	Penmann-Montheith type.
Mortality and litter	Probability of death of an individual tree and stand level self-thinning, organ specific turnover rates for foliage, branches, course roots and fine roots.
Seasonality	Temperature control photosynthetic capacity, respiration and phenology.
Soil processes	
Temperature	Soil temperature controlled by radiation balance and physical properties of soil.
Water	Soil moisture controlled by precipitation, evapotranspiration and water outflow.
Nitrogen	Available nitrogen controlled by litter fall, decomposition of litter and humus and uptake of nitrogen by trees.
Carbon	Dynamics controlled by heterotrophic losses under the control of soil moisture and temperature and quality of litter.
Main model outputs	
Water balance	Precipitation, evaporation, transpiration, runoff (surface and groundwater), available soil water, snow cover.
Nitrogen cycle	Uptake, deposition, litter fall, decomposition, available nitrogen.
Carbon balance	Gross primary production, autotrophic respiration, heterotrophic respiration, carbon in trees and soil.
Structure/properties of trees/stand, harvested biomass/timber	Trees and stand structure. Harvested biomass and timber (logs, pulp and energy biomass), carbon sequestration.

The decay of litter and humus releases CO_2 and nitrogen. The main model output represents the water and carbon balances, the structure and properties of tree stand, and the amount of biomass and timber harvest and the sequestration of carbon during the selected rotation. The water balance involves the effects of precipitation, evaporation, transpiration, runoff and availability of water in the rooting zone for the uptake of trees. The carbon balance is indicated, for example, by gross primary production, autotrophic and heterotrophic respiration and the amount of carbon in trees and soil. The carbon balance is further related to the growth of biomass and wood, which are harvested in cuttings in the form of biomass and timber.

13.2 Simulation of Changes in Climate and Weather

In climatic statistics, the climate is widely described using the long term averages (e.g., 30 years) of a set of climate parameters, which vary in time and space. Climatic variation on the daily and local scale refers to weather, which directly or indirectly affects the climatic and edaphic properties of a site and subsequently the physiology and ecology of trees. In this context, climate change refers to any systematic change in climate over time, whether due to natural variability or a result of human activity (IPCC 2007). Short-term variability (hourly, daily) is especially important in order to meet the response constants of physiological and ecological processes behind the growth and development of trees.

Figure 13.2 outlines the weather generator built by Strandman et al. (1993) for the current ecosystem model, utilizing weather statistics on a monthly basis. Under climate change, a trend-like variability in the weather pattern is assumed. For example, the growing season may become systematically longer, but not excluding the annual, seasonal and diurnal variability characterizing climate and weather. Even under climate change, there is a pronounced seasonal component in the annual weather pattern. Diurnal variability further specifies the seasonal weather pattern with regard to short-term variability, with a clear autocorrelation between consecutive moments. There is also much random variability associated with the seasonal and diurnal components of the annual weather patterns (Fig. 13.2, left; Box 13.1).

In simulations, the climate and weather are given at the scale of points representing a stand/community of trees occupying a site. On a larger spatial scale (e.g., region), the points form a grid, which facilitates the extrapolation of model output outside the points. The points, with coordinates (x, y, z), also represent spatially differentiated information about the measured values of properties of tree stands/communities; for example, tree species composition, diameter distribution, spacing, etc., provided by forest inventories. The properties of climate and soil are also given in the point

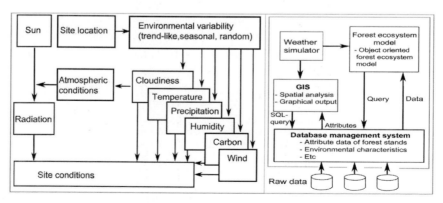

Fig. 13.2 Left: Outline used in calculating the weather pattern (Strandman et al. 1993). Permission of Elsevier. Right: Schematic presentations of interfacing ecological models with a data management system and geographical information system (Kellomäki et al. 1993).

coordinate thus making it possible to assess climate/forest interactions under climate change over selected regions (Fig. 13.2, right).

Box 13.1 Simulating short-term variability of weather factors

The weather simulator includes: (i) linear or non-linear long-term trend in the changing weather patterns; (ii) seasonal variability of annual weather patterns; (iii) diurnal variability in seasonal weather patterns; (iv) random variability in seasonal and diurnal weather patterns; and (v) autocorrelation among weather factors as in temperature, precipitation and radiation (Strandman et al. 1993).
Hourly temperature values (hT (h), °C) are related to daily mean temperature (dT (d), °C):

$$hT(h) = dT(d) + (0.5 \times a_{dT}) \times \sin((h-6) \times 15) \qquad (13.2)$$

where a_{dT} is the monthly amplitude of daily temperature. Daily mean temperature is calculated based on the mean monthly values (Richardson 1981):

$$dT(d) = \mu_{mT} + \rho_{mT} \times (dT(d-1) - \mu_{mT}) + \sigma_{mT} \times n_i \times (1 - \rho^2_{mT})^{0.5} \qquad (13.3)$$

where μ_{mT} is the monthly mean temperature (Fig. 13.3). Precipitation ((dprecip(d), mm) is based on daily cloudiness:

$$dprecip(d) = \frac{0.091}{A(m)} \times (e^{0.935 \times dcould(d)} - 1) \qquad (13.4)$$

where the coefficients 0.091 and 0.935 were estimated from the weather statistics. The values of A(m) are:

$$A(m) = \frac{1}{mnumber(m)} \times \frac{monthly\ mean\ precipitation(m)}{monthly\ mean\ cloudiness(m)} \qquad (13.5)$$

where mnumber(m) is the number of days in a month (Fig. 13.3). Total hourly radiation (RADCLR(h), W m^{-2}, wave length < 2,500 nm) above clouds is (Gates 1980, pp. 111–117):

$$RADCLR(t) = S_0 \times \tau^{th} \times \sin(alt) + S_0 \times (0.271 - 0.294 \times \tau^{th}) \times \sin(alt) \qquad (13.6)$$

where S_0 is the solar constant (1,360 W m^{-2}), τ is the transmission [dimensionless] of the atmosphere, th is the thickness of the atmosphere (i.e., th ≈ 1/sin(alt), where alt is the altitude of sun). Clouds reduce radiation and split it into direct (cdirrad(h)) and diffuse (cdifrad(h)) radiation under a partly cloudy sky and totally diffuse radiation under a fully cloudy sky:

$$cdirrad(h) = RADTOT(h) \times FDIFF(h) \qquad (13.7)$$

$$cdifrad(h) = RADTOT(h) \times (1 - FDIFF(h)) \qquad (13.8)$$

where FDIFF(h) = (1-hcloud(h)) × 0.77. Consequently, the hourly total radiation below clouds (RADTOT(h)) is (Lumb 1963):

$$RADTOT(h) = RADCLR(h) \times (1 - 0.71 \times hcloud(h)) \qquad (13.9)$$

Part of the radiation is reflected from the canopy, with the consequence that hourly net radiation (cnetrad(h)) is:

$$cnetrad(h) = (1-a) \times cshortrad(h) + clongrad(h) \qquad (13.10)$$

where a is albedo, cshortrad (h) is the short-wave (300–2,500 nm) and clongrad(t) is the long-wave (wave length > 2500 nm) radiation. The short- and long-wave radiations are:

$$cshortrad(h) = cdirrad(h) + cdifrad(h) \qquad (13.11)$$

$$clongrad(h) = \sigma \times (hT(h) + 273.15)^4 \times (r_1 + r_2 \times \sqrt{e(h)}) \times (r_3 + r_4 \times nsun(h)) \qquad (13.12)$$

where σ is the Stefan-Boltzman constant, hT(h) is the hourly temperature [°C], e(h) is the vapor pressure of air, nsun(h) is the relative duration of sunshine (nsun(h) = 1- hcloud(h)), and $r_1...r_3$ are dimensionless parameters (Jansson 1991a,b) (Fig. 13.3).

Box 13.1 contd....

Box 13.1 contd....

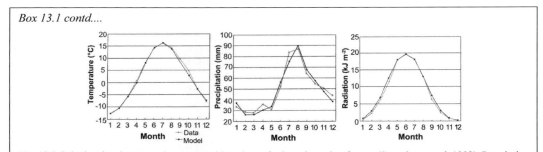

Fig. 13.3 Calculated and measured mean monthly values of selected weather factors (Strandman et al. 1993). Permission of Elsevier. Left: Temperature, middle: Precipitation, and right: Radiation. Calculated and measured values of temperature and precipitation are for Joensuu airport (62° 30' N, 29° 38' E) and radiation for Luonetjärvi airport (62°24'N, 25°40'E).

13.3 Organic Matter on Soil Surface and in Soil Profile

The soil surface is the boundary between the atmosphere and soil system. It may be mineral soil alone or covered by an organic layer (litter and humus), which is assumed to be too coarse to store heat and water itself. Water and heat infiltrate through the organic layer into the mineral soil, where they are driven down by the moisture and temperature gradients between the soil surface and underlying soil layers. The layer of litter on the soil surface includes litter in varying phases of decay with an identifiable origin. Further decay produces humus, the origin of which is no longer identifiable. A nitrogen content in humus is greater than the value specific to the litter from each organ (Fig. 13.4). In general, the border between litter/humus and the underlying mineral soil in boreal upland sites is clear, but infiltrating water transfers humus particles deeper into the soil profile. Humus in soil profile origins also from root litter, especially from fine roots, which are an important carbon sink of boreal trees (Högberg et al. 2001). Total Soil Organic Matter (SOM) is the combined amount of litter and humus on the mineral soil and humus in the soil profile, representing partly historic accumulation of carbon compounds in the soil.

Litter is dead mass falling annually (litter cohort) on a soil surface from the dying foliage, branches and stems or bound in soil from coarse and fine roots. In the current model, the weight loss in decaying litter is calculated on an annual basis (Fig. 13.4). The weight loss (%) is dependent on annual evaporation and ratio between the lignin (L) and nitrogen (N) content (ratio L/N) in decaying litter. When a whole tree dies, the organs of the tree, including the stem, are converted to litter and decay. The decomposition of humus is also dependent on annual evaporation and the contents of carbon (C) and nitrogen (C/N ratio) in humus. The weight loss of litter and humus is converted to CO_2, which is emitted into the atmosphere. The further decay of humus makes nitrogen available for uptake and growth (Box 13.2).

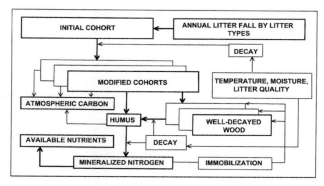

Fig. 13.4 Model for the decay of soil organic matter (Kellomäki et al. 1993).

Box 13.2 Outlines of decay of soil organic matter

The current weight ($W_{litter}(t)$, kg ha^{-1}) of litter and the weight of nitrogen in the litter are:

$$W_{litter}(t) = W_{litter}(t-1) - f_{litter}(t) \times W_{litter/orig}(t-1) \qquad (13.13)$$

$$W_{litter/N}(t) = W_{litter/N}(t-1) - N_{imm/year}(t) \times f_{litter}(t) \times W_{litter}(t-1) \qquad (13.14)$$

where $f_{litter}(t)$ is the yearly fraction of weight loss, $W_{litter/orig}(t)$ is the original weight of litter, $W_{litter/N}(t)$ (kg ha^{-1}) is the weight of nitrogen and $N_{imm/year}$ is the amount of nitrogen [kg ha^{-1}] immobilized in decay. The weight loss of the litter (W_{Loss}) is the function of evaporation at the site and the lignin and nitrogen contents (ratio lignin/nitrogen) of the litter (Fig. 13.5):

$$W_{Loss} = A - B \times (L/N) \qquad (13.15)$$

where A and B are variables dependent on evaporation.
Carbon dioxide ($C_{litter}(t)$, kg ha^{-1}yr^{-1}) is emitted according to the weight loss of litter, making nitrogen available ($N_{avail}(t)$, kg ha^{-1}) for uptake:

$$C_{litter}(t) = 0.48 \times f_{litter}(t) \times W_{litter}(t) \qquad (13.16)$$

$$N_{avail}(t) = \max(0, N_{\min}(t) - N_{imm}(t)) \qquad (13.17)$$

where $N_{\min}(t)$ is the amount of nitrogen mineralized per a year. The nitrogen released is that not immobilized in the decay of litter. The nitrogen released from humus ($N_{\min/humus}$, kg ha^{-1}) at the same time as carbon emission ($C_{year/hum}(t)$, kg ha^{-1} yr^{-1}) is:

$$N_{\min/humus}(t) = W_{hum}(t) \times \frac{a \times (C/N)}{b + (C/N)} \times f(E(t)) \qquad (13.18)$$

$$C_{hum}(t) = 0.48 \times \frac{N_{\min/hum}(t-1)}{N_{hum}(t)} \qquad (13.19)$$

where is $W_{hum}(t)$ is the amount of humus [kg ha^{-1}], and C/N is the ratio between the carbon and nitrogen content in humus. Evaporation drives a function ($f(E(t))$) scaling mineralization to the prevailing evaporation [mm yr^{-1}] in the site.

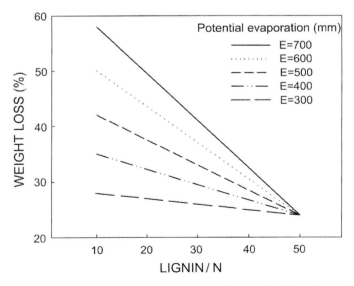

Fig. 13.5 Weight loss of decaying litter as a function of annual potential evaporation (E, mm) at a site and lignin and nitrogen content (ratio lignin/nitrogen) of litter, based on Meentemayer (1978), Meentemayer and Berg (1986) and Pastor and Post (1986).

13.4 Water in Canopy and Soil

In forest with standing trees, precipitation (water/snow) falls on soil through the canopy, where some precipitation is intercepted and evaporated (Fig. 13.6). Water pooling on the soil surface represents direct precipitation, throughfall and melting snow. Water on the surface may flow in surface flow, infiltrate the soil profile and flow more deeply to ground water outside the rooting zone, or evaporate from the surface pool. Water in the rooting zone is further taken up by trees and lost to the atmosphere in transpiration, which is controlled by the properties of foliage and atmosphere. In the same way, the evaporation rate of water from the wet canopy is affected by the properties of foliage and atmosphere, as is the case for evaporation from the ground surface.

Evaporation from a wet canopy (or from canopy pool) is dependent on the balance between incoming precipitation (P, mm) and intercepted water evaporated from canopy surfaces, and water infiltrated through the canopy onto soil. Water intercepted on the canopy surfaces (AWV, mm) is equal to the interception (INTERCEPT, mm) minus evaporation (EVA, mm): AWV = INTERCEPT− EVA. The daily climatic potential of the evaporative water loss (PEV, mm d^{-1}) from the wet canopy is calculated using the Penman-Monteith equation, applying the big leaf approach (i.e., foliage in different layers is pooled in one homogenous layer) (Jarvis and McNaughton 1986, Monteith and Unsworth 1990):

$$PEV = \frac{s \times R_{nc} + \rho_a \times C_p \times (e_s(T_a) - e_a)/r_{ac}}{s + \gamma \times (1 + r_{sc}/r_{ac}) \times Hv} \qquad (13.20)$$

where s is the derivate of the slope of saturation vapor pressure curve [Pa °C^{-1}], R_{nc} [W m^{-2}] is the radiation intercepted in foliage (the difference between the radiation above the canopy and the radiation on the soil surface below the canopy), ρ_a is the air density [1.2220

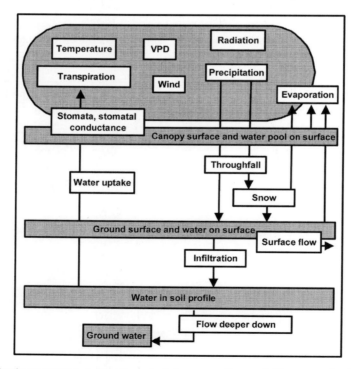

Fig. 13.6 Outline of main processes controlling the water balance on a site occupied by trees under the control of climatic factors and properties of soil and the tree stand, based on Kellomäki and Väisänen (1997).

kg m⁻³], C_p is the specific heat capacity of air at a constant pressure [1004.0 J kg⁻¹ °C⁻¹], T_a is air temperature [°C], e_s the saturated vapor pressure, e_a is the current vapor pressure, r_{ac} [s m⁻¹] is the aerodynamic resistance of foliage, Hv is the vaporization heat of water [2,353,000 J kg⁻¹], r_{sc} is the surface resistance [s m⁻¹] when intercepted water occur, and γ is the psychrometric constant [66.0 Pa °C⁻¹].

Water in soil represents water on the soil surface and water in the soil profile down to a given depth through the rooting zone. On a daily basis, water in the surface pool (W_{surf}, mm d⁻¹) is the balance between the incoming and outgoing water flows:

$$W_{surf}(t) = W_{surf}(t-1) + (W_{in}(t) - E_{ground}(t) - W_{runoff}(t) - W_{insoil}(t)) \tag{13.21}$$

where W_{in} [mm d⁻¹] is the incoming water, E_{ground} [mm d⁻¹] is the evaporation from the soil surface pool, W_{runoff} [mm d⁻¹] is the runoff from the surface pool and W_{insoil} [mm d⁻¹] is the infiltrating of water from the surface pool to the soil profile (Jansson 1991a,b).

Potential evaporation from the soil surface is dependent on the microclimate at the surface and the properties of the surface. Under boreal conditions, water on the soil surface is further related to the snow accumulation in winter, and snow melting in winter and spring providing water for the early growing season. The dynamics of snow cover and soil frost are related to the weather and soil factors, and also to the dynamics of trees occupying the site (Fig. 13.7). The canopy cover controls the interception of snow on the canopy, where water in the snow is evaporated, thus reducing water entering the ground in the form of snow (Box 13.3).

Box 13.3 Snow on soil surface

Precipitation is only water if the air temperature > +1°C and only snow if the air temperature is < –1°C, otherwise (air temperature in the range from –1°C to +1°C) precipitation is a mixture of water and snow (Strandman et al. 1993). Snow is accumulated on the soil surface if the air temperature is < 0°C. The accumulation of snow is controlled by snow fall and the hydraulic and thermal processes related to the properties of snow (e.g., old or new snow, density of snow), which affect the water content in snow (L (t), mm) (Fig. 13.7):

$$L(t) = L(t-1) + (RS(t) + D(t)) \times s \tag{13.22}$$

where RS (t) is the daily snow fall [mm d⁻¹], D(t) is the daily value of the melting/freezing function [mm d⁻¹] and s the integration step. Melting/freezing is related to the air temperature, the amount of daily short wave radiation [W m⁻²], the heat flow into the soil surface [J m⁻² d⁻¹], the latent heat of melting and the age of the snow.

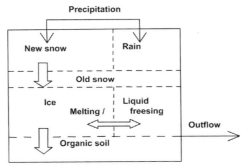

Fig. 13.7 Process for calculating snow dynamics (Wang and Kellomäki 2003).

The soil surface forms the upper boundary for water in soil, and the lower boundary is the lowest soil layer included in the calculations of water in the soil profile (Jansson 1991a,b):

$$\frac{\partial W}{\partial t} = \frac{\partial}{\partial z}\left[k_w \times \left(\frac{\partial p}{\partial z}\right)\right] + s_w \tag{13.23}$$

where W is the water content [m³ m⁻³], z is depth [m] in the soil profile, p is water tension [Pa], k_w is hydraulic conductivity [mm day⁻¹] and S_w is the sink/source term for water [mm day⁻¹]. Water content indicates the volume of water per unit volume of soil [m³ m⁻³]. This can further be translated to the water tension [Pa] binding water in soil particles. The water available to plants (W_a) is defined by the field capacity and wilting point specific to different soil types:

$$W_a = \frac{W_{soil} - W_w}{W_x - W_w} \qquad (13.24)$$

The numerator indicates the actual extractable water defined by the difference between the volumetric water content in the rooting zone (W_{soil}) and the volumetric water content at the wilting point (W_w), while the denominator defines the maximum water content at the field capacity (W_x). Figure 13.8 shows a simulated example of the annual water content under boreal conditions with and without tree stands.

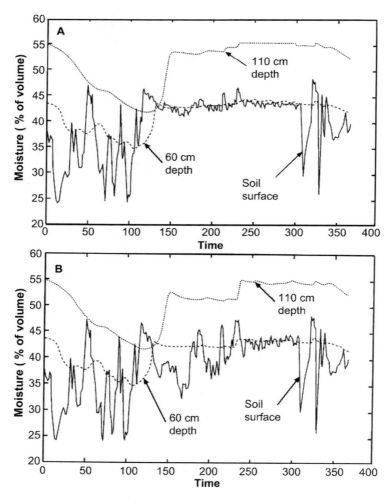

Fig. 13.8 Simulated values of soil moisture in selected layers over a year for the middle boreal zone (62° N) at a forest site of *Myrtillus* type under the current climate. Above: Without tree stand, and Below: With Scots pine stand. The stand structure was characterized by the density of 3,000 trees ha⁻¹, stem wood stocking 60 m³ ha⁻¹ and basal area 24 m² ha⁻¹, mean height 10 m, mean diameter 10 cm). In the simulation, a process-based model (FinnFor) was used (Kellomäki and Väisänen 1997).

Transpiration is the water lost in evaporation through stomata. The big leaf model (Jarvis and McNaughton 1986) is commonly used in calculating transpiration (g H_2O m^{-2} s^{-1}) from the whole crown. In doing this, foliage in different crown layers is pooled in a single homogenous layer with the cumulative effects of single-leaf conductance of stomata. Water lost through stomata may be calculated with the Penman-Monteith equation, assuming that the ambient vapor pressure deficit and temperature are the same through the whole canopy, and water loss in transpiration occurs only through stomata (McMurtrie et al. 1990). Stomatal conductance is a function of photon flux density, air temperature, leaf-to-air vapor pressure difference, CO_2 concentration, soil water potential and soil temperature in a Jarvis-type model (Jarvis 1976; Chapter 8):

$$\begin{aligned} g_s &= g_{s\,max}, \text{if no factor reduces transpiration} \\ g_{s\,min} &< g_s < g_{s\,max}, \text{if some factor partly reduces transpiration} \\ g_s &= g_{s\,min}, \text{if some factor fully reduces transpiration} \end{aligned} \qquad (13.25)$$

where g_{smax} is maximum and g_{smin} minimum stomatal conductance.

13.5 Heat and Soil Temperature

The properties of the soil surface (e.g., surface of mineral soil with/without organic layer (litter, humus)) affect heat transfer into the soil profile. Soil temperature further determines whether the soil is frozen or not, with the consequent effect on water conditions in the soil. The properties of tree stands, especially the canopy, affect radiation and water on the soil surface and affect the soil temperature and moisture of the surface, with feedback to the growth and development of trees on the site. Temperature on the soil surface equals the hourly air temperature, if the depth of snow covers is < 0.1 mm; otherwise the soil surface temperature (T_s, °C) is a function of the heat flow through snow to the soil surface (Jansson 1991a,b):

$$T_s = \frac{T_1 + a \times T_a}{1 + a} \qquad (13.26)$$

where T_1 is the temperature [°C] of the topmost layer of inorganic soil, T_a is the air temperature [°C], a is a factor related to the thickness of the topmost layer of inorganic soil, the thickness of snow cover, and the thermal conductivity of the topmost layer of inorganic soil.

The surface temperature is used in calculating the downward heat flow based on heat conduction and convection (Jansson 1990a,b). The soil surface forms the upper boundary conditions, and the lower boundary is the lowest soil layer included in the calculations:

$$\frac{\partial(C_{soil} \times T)}{\partial t} - L_{ice} \times \rho_{ice} \times \frac{\partial W_{ice}}{\partial t} = \frac{\partial}{\partial z}\left(k_h \times \frac{\partial T}{\partial z}\right) - C_w \times \frac{\partial(T \times q_w)}{\partial z} + S_h \qquad (13.27)$$

where C_{soil} is the heat capacity of the inorganic soil layer [J m^{-3} °C^{-1}], L_{ice} is the latent heat of freezing water [J kg^{-1}], ρ_{ice} is the density of ice, W_{ice} is the volumetric water content of ice, C_w is the heat capacity of water [J m^{-3} °C^{-1}], k_h is thermal conductivity [W m^{-1} °C^{-1}], q_w is water flow [mm day^{-1}], s_h is the sink/source term for heat [mm day^{-1}], and T is temperature [°C] of the inorganic layer (Fig. 13.9).

Soil temperature is given layer by layer through the soil profile. The processes in the calculations are linked in such a way that the water in the soil influences soil temperature through its heat content and heat capacity relative to the physical properties of soil. Similarly, the volumetric water content depends on the freezing temperature and evaporation of water from the soil. Figure 13.10 shows that there is a close correlation between the measured and modeled timing of freezing, but there is a wide variability in the absolute depth of soil frost depending on soil type, accumulation of snow and local conditions. The temperature on the soil surface may exceed 0°C during warm spells, however, in late winter or early spring the soil remains frozen deeper below the unfrozen soil surface. In autumn, the soil remains unfrozen deeper in the soil profile than in the uppermost soil layer, demonstrating the delay in thermal conditions in different parts of the soil profile.

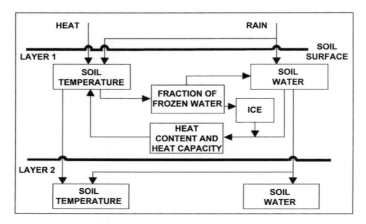

Fig. 13.9 Connections between temperature and water in different layers of a soil profile (Kellomäki and Väisänen 1997). Courtesy of the Finnish Society of Forest Science and the Finnish Forest Research Institute.

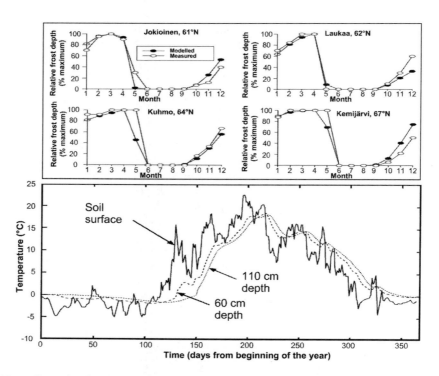

Fig. 13.10 Above: Comparison between the measured and simulated timing of soil freezing throughout Finland. The relative depths of soil frost for measured and simulated values represent the monthly mean values of the depth of soil frost related to the mean maximum depth of the year in the period 1971–2000 (Kellomäki et al. 2010). Below: Simulated values of soil temperature on soil surface and in selected layers in soil profile over a year for middle boreal zone (62° N) at a forest site without tree cover. Figures above: Courtesy of the Finnish Society of forest Science and the Finnish Forest Research Institute.

13.6 Uptake of Carbon and Nitrogen

Simulations are launched using initial stands, which include one or more tree cohorts. Photosynthetic rate of object tree (A_n, μmol m^{-2} s^{-1}) is calculated on an hourly basis using the biochemical model of Farquhar et al. (1980). The total photosynthesis (P_{tot}) in the crown of object tree is the sum of photosynthesis over crown layers:

$$P_{tot} = \sum_{l=1}^{n} A_{nl} = \sum_{l=1}^{n} f(properties\ of\ object\ tree) \times f(climatic\ factors) \times f(edaphic\ factors) \quad (13.28)$$

where l in the number of crown layers [l = 1…n] from the stem apex.

Regardless of tree species, the crown is ellipsoid shaped over the life time of the tree. The foliage area is distributed uniformly in the crown, divided in layers. In Scots pine and Norway spruce, the foliage is further divided into needle age classes. Any tree cohort is shaded by the surrounding tree cohorts, which reduces the direct and diffuse radiation incoming to the crown surface. The internal shading of the crown further reduces the radiation on a crown layer, which is the sum of direct and diffuse radiation. The calculation of radiation on the crown layer is based on: (i) the probability of direct and diffuse radiation passing through the crown of other trees and reaching the trees in a particular cohort; (ii) the length of the path through tree crowns in other cohorts; (iii) the thickness of crown layers; and (iv) the shadow area that the layer produces on a particular layer below (Oker-Blom 1985, 1986) (Fig. 13.11) (Box 13.4).

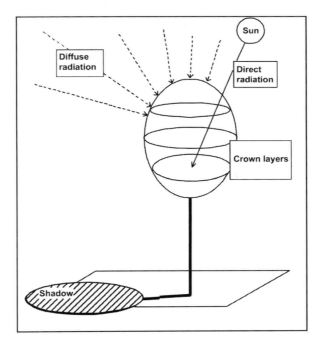

Fig. 13.11 Performance of direct and diffuse radiation in the tree crown, divided into layers (Kellomäki and Väisänen 1997). Permission of Elsevier.

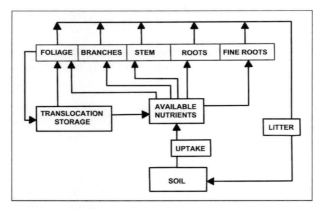

Fig. 13.12 Outline of the calculation of the nutrient cycle in tree stands (Kellomäki and Väisänen 1997). Permission of Elsevier.

Box 13.4 Radiation in tree crown

The mean direct irradiation I_{ID} [W m^{-2}] on the surface of a leaf/needle at the point (x, y, z) is:

$$I_{ID} = k \times I_D \times p_o \qquad (13.29)$$

where I_D is the direct radiation above the canopy, p_o the transmission fraction (gap probability) for direct radiation and k is the light extinction coefficient [dimensionless]. Similarly, the mean diffuse irradiation (I_{id}) is:

$$I_{id} = 2 \times k \times I_d \int_0^{\pi/2} \exp(-\lambda \times T(z,\alpha)) \times \cos\alpha \times \frac{1}{C(z)} \int_{C(z)} \exp(-k \times LAD \times t(\alpha,x,y,z)) dx dy) d\alpha \qquad (13.30)$$

where I_d is the diffuse radiation above the canopy, crown shadow area (T (z, α), where α is the altitude of sun in radians), the horizontal cross-section of the crown at the depth [m] z from stem apex, and the foliage area density (LAD (α, x, y, z), m^2 m^{-3}).

The transmission fraction is a function of stand density (λ, trees ha^{-1}), foliage area density, the shadow area [m^2] cast by a crown layer (T (z, α)) on the horizontal plane at the depth z [m] from the stem apex, and the length of path (t (α, x, y, z,)) in the crown:

$$p_o = \exp(-\lambda T(z,\alpha))\exp(-kLADt(\alpha,x,y,z)) \qquad (13.31)$$

where the first exponential refers to the probability of a gap through the crowns of neighboring trees and the second exponential to the gap probability in the crown of the object tree. The shadow area is the function of the leaf area density and the cross-section of the crown (C(z)):

$$T(z,\alpha) = \int_0^z \int_{C(z)} \exp(k \times LAD \times t(\alpha,x,y,z) \times k \times LAD/\sin\alpha) dx dy dz \qquad (13.32)$$

The length of the path [m] in the crown is:

$$t(\alpha,x,y,z) = \frac{B + \sqrt{B^2 - A \times C}}{A} \qquad (13.33)$$

where

$$A = (h/2)^2 \times (\cos\alpha)^2 + r^2 \times (\sin\alpha)^2 \qquad (13.34)$$

$$B = r^2 \times \sin(\alpha) \times ((h/2) - z) - (h/2)^2 \times x \times \cos\alpha \qquad (13.35)$$

$$C = (h/2)^2 \times (x^2 + y^2) + r^2 \times z \times (z - h) \qquad (13.36)$$

and h [m] is the length of the crown and r is the radius of the crown [m].

Photosynthetic rate is related to the nitrogen content in foliage affected by two sources of nitrogen (Fig. 13.12); i.e., that taken up from the soil and that translocated from dying foliage and other tissues. Nitrogen fraction cN (k, l, m, t) indicates the fraction [0...1] in the tree k, at the crown layer l, and in the foliage age class number m at the moment t:

$$cN(k,l,m,t) = \frac{N(k,l,m,t)}{Wn(k,l,m,t)} \tag{13.37}$$

where Wn indicates the respective foliage mass [kg]. The amount of nitrogen [kg] is:

$$N(k,l,m,t) = N(k,l,m,t-1) + NU(k,l,m,t) - NL(k,l,m,t) \tag{13.38}$$

where NU is the uptake [kg] and NL the removal [kg] of nitrogen. Nitrogen is removed from foliage along with litter: $NL = cNL \times N(k,l,m-1,t-1)$ where cNL is nitrogen fraction in litter. A new foliage age class receives nitrogen through uptake and translocation:

$$N(k,l,m,t) = trN(k,l,m,t) + upN(k,l,m,t), \tag{13.39}$$

where trN is the amount of nitrogen obtained through translocation and upN is the amount of nitrogen obtained in uptake.

13.7 Growth and Development of Trees

Simulations are initialized by describing a tree stand compiled by tree cohorts. Based on Marklund (1987, 1988), the initial mass ((M(D) or M (D, H), kg) of foliage, branches, stem biomass, coarse roots and fine roots in the object tree is calculated as a function of diameter (D, cm) or diameter and height (H, m) of tree. The annual gross photosynthesis in the object tree k in the year t is allocated to different tree organs i for calculating growth, which is the net amount of carbon bound in organs after removing the carbon used in maintenance (R_m (k, i, t)) and growth (R_g (k, i, t)) respiration:

$$\Delta W(k,i,t) = a(k,i,t) \times P_{tot}(k,t) - (R_m(k,i,t) + R_g(k,i,t)) \tag{13.40}$$

where P_{tot} (k, t) is the annual gross photosynthesis, a(k,i,j) is the allocation of photosynthesis and ΔW(k,i,j) is the growth of organs shown in Fig. 13.13.

The diameter and height of a tree is related to the stem volume, which is calculated by determining the volume of annual rings over the stem. The mean volume of an annual ring over stem is obtained by converting the mass of annual stem growth (ΔM (k, t), kg) to the volume using the density of wood. Based on Hakkila (1979), the density of wood (ρ (k, t), kg m^{-3}) is a function of the annual stem growth and the stem mass (M (k, t), kg) at the year t:

$$\rho(k,t) = d(k) \times \Delta M(k,t)^a \times M(k,t)^b \tag{13.41}$$

where d(k) is the density of juvenile wood specific for tree species k, with the value 305 kg m^{-3} for Scots pine, 275 kg m^{-3} for Norway spruce and 325 kg m^{-3} for birch and a and b parameters (Fig. 13.14).

The height (H (k, t), m) and the diameter (D(k, t)) of tree k at the year t was calculated as a function (f1, f2) of stem wood volume (V (k, t), m^3) and the initial height H(k,0) and diameter (D(k,0)) of tree:

$$H(k,t) = H(k,0) + f1(V(k,t) \times f(TS(r))) \tag{13.42}$$

$$D(k,t) = D(k,0) + f2(V(k,t)) \tag{13.43}$$

In Equations (13.42), TS(r) is the long-term site-specific current temperature sum (a +5°C threshold) scaling the height to meet the ecotype differences in D/H ratio across the south-north temperature gradient through the whole country (Kellomäki et al. 2008). The stem volume V (k, t) of tree k in the year t since the start of simulation is:

$$V(k,t) = V(k,0) + \sum_{t=1}^{n} \Delta V(k,t) = V(k,0) + \sum_{t=1}^{n} \left[\frac{\Delta M(k,t)}{\rho(k,t)} \right] \quad (13.44)$$

where V(k,0) is the initial volume of stem wood at launching the simulation and ΔV (k, t) is the volume growth of tree k in the year t, ΔM (k, t) is the mass growth of stem, and n is the years since the start of growth.

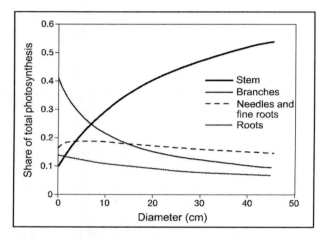

Fig. 13.13 Fraction of annual gross photosynthesis allocated to different organs of Scots pine used in simulations in different developmental phases of trees indicated by the stem breast height diameter, based on Vanninen and Mäkelä (2005). Optionally, the constant allocations with the fractions of gross photosynthesis for the foliage 0.18, branches 0.24, stem 0.35, course roots 0.05 and fine roots 0.18 give fairly similar results on the long-term growth and development of object trees.

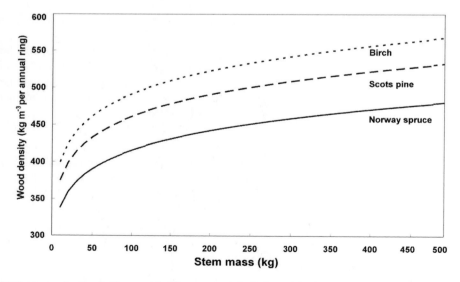

Fig. 13.14 Mean annual values (or the wood density per annual ring) of wood density in Scots pine, Norway spruce and birch as a function of stem wood mass and stem wood growth (Seppo Kellomäki, University of Eastern Finland, unpublished).

13.8 Mortality and Litter Fall

Mortality involves the death of living tree organs that form litter (litter fall) or the death of a whole tree, with the immediate death of all its organs. In living coniferous trees, non-woody litter includes mainly dead needles and fine roots. The needle mass born in a given year reduces depending on the survival rate related to the needle age (Muukkonen and Lehtonen 2004, Muukkonen 2005). The mean maximum survival in Scots pine needles is 4–6 years and in Norway spruce 8–12 years, longer on poor and northern sites than on southern and fertile sites. The mortality of fine roots may be related to the mortality of needles using an allometric approach and the life span of foliage (Vanninen and Mäkelä 2005, Franklin et al. 2012).

In deciduous trees, the built-up and senescence of the whole foliage are repeated every year. Calculation of the built-up of deciduous foliage is based on the temperature sum (TS, d.d.). The built-up of foliage (budburst) is assumed to occur both in the south and north whenever the springtime value of TS exceeds the value 36 d.d. (i.e., a = 36 in Equation (13.45). Similarly, the build-up of foliage is complete, whenever the value of TS exceeds the value of b = 865 d.d. for the south (< 64° N), and 600 d.d. for the north (> 64° N) (Raulo and Leikola 1974). The percentage share (SH_{up}, %) of the maximum foliage during the built-up phase is (Kellomäki et al. 2001a,b):

$$SH_{up} = \sum_{1}^{n}\left(\left(\frac{TS-a}{(b-a)/2}\right)\times 100\right) \qquad (13.45)$$

where n is the number of days with the daily mean temperature ≥ 5°C (Fig. 13.15). Similarly, calculation of the senescence of foliage is based on temperature (T) conditions and further day length (DL) (Kellomäki et al. 2001b):

$$SH_{down} = \min\left(SH_{down}(T), SH_{down}(DL)\right) \qquad (13.46)$$

where SH_{down} is the percentage of senescence of foliage, and $SH_{down}(T)$ and $SH_{down}(DL)$ are the effects of temperature and day length on senescence (Fig. 13.15).

The dynamics of the woody parts in branches, stems and coarse roots are characterized by the annual addition of a new annual ring of sapwood on wood accumulated in previous years. At the same time, the growth rings born in previous years are dying, and sapwood is converted to heartwood with no physiological function, but remaining attached to the organ structure.

The mechanism of the mortality of whole trees is poorly known, except where death is due to, e.g., wind, snow or fire damage. However, the monitoring of permanent sample plots with subsequent growth and yield studies provides information to model the effects of mortality of whole trees on the dynamics of forest ecosystems. Mazziotta et al. (2014) used this information to compare the success of the modeling of mortality in Norway spruce, Scots pine and birch (Box 13.5). Over a 100-year period, 20–40% of the tree biomass may die, depending on tree species, site type and density of initial stand (Table 13.2). The models used to make parallel simulations show similar trends, with the highest mortality in dense birch stands on fertile sites. A similar trend held for Norway spruce and Scots pine.

13.9 Management

Natural regeneration and planting

Under boreal conditions, regeneration based on natural seeding is directly dependent on temperature controlling flowering and formation and maturing of seeds with impacts on the quantity and quality

182 Managing Boreal Forests in the Context of Climate Change

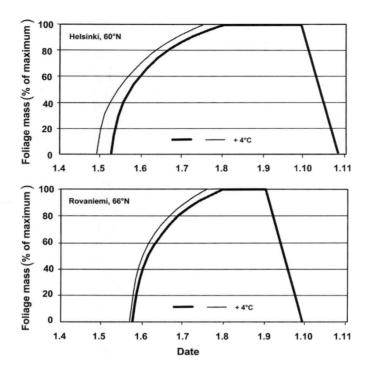

Fig. 13.15 Build-up and senescence of birch foliage for southern (Helsinki, 60° N) and northern Finland (Rovaniemi, 66° N) under current temperature (thick line) and elevated temperature (thin line for current temperature +4°C) (Kellomäki et al. 2001b). Permission of Wiley & Sons.

Box 13.5 Calculating mortality with two separate models for Table 13.2

The death of trees is determined (i) purely by random reasons (e.g., wind, snow) or (ii) by crowding with the consequent reduction in growth and the increase in the probability of a tree to die at a given moment. In the case (ii), trees in each cohort die if the uniform random number [0…1] has the value < L:

$$L = \frac{BA \times (0.5 \times \rho)^{0.7}}{610.747 \times AgeMax} \qquad (13.47)$$

where BA [m² ha⁻¹] is the basal area of trees in the stand, ρ is the density of trees in the stand [trees ha⁻¹] and AgeMax is the maximum age of trees, with values 350 years for Scots pine, 180 years for Norway spruce and 150 years for birch (Koivisto 1959, Nikolov and Helmisaari 1992). Table 13.2 also utilizes the results of the growth and yield model Motti (Hynynen et al. 2002), in which the mortality of trees is calculated partly based on the inverse relationship between the stand density and the mean volume of trees following Reineke (1933). This mortality model is optionally used in the current ecosystem model.

Box 13.5 contd....

Box 13.5 contd....

Table 13.2 Dead wood share (%) of total production calculated using the results of the Sima and Motti models (Mazziotta et al. 2014) compared to the values from growth and yield studies (Koivisto 1959). The simulations are for southern (N 62°) and in northern boreal forests conditions (N 66°) as a function of tree species and site type.

Species and site type	Initial stand density, trees ha⁻¹	Southern boreal Sima model %	Motti model %	Growth and yield studies, %	Northern boreal Sima model %	Motti model %	Growth and yield table, %
Norway spruce • MT	1200	21.9	5.5		45.4	5.5	-
	2400	28.1	21.6	22	48.3	12.8	
	4800	33.6	34		54.3	23.3	
• OMT	1200	23.9	10.5		50.4	4.6	-
	2400	30.5	24.9	16.6	55.9	13.5	
	4800	36.2	34		58.5	26.4	
Scots pine • VT	1200	18.9	15.6		24.9	13.7	
	2400	26.8	26.8	31	31.3	23.8	35.7
	4800	35.8	36		41	32.1	
• MT	1200	28.7	24.7		41.3	15	
	2400	37.8	36.5	29.2	45.7	31.8	41.2
	4800	40.3	44.8		49.6	40.1	
• OMT	1200	27.3	19.9		-	-	-
	2400	34.2	33.1	27.6			
	4800	40.6	39.7				
Birch • MT	1200	39.6	33.7		44.4	7.7	-
	2400	38.5	43.2	32.5	41.8	17.4	
	4800	38.6	47.5		46.7	26.5	
• OMT	1200	44.8	39.6		55	9.8	-
	2400	43	47.9	35.5	49.5	23.2	
	4800	44	52.2		51.7	32	

of seed crop (Sarvas 1962, Pukkala 1985a,b) (Fig. 13.16). Germination is further affected by seed viability, which reduces as a function of the time since the dissemination of seeds; i.e., the main part of seeds of boreal trees lose their capacity to germinate during the summer following dissemination in spring (Yli-Vakkuri 1961a,b). Germination is dependent on temperature and moisture on surface soil. Moisture is assumed to increase linearly from the wilting point to field capacity, representing the water content in surface soil (Satoo 1966). Finally, climatic and edaphic conditions near the soil surface and in soil affect growth and development of established seedlings.

Over the regeneration period (the time from the regeneration cut to the time when a seedling population fulfills the given criteria), the seed crop is repeated annually, but only a few germinated seeds from each crop form established seedlings. Thus, the density of a seedling stand at a given moment represents the accumulation of seedlings over the years:

$$Density\ of\ seedlings(n) = \sum_{t=0}^{n} a(t+1) \times Established\ seedlings(t) \tag{13.48}$$

where t is the year for the establishment of a seedling cohort (t = 0), n is the end year of regeneration period, and a(t + 1) is the fraction of seedlings born in the year t surviving to the year t + 1. Following

Pukkala (1985a,b), the establishment of seedlings from a given seed crop is influenced by several random processes comprising the main phases of the regeneration process:

$$Established\ seedlings(t) = Seeds(t) \times Surf(t) \times Fulls(t) \times Mats(t) \times Gers(t) \times Dists(t) \qquad (13.49)$$

where Established seedlings(t) is the number of seedlings [seedlings m^{-2}] born in the year t, Seeds(t) is the seed crop [seeds m^{-2}], Surf(t) is the fraction of stockable soil [0 – 1], Fulls(t) is the fraction of full seeds [0 – 1], Mats(t) is the fraction of mature seeds [0 – 1], Gers(t) is the fraction of germinated seeds [0 – 1], and Dists(t)I is the fraction [0–1] of seeds destroyed by herbivores and pests. The probability, that a seedling in a cohort will survive to next year (a(t + 1)), increases with the age (and growth) of the seedlings.

Regeneration based on planting uses seedlings raised in nurseries, whereas direct seeding refers to sowing seeds in the area to be regenerated. When initializing the simulation, the seedlings to be planted are defined by species, number (stand density) and size (diameter, height) in the cohorts forming the initial stand. In climatically homogeneous areas, site fertility (site type) and soil texture are used in selecting suitable tree species for the site (Table 13.3).

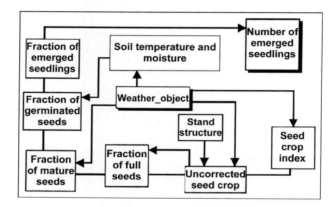

Fig. 13.16 Main factors of parent trees and weather affecting the success of natural regeneration, based on Pukkala (1985a).

Table 13.3 Recommended use of tree species suitable for in regeneration considering site fertility and soil texture in Finland.

Site fertility and soil texture	Tree species, first choice	Tree species, second choice
High site fertility (*Oxalis–Myrtillus* type, OMT)		
• Fine texture	*Betula pendula*	*Picea abies*
• Coarse texture	*Picea abies*	*Betula pendula*
Intermediate site fertility (*Myrtillus* type)		
• Fine texture	*Picea abies, Betula pendula*	*Betula pendula, Picea abies*
• Coarse texture	*Pinus sylvestris*	*Picea abies, Betula pendula*
Low site fertility (*Vaccium, Calluna* and *Cladonia* types)		
• Fine texture	*Pinus sylvestris*	*Pinus sylvestris* or related
• Coarse texture	*Pinus sylvestris*	*Pinus sylvestris* or related

Pre-commercial and commercial thinning, rotation length

Pre-commercial thinning refers to the removal of trees not large enough to produce pulp wood or saw logs but usable for energy biomass. Pre-commercial thinning is used to control competition between trees and provide more space for those remaining in order for them to maintain a fast growth. This holds also for commercial thinning, where trees large enough for pulp wood alone, or pulp wood and saw logs, are removed. The total removal in a commercial thinning is given in terms of reducing the basal area, which is converted into the number of trees to be removed from each tree cohort. The thinning may be done from above or below. In the former case, dominant and co-dominant trees representing the upper quartile of diameter distribution are mainly removed, and in the latter case suppressed and intermediate trees representing the lower quartile of diameter distribution are removed.

In commercial thinning, intensity and timing are related to the basal area and the dominant height of trees (Fig. 13.17). Trees removed in thinning and the final cut can be converted to saw logs, pulp wood and energy biomass. The concept of the saw log involves the butt of the stem with a given minimum diameter at the top of the log, and pulp wood is the remaining parts of the stem, with a given minimum diameter. The top part of the stem with a diameter smaller than that for pulp wood is logging residue, which also includes foliage, branches and stumps with coarse roots.

In stand-based management, the management cycle or rotation is terminated in final cut related to the maturity of trees. In general, rotation length is related to the growth of trees, and it can be defined in biological and/or economic terms. In biological terms, the optimum rotation length is the point where the slope of mean annual growth is equal to zero (Fig. 13.18). This is equivalent to the intersection of mean annual growth and the current annual growth, beyond which mean annual growth reduces. In economic terms, the optimal rotation length indicates the time from regeneration to terminal cut, generating the maximum revenue or economic yield. The optimal rotation length is that maximizes the Net Present Value (NPV) considering the costs (PVC) and income (PVR) from management and harvest: NPV = PVR–PVC.

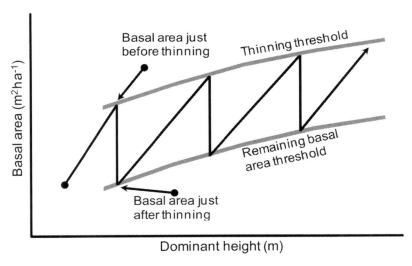

Fig. 13.17 Timing and intensity of thinning are determined as a function of the dominant height and basal area of tree stands. The dominant height is the mean height of the 100 thickest trees in the stand. The thinning threshold indicates the basal area at a given dominant height which triggers thinning. The remaining basal area indicates the remaining stocking, which is high enough to sustain productivity. A higher thinning threshold will delay thinning, as does a lower limit for the remaining basal area. Later in this book, the label BT (0,0) indicates current thinning rules, and BT (15,0) indicates a thinning regime with a 15% increase in the upper limit of stocking and 0% change in the remaining stocking compared to that under basic thinning (BT (0,0)), while UT (0,0) indicates no thinning before the terminal cut.

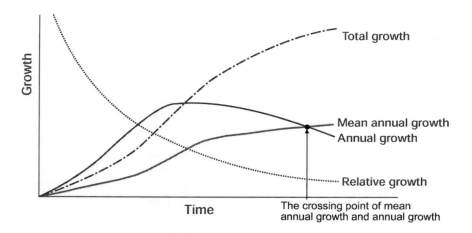

Fig. 13.18 Schematic presentation of the course of annual growth, total growth and relative growth. Current growth refers to the annual growth [e.g., m³ ha⁻¹ yr⁻¹], and mean annual growth to the mean growth during a specific period [e.g., m³ ha⁻¹ yr⁻¹]. Total growth is the sum of annual values of growth over the selected time period [e.g., m³ ha⁻¹ per period], and relative growth indicates the percentage value of current growth in relation to the current stocking of trees [%].

Fertilization

Nitrogen fertilization is an effective way to increase forest growth under boreal conditions, where short supply of nitrogen substantially limits forest growth (Jarvis and Linder 2000). Growth enhancement is mediated through the nitrogen content in foliage and the subsequent increase of the photosynthetic rate. In modeling, the increase and duration of nitrogen content in foliage is related to the nitrogen added in fertilizing events (Fig. 13.19). According to Jonsson (1978), the nitrogen content in the foliage prior to fertilizing is doubled if the maximum addition of 600 kg N ha⁻¹ is used. The fertilizing effect reduces over time and disappears finally:

$$d(t) = \left(1 + \frac{F(t)}{F(t_{max})}\right) \times \frac{f(NT)}{100} \tag{13.50}$$

where d(t) is the reduction of nitrogen in foliage, t is the time in years since fertilizing, and F(t) is the fraction of fertilizer affecting nitrogen content in foliage. The function f(NT) scales the increase (0–100%) of nitrogen content in foliage as a function of nitrogen addition (NT, kg ha⁻¹) in single fertilizing events (Routa et al. 2011):

$$f(NT) = 100 \times z(NT) / z(600) \tag{13.51}$$

$$z(NT) = a \times NT^b \tag{13.52}$$

where 600 kg N ha⁻¹ is the maximum addition of nitrogen in a single application and a and b are parameters.

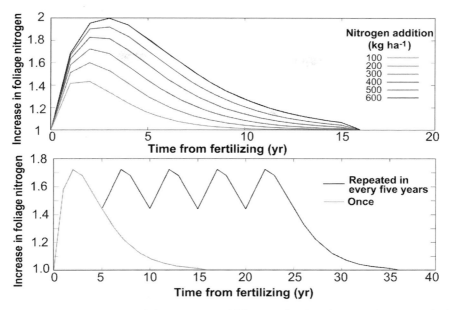

Fig. 13.19 Above: Increase and duration of nitrogen content of foliage as a function of amount of nitrogen added in a single event, and Below: The same as above, but under repeated fertilization (Routa et al. 2011). Permission of Oxford University Press.

13.10 Model Performance and Management Applications

Modeled photosynthesis vs. measurements

Following Swartzman and Kaluzny (1987), Table 13.4 summarizes methods for evaluation of the model. In validation, the model output is compared to the measured values in order to identify whether the model is able to replicate the system behavior. Sensitivity analysis is further used to evaluate whether the data used in parameter estimation is sound. Simulation experiments include the application of the model in varying simulation situations in order to understand ecosystem performance. These simulations may also include simulations where the model ecosystem is disturbed (e.g., thinning interventions) in order to identify if the model dynamics response in a correct way to altered ecosystem structure. The recalibration of the model to new environmental

Table 13.4 Outline of analyzing model performance, based on Swartzman and Kaluzny (1987).

Methods to analyze model performance	Objectives in analyzing model performance
Validation, sensitivity analysis, identifying key processes with main impacts on model performance	• Comparison of model performance against measured performance. • Identifying research needed to improve the model structure to meet simulation needs. • Identifying key parameters and needs to experiments and measurements for parameter estimation.
Simulation experiments, calibration of model to new sites and simulation layouts, management experiments	• Understanding model performance. • Generality of model beyond particular sites and simulation layouts. • Management impacts on model dynamics and performance against measured performance.

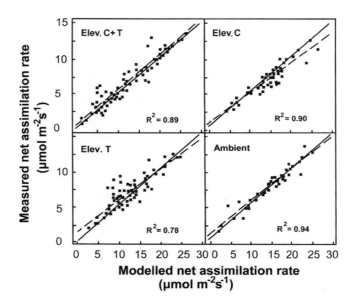

Fig. 13.20 Modelled values of net assimilation rate against the measured values for Scots pine grown in elevated CO_2 (Elev. C), temperature (T) and combined both (Elev. C + T) and in ambient conditions (Wang et al. 1996). Permission of Elsevier. Solid lines represent 1:1 line, and the dashed lines represent the regression between the modelled and measured values.

conditions indicates further the generality of the model, and the need to develop model structure and parameterization. In each case, the performance of the model is considered against modeling objectives.

Figure 13.20 shows the relationship between the measured and modeled values of net assimilation of Scots pine grown under elevated CO_2 and temperature using the biochemical photosynthetic model of Farquhar et al. (1980). The input variables in the simulations were photon flux density, intercellular concentration of CO_2 and needle temperature (Wang et al. 1996). Only the measurements representing water vapor deficit < 0.6 kPa were used in the comparison. The data used to parameterize the model was from July–August 1993, and the measurements were from July 1994. The measured and simulated values agreed with each other, especially under ambient conditions. The simulated values for elevated CO_2 were slightly lower than those measured, whereas the situation was opposite for elevated temperature alone or combined with elevated CO_2.

Modeled net ecosystem exchange vs. measurements

Kramer et al. (2002) compared the capacity of several ecosystem models including the current one (FinnFor), to simulate the net ecosystem exchange (daily net change of carbon between ecosystem and atmosphere per ground area (NEE, g C m^{-2} d^{-1})) of a Scots pine dominated ecosystem in the middle boreal conditions. The measurements represented comprehensive monitoring of the meteorological and soil conditions, meeting the requirement to link the variability of CO_2 flux to the prevailing conditions. Figure 13.21 shows that the values simulated with the current and other models agreed with the measured values: R^2 value was 0.86 with the mean squared value (MSE) 0.62 and the mean systematic error (MSE$_s$) 0.15. Kramer et al. (2002) concluded that the accuracy of all the models was high, and the models were realistic, representing causal links to such drivers as radiation, temperature and soil moisture. Kramer et al. (2002) further demonstrated that the measured and simulated values of gross primary production [g C m^{-2} d^{-1}] and ecosystem

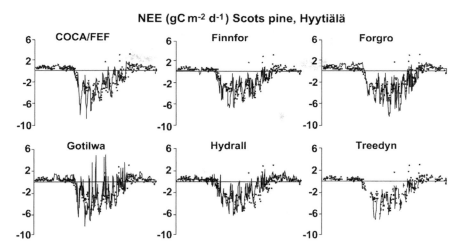

Fig. 13.21 Seasonal patterns of the measured and simulated net ecosystem exchange (NEE) of Scots pine at a middle boreal site in western Finland (61°51'N, 23°18'E, 170 m above sea level) using several process-based ecosystem models, including the FinnFor model (Kramer et al. 2002). Permission of Wiley & Sons.

respiration (combined values of autotrophic and heterotrophic respiration, g C m^{-2} d^{-1}) agreed with the measured values: R^2 values varied for gross primary production in the range 0.85–0.93 (MSE 0.75–2.30, MSE$_s$ 0.03–1.30), whereas R^2 values for ecosystem respiration were in the range 0.83–0.93 (MSE 0.26–2.67, MSE$_s$ 0.02–2.53).

Modeled growth vs. measurements

Ge (2011) and Ge et al. (2010, 2011a,b) studied further the performance of the current model (FinnFor) by comparing the simulated mean values of volume growth of single trees against the measured value of the same trees over a 10-year period (1985–1995). In total, 1191 sample trees were selected from the permanent sample plots of the Finnish Forest Research Institute (Metla) throughout Finland. The sample trees were those whose presence in the sample plots was identified in both 1985 and 1995. The structure (site type, stand density, tree species, diameter and height of object trees of cohorts) of the stand in 1985 was used as input for the simulation. Climate for the plots represented the weather records for the period 1985–1995 in the closest weather monitoring site on a monthly basis, but transformed to hourly values with the weather simulator (Strandman et al. 1993). Figure 13.22 shows that the calculated mean volume growth of single trees per plots over the 10 years agreed with those measured, regardless of tree species. The simulations were further run across the forest administrative centers and compared to the measured mean growth for the centers throughout the country. The performance of the model was reasonable when considering the large geographical distribution of the sample plots and their wide coverage of site types.

Model comparisons

Matala et al. (2003) further analyzed the performance of the current model (FinnFor) against measured growth in stands of Scots pine, Norway spruce and birch located in southern and central Finland (60°–62° N). The same simulations were run with the Motti model, which is a growth and yield model utilizing diameter, height and site type (site fertility) as the drivers behind growth (Hynynen et al. 2002). Both models used the same submodel for mortality, but otherwise they

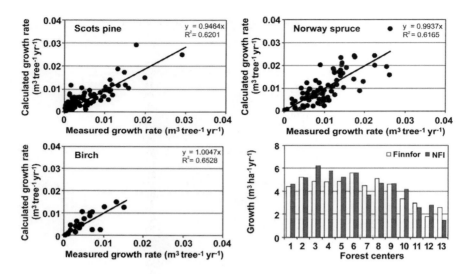

Fig. 13.22 Relationship between the measured and simulated volume growth in tree stand per year for Scots pine (upper left), Norway spruce (upper right) and birch (lower left), and the relationship between measured and calculated volume growth for different forest centers (lower right) indicated by the numbers on the *x*-axis. The numbers 1–10 indicate the forest centers in southern and middle boreal conditions (below 63 °N), while the numbers 11–13 indicate the forest centers in northern (above 63 °N) boreal conditions (Ge et al. 2010, 2011a,b). Permission of Springer.

were fully independent of each other. Parameterization of the Motti model is based on the Finnish National Forest Inventory data including several tens of thousands of inventory plots throughout Finland (60°–70° N).

Both models simulated the effects of within-stand competition and position on growth in a similar way (Fig. 13.23). The Motti model was less sensitive to the initial stand conditions and management than the FinnFor model, but the models agreed in their dynamics (Matala et al. 2003). Both models simulated the development of Scots pine stands in a similar way. The density in unmanaged and thinned managed stands reduced in an equal manner, even though the FinnFor model gave a slightly smaller basal area in unmanaged stands than the Motti model. Stem wood stocking was performed in the same way in unmanaged stands, but in managed stands it remained smaller in the Motti than in the FinnFor model, especially in mature stands. These differences were repeated in the total growth (total yield), but again both models gave fairly similar results, with higher total growth in unmanaged than in managed stands. In the FinnFor model, the height growth was slight greater than in the Motti model with the consequence that the stem form in the FinnFor model was slightly slenderer than in the Motti model.

Matala et al. (2003) also used the VG analysis (Mäkelä et al. 2000, Sievänen et al. 2000) in studying the performance of the FinnFor model. In the VG analysis, the rate of volume growth of individual trees/cohort (ΔV) was divided by the maximum growth rate (ΔV^*) in the same stand. The values of $\Delta V/\Delta V^*$ were further related to the relative diameter (d/d_{max}) and relative height (h/h_{max}), where d is diameter and h is height of individual trees/cohorts and d_{max} and h_{max} are the maximum values of individual trees/cohorts in the stand, respectively. Figure 13.24 shows the VGs graphs for unmanaged and managed (thinned) Scots pine stands at the ages of 39 and 79 years. In general, the graphs are slightly convex in shape, thus indicating that the volume growth is dependent on tree size in both unmanaged and managed stands regardless of the model. The shapes, slopes and locations of graphs for the unmanaged stand were similar for both models. However, the Motti model seemed to produce a narrower diameter and height distribution than did the FinnFor model. Regardless of the model, the total growth in the stand seemed, however, to be allocated for individual trees in the same way.

Modeling Eco-physiological Responses of Boreal Forest to Climate Change   191

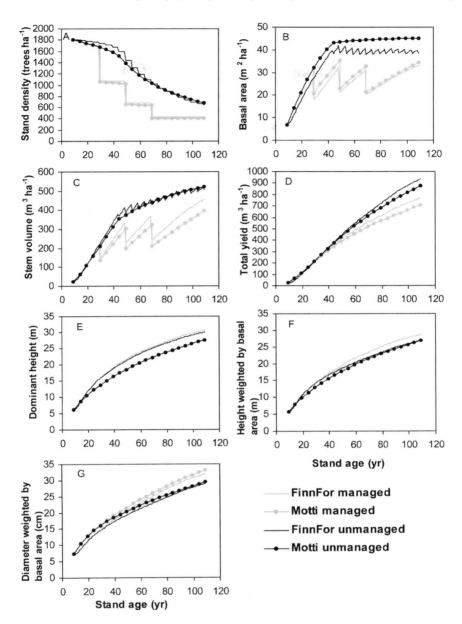

Fig. 13.23 Development of stand density, basal area, stem wood volume, total growth, dominant height, mean height and mean diameter over time in unmanaged and managed (thinned) Scots pine stands as simulated by the Motti and FinnFor models. In both cases, the simulations were done for a site of the *Myrtillus* type at latitude 63° N applying the initial stand density of 2,500 seedling per hectare (Matala et al. 2003). Permission of Elsevier.

Matala et al. (2006) also analyzed the performance of the FinnFor model by simulating the growth of Scots pine and Norway spruce over sites of varying fertility through Finland. The site fertility spanned from poor (*Vaccinium* type, VT) to rich sites (*Oxalis-Myrtillys* type, OMT) combined with a temperature sum from 800 d.d. (66° N) to 1,300 d.d. (60° N). The effect of site type was introduced to the model using the species-specific nitrogen content of foliage as a function of site type and temperature sum. The nitrogen values were obtained from the database of the Pan-

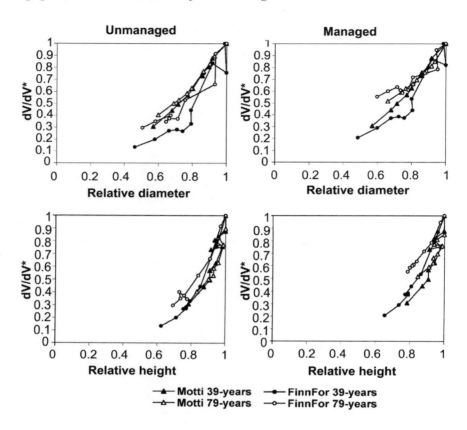

Fig. 13.24 Relative volume growth rate of individual Scots pines/cohorts at young (age 39 yr) and mature (age 79 yr) phases plotted against the relative diameter and height of trees as simulated by the Motti and FinnFor models (Matala et al. 2003). Permission of Elsevier.

European Programme for Intensive and Continuous Monitoring of Forest Ecosystems (Level II). The nitrogen content increased along with increasing temperature sum from north to south, and it was higher on rich sites than on poor sites, regardless of tree species. The calculated values of total stem wood growth were compared with the values simulated with the Motti model made responsive to climate change (Matala et al. 2005). The calculations were run over a 100-year period for the current climate and the changing climate, which represented the current CO_2 + 350 ppm and the current temperature + 4°C at the annual basis.

Figure 13.25 shows that the total yield obtained with both models was greater on a site of higher fertility regardless of the temperature sum. Under the elevating CO_2 and temperature, growth of Scots pine increased more in the FinnFor simulations than in the Motti simulations. In the FinnFor case, the growth increase was 39% on the *Oxalis-Myrtillus* site and 43% on the *Vaccinium* site, whereas in the Motti case the growth increase was 23 and 27% assuming the 1,130 d.d. on both sites. In the Motti simulations, the total growth increased more in the north than in the south; i.e., the growth increase on the *Myrtillis* sites was 31% for a temperature sum of 800 d.d., and 23% for a temperature sum of 1,300 d.d. Similarly, the FinnFor simulations yielded a higher growth increase, 65% for temperature sum 800 d.d. and 37% for temperature sum of 1,300 d.d. The pattern was similar in relation to the site fertility and temperature sum for Norway spruce and birch, but the values of total growth were species-specific, as was the case for the increase of growth due to elevating CO_2 and temperature.

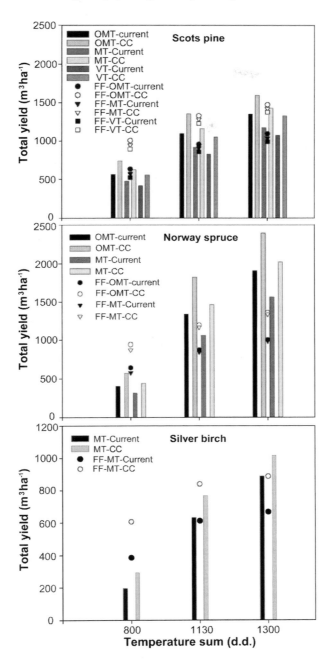

Fig. 13.25 Total yield of Scots pine (upper), Norway spruce (middle) and silver birch (lower) stands over a 100-year period for fertile (OMT) and medium fertile (MT) site types and three locations indicated by the current temperature sum at the x-axis. Two climate scenarios (current = current climatic conditions; CC = climate change with the gradual elevation of T and CO_2 by +4°C and +350 ppm from current levels) were used in the simulations. Bars show the values based on the Motti model and the dots (FF) those based on the FinnFor model (Matala et al. 2006). Permission of Elsevier.

Sensitivity to climate change and variability in weather

Briceño-Elizondo et al. (2006) used the FinnFor model to study the sensitivity of tree growth to changes in temperature, precipitation, atmospheric CO_2 and nitrogen content in foliage. Simulations were run for a southern (62° N) and northern boreal site (66° N), both of medium fertility (*Myrtillus* type, MT) with a high water retention capacity suitable for Scots pine, Norway spruce and birch. The density of initial stands was 1,600–1,880 trees per hectare, with the mean diameter 12–15 cm and the mean height 10–15 m. The simulation started 15 years prior to the culmination of growth in the tree stand and terminated 15 years after the growth culmination. In the simulations, the current climate was used, but the current temperature and precipitation conditions (1961–1990) were changed systematically in order to identify the model sensitivity to the variability in temperature and precipitation. In the former case, the daily mean temperature was increased by +1, +2, +3, +4, and +5°C. Similarly, the precipitation was increased/decreased by +20, +10, –10, and –20%. During the simulations, stable CO_2 concentrations of 350 was used. The relative change (%) in growth was:

$$\Delta Growth = 100 \times \frac{Growth(T+\Delta T, P+\Delta P) - Growth(T,P)}{Growth(T,P)} \quad (13.53)$$

where T and P indicate the current temperature and precipitation and ΔT and ΔP the change in temperature and precipitation.

In the north, warming substantially enhanced growth of Scots pine regardless of precipitation, whereas in the south the response to temperature increase was smaller and even decreasing under reduced precipitation (Fig. 13.26). Similarly, the growth of Norway spruce reduced in the south, and elevating temperature increased growth only under substantially increased precipitation. The stronger temperature response of Norway spruce at low precipitation suggested that temperature-induced water limitation is likely to reduce the productivity of Norway spruce. Conversely, the temperature increase in the north clearly increased growth of Norway spruce even under reducing precipitation. The response of birch to the changes in temperature and precipitation were close to that of Scots pine. In this case, the large temperature increase combined with a reduction in precipitation reduced growth in the south, whereas in the north a growth increase was clear.

Sensitivity to management

Briceño-Elizondo et al. (2006) studied further how thinning interventions affect: (i) the site conditions in short term and the subsequent changes in photosynthesis and growth; and (ii) the long-term growth and timber yield. Regardless of the site, thinning interventions increased the amount of precipitation per remaining trees, with a consequent increase of soil moisture (Fig. 13.27). The increase was greater in the north than in the south, whereas the increase in available nitrogen per remaining tree was larger in the south. Thinning thus increased the amount water and nitrogen available for the remaining tree along with increasing thinning intensity, with the enhancement of photosynthetic production and the growth of mass per remaining trees as may be expected. However, thinning reduced the total growth of tree stands due to the reduced total stocking.

Over the 100-year rotation, the growth and timber yield were dependent on the thinning regime shown in Fig. 13.28. Basic thinning (BT(0,0)) followed the thinning rules currently recommended for the sites of *Myrtillus* type (MT) in the middle boreal forests. On average for thinning regimes,

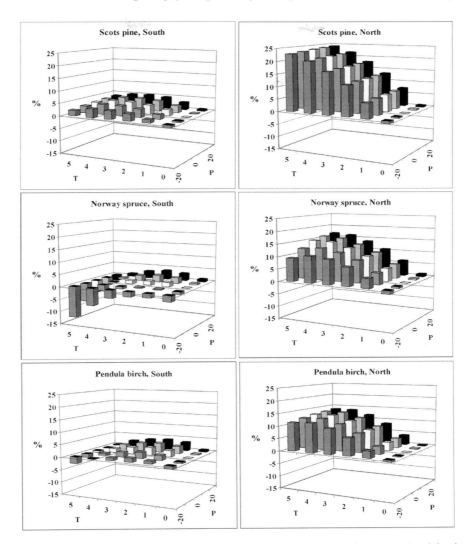

Fig. 13.26 Relative effects (%) of the elevation of temperature (T, °C from the mean annual current one) and the change of precipitation (P, % of the mean annual current one) on growth Scots pine, Norway spruce and birch in southern (62° N) and northern (66° N) boreal conditions (Briceño-Elizondo et al. 2006). Permission of Elsevier. For details of the simulations, see the text.

climate change increased the mean growth of Scots pine by 28% in the south and 53% in the north. For Norway spruce, the growth increase was smaller than that for Scots pine: 23% in the south and 30% in the north. The growth response of birch was smaller than that of conifers in the south (21%), and in the north it was nearly the same (33%) as that of Norway spruce. Over the thinning regimes, climatic warming increased the mean timber yield for Scots pine by 26% in the south and 50% in the north. Similarly, climate change increased Norway spruce timber, but the increase was somewhat smaller: 23% in the south and 30% in the north. The increase of birch timber was 20% in the south and 33% in the north.

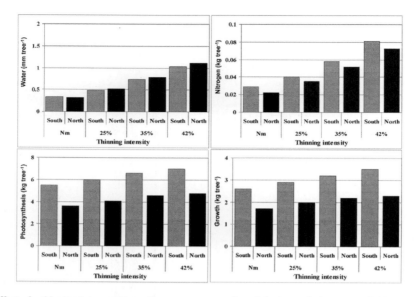

Fig. 13.27 Effect of a thinning intervention on the mean amount of precipitation and nitrogen available per tree in a Scots pine population during 30 years of simulation, with the consequent changes in photosynthetic production and growth in the south (62° N) and north (66° N) (Briceño-Elizondo et al. 2006). Permission of Elsevier. Thinning was performed 15 years prior to the culmination of thinning, and the values are the cumulated values over 30 years since thinning. Thinning options were no thinning (Nm) and the reduction of the basal area by 25%, 35% and 42% from below per thinning. In the simulation, the current climate was used, with the mean annual temperature 3.5°C in the south and 1°C in the north. Similarly, the precipitation was 480 mm and 450 mm in the south and north, respectively. The constant atmospheric CO_2 of 350 ppm was used in the simulations.

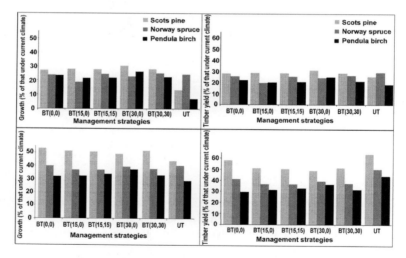

Fig. 13.28 Left: Warming-induced increase of growth (%) in the south (left upper, 62° N) and north (left lower, 66° N). Right: Percentage increase of timber yield for the south (right upper) and north (right lower) for Scots pine, Norway spruce and birch as a function of thinning regime (Briceño-Elizondo et al. 2006). Permission of Elsevier. The climatic warming represented the mean of HdCM2 and ECHAM4 climate change scenarios based on the IS92a emission scenario, with the gradual increase of atmospheric CO_2 from 350 ppm to 653 ppm in 100 years. At the same time, the mean annual temperature increased in the south from 3.5°C to 7°C and in the north from 1 to 5°C. Similarly, the precipitation increased from 480 to 570 mm and from 450 to 600 mm, respectively. Legend for thinning: BT (0,0) indicates current thinning rules (basic thinning), and BT(15,0) indicates a thinning regime with a 15% increase in the upper limit of stocking and 0% change in the remaining stocking compared to that under basic thinning (BT(0,0)), while UT(0,0) indicates no thinning before the terminal cut. For thinning and labels, see further Fig. 13.17, Chapter 13.

13.11 Concluding Remarks

Boreal forests grow and develop slowly, which limits the use of experimental studies to answering the questions: (i) how might climate change affect the long-term dynamics of boreal forests; and (ii) how are forests responding to management under climate change. Such information is of prime importance in developing proper management to sustain forestry to meet climate change and utilizing forest in timber and biomass production and in storing carbon for mitigating climate change. Growth and yield models based on inventory data may be done responsive to climate change as demonstrated by Matala et al. (2003) and used to outline the growth and development of forests under climate change. Such hybrid models provide a solid basis for decision-making for predictions and sensitivity analysis, if the assumptions behind the eco-physiological properties included in growth and yield models are justified. Process-based models with key physiological and ecological processes in ecosystem dynamics may be equally used for developing management practices to meet climate warming in a sustainable way. Successful modeling also requires that it is possible to estimate the values of the necessary parameters, and that the model is valid for simulating the dynamics of the forest ecosystem; i.e., the model includes the important factors needed to explain the performance of the phenomenon as regards the selected approach to the study problem.

References

Briceño-Elizondo, E., J. Garcia-Gonzalo, H. Peltola, J. Matala and S. Kellomäki. 2006. Sensitivity of growth of Scots pine, Norway spruce and silver birch to climate change and forest management in boreal conditions. Forest Ecology and Management 232: 152–167.

Farquhar, G. D., S. von Caemmerer and J. A. Berry. 1980. A biochemical model of photosynthetic assimilation in leaves of C_3 species. Planta 149: 67–90.

Franklin, O., J. Johansson, R. C. Dewar, U. Dieckmann, R. E. McMurtrie, Å. Brännström et al. 2012. Modeling carbon allocation in trees: a search for principles. Tree Physiology 32: 648–666.

Garcia–Gonzalo, J., M. J. Lexer, H. Peltola and S. Kellomäki. 2008. Designing a forested landscape in Finland under different climate scenarios. In: K. von Gadow and T. Pukkala (eds.). Designing Green Landscapes. Volume 15. Springer Science + Media.

Gates, D. M. 1980. Biophysical Ecology. Springer-Verlag, New York, USA.

Ge, Z. -M., X. Zhou, S. Kellomäki, K. -Y. Wang, H. Peltola, H. Väisänen and H. Strandman. 2010. Effects of changing climate on water and nitrogen availability with implications on the productivity of Norway spruce stands in southern Finland. Ecological Modelling 221(13–14): 1731–1743.

Ge, Z. -M. 2011. Effects of climate change and management on growth of Norway spruce in boreal conditions—an approach based on ecosystem model. Ph. D. Thesis, University of Eastern Finland, Finland.

Ge, Z. -M., S. Kellomäki, H. Peltola, X. Zhou, K. -Y. Wang and H. Väisänen. 2011a. Effects of varying thinning regimes on carbon uptake, total stem wood growth, and timber production in Norway spruce (*Picea abies*) stands in southern Finland under the changing climate. Annals Forest Science 68: 371–383.

Ge, Z. -M., S. Kellomäki, X. Zhou, K. -Y. Wang and H. Peltola. 2011b. Evaluation of carbon exchange in a boreal coniferous stand over a 10-year period: An integrated analysis based on ecosystem model simulations and eddy covariance measurements. Agricultural and Forest Meteorology 151: 191–203.

Hakkila, P. 1979. Wood density survey and dry weight tables of pine, spruce and birch stems in Finland. Communicationes Instituti Forestalis Fenniae 96(3): 1–59.

Högberg, P., A. Nordgren, N. Buchmann, A. F. Taylor, A. Ekblad, M. N. Högberg et al. 2001. Large-scale forest girdling shows that current photosynthesis drives soil respiration. Nature 411: 789–792.

Hynynen, J., R. Ojansuu, H. Hökkä, H. Salminen, J. Siipilehto and P. Haapala. 2002. Models for predicting stand development in MELA system. Finnish Forest Research Institute. Research Paper 835: 1–116.

IPCC. 2007. Climate Change 2007: The Physical Science Basis. Contribution of Working Group I to the Fourth Assessment Report of the Intergovernmental Panel on Climate Change. Cambridge University Press, Cambridge, UK.

Jansson, P. E. 1991a. Simulation model for soil water and heat conditions. Sveriges Lantbruksuniversitet. Uppsala. Raport 165: 1–72.

Jansson, P. E. 1991b. Simulation model user's manual. Sveriges Lantbruksuniversitet. Uppsala. Avdelningsmeddelande 91/7: 1–59.

Jarvis, P. G. 1976. The interpretation of the variations in leaf water potential and stomatal conductance found in canopies in the field. Philosophical Transactions of the Royal Society B 273: 593–610.

Jarvis, P. G. and K. G. McNaughton. 1986. Stomatal control of transpiration–scaling up from leaf to region. Advances in Ecological Research 15: 1–49.
Jarvis, P. G. and S. Linder. 2000. Constraints to growth of boreal forests. Nature 405: 904–905.
Jonsson, S. 1978. Resultat från en tioårig försökserie med hög kvävegivor. Föreningen Skogsädling Institutet Skogsförbättring. Årsbok 1977.
Kellomäki, S., H. Väisänen and H. Strandman. 1993. FinnFor: a model for calculating the response of the boreal forest ecosystem to changing climate. University of Joensuu, Faculty of Forestry. Research Notes 6: 1–120.
Kellomäki, S. and H. Väisänen. 1997. Modelling the dynamics of the boreal forest ecosystems for climate change studies in the boreal conditions. Ecological Modelling 97(1,2): 121–140.
Kellomäki, S., I. Rouvinen, H. Peltola and H. Strandman. 2001a. Density of foliage mass and area in the boreal forest cover in Finland, with applications to estimation of monoterpene and isoprene emissions. Atmospheric Environment 35: 1491–1503.
Kellomäki, S., I. Rouvinen, H. Peltola, H. Strandman and R. Steinbrecher. 2001b. Impact of global warming on the tree species composition of boreal forests in Finland and effects on emissions of isoprenoids. Global Change Biology 7: 531–544.
Kellomäki, S., H. Peltola, T. Nuutinen, K. T. Korhonen and H. Strandman. 2008. Sensitivity of managed boreal forests in Finland to climate change, with implications for adaptive management. Philosophical Transactions of the Royal Society B363: 2341–2351.
Kellomäki, S., M. Maajärvi, H. Strandman, A. Kilpeläinen and H. Peltola. 2010. Model computations on the climate change effects on snow cover, soil moisture and soil frost in the boreal conditions over Finland. Silva Fennica 44(2): 213–233.
Koivisto, P. 1959. Growth and yield tables. Metsäntutkimuslaitoksen julkaisuja 51(8): 1–49.
Kramer, K., I. Leinonen, H. H. Bartelink, P. Berbigier, M. Borghetti, C. Bernhofer et al. 2002. Evaluation of six process-based forest growth models using eddy-covariance measurements of CO_2 and H_2O fluxes at six forest sites in Europe. Global Change Biology 8: 213–230.
Lumb, F. E. 1963. The influence of clouds on hourly amounts of total solar radiation at the sea level. Quaternary Journal of Royal Meteorological Society 90: 43–56.
Mäkelä, A., R. Sievänen, M. Lindner and P. Lasch. 2000. Application of volume growth and survival graphs in evaluation of four process-based forest growth models. Tree Physiology 20: 347–355.
Marklund, L. G. 1987. Biomass functions for Norway spruce (*Picea abies* (L.) Karst.) in Sweden. Sveriges Lantbruksuniversitet, Institutionen för Skogstaxering, Rapport 43: 1–127.
Marklund, L. G. 1988. Biomassafunktioner för tall, gran och björk i Sverige. Sveriges Lantbruksuniversitet, Institutionen för Skogstaxering, Rapport 45: 1–73.
Matala, J., J. Hynynen, J. Miina, R. Ojansuu, H. Peltola, R. Sievänen et al. 2003. Comparison of a physiological model and a statistical model for prediction of growth and yield in boreal forests. Ecological Modelling 161: 95–116.
Matala, J., R. Ojansuu, H. Peltola, R. Sievänen and S. Kellomäki. 2005. Introducing effects of temperature and CO_2 elevation on tree growth into a statistical growth and yield model. Ecological Modelling 181(2-3): 173–190.
Matala, J., R. Ojansuu, H. Peltola, H. Raitio and S. Kellomäki. 2006. Modelling the response of tree growth to temperature and CO_2 elevation as related to the fertility and current temperature sum of a site. Ecological Modelling 199(1): 39–52.
Mazziotta, A., M. Mönkkönen, H. Strandman, J. Routa, O. -P. Tikkanen and S. Kellomäki. 2014. Modeling the effects of climate change and management on the dead wood dynamics in boreal forest plantations. European Journal of Forest Research 133: 405–421.
McMurtrie, R. E., D. A. Rook and F. M. Kelliher. 1990. Modelling the yield of *Pinus radiata* on a site limited by water and nitrogen. Forest Ecology and Management 30: 381–418.
Meentemeyer, V. 1978. Macroclimate and lignin control of litter decomposition rates. Ecology 59: 465–472.
Meentemeyer, V. and B. Berg. 1986. Regional variation in rate of mass loss of *Pinus sylvestris* needle litter in Swedish pine forests as influenced by climate and litter quality. Scandinavian Journal of Forest Research 1: 167–180.
Monteith, J. L. and M. H. Unsworth. 1990. Principles of Environmental Physics. Edward Arnold, London UK.
Muukkonen, P. and A. Lehtonen. 2004. Needle and branch biomass turnover rates of Norway spruce (*Picea abies*). Canadian Journal of Forest Research 34: 2517–2527.
Muukkonen, P. 2005. Needle biomass turnover rates of Scots pine (*Pinus sylvestris* L.) derived from needle-shed dynamics. Trees 19: 273–297.
Näslund, M. 1936. Skogsförsöksanstaltens gallringsförsök i tallskog. Primärbearbetning. Meddelanden från Statens Skogsförsköksanstalt. Häfte 29: 1–179.
Nikolov, N. and H. Helmisaari. 1992. Silvics of the circumpolar forests tree species. pp. 13–84. *In*: H. H. Shugart, R. Leemans and G. B. Bonan (eds.). A System Analysis of the Global Boreal Forest. Cambridge University Press, New York, USA.
Oker-Blom, P. 1985. Photosynthesis of Scots pine shoot: simulation of the irradiance distribution and photosynthesis of a shoot in different radiation fields. Agricultural and Forest Meteorology 34(1): 31–40.
Oker-Blom, P. 1986. Photosynthetic radiation regime and canopy structure in modelled stands. Acta Forestalia Fennica 197(1): 1–44.
Pastor, J. and W. Post. 1986. Influence of climate, soil moisture, and succession on forest carbon and nitrogen cycles. Biogeochemistry 2: 3–27.
Pukkala, T. 1987a. Kuusen ja männyn siemensadon ennustemalli. Abstract: A model for predicting the seed crop of *Picea abies* and *Pinus sylvestris*. Silva Fennica 21: 135–144.

Pukkala, T. 1987b. Siementuotannon vaikutus kuusen ja männyn vuotuiseen kasvuun. Abstract: Effect of seed production on the annual growth of *Picea abies* and *Pinus sylvestris*. Silva Fennica 21: 145–158.
Raulo, J. and M. Leikola. 1974. Studies on annual height growth of trees. Communications Instituti Forestalis Fenniea 81(2): 1–19.
Reineke, H. 1933. Perfecting a stand-density index for evenaged forests. Journal of Agricultural Research 36: 627–638.
Richardson, C. W. 1981. Stochastic simulation of daily precipitation, temperature and solar radiation. Water Resource Research 17: 182–190.
Routa, J., S. Kellomäki, H. Peltola and A. Asikainen. 2011. Impacts of thinning and fertilization on timber and energy wood production in Norway spruce and Scots pine: scenario analyses based on ecosystem model simulations. Forestry 84(2): 159–175.
Sarvas, R. 1962. Investigations on the flowering and seed crop of *Pinus silvestris*. Communicationes Instituti Forestalie Fenniae 53(4): 1–198.
Satoo, T. 1966. Variation in response of conifer seed germination to soil moisture conditions. Tokyo University Forest. Miscellaneous Information 16: 17–20.
Sievänen, R., M. Lindner, A. Mäkelä and P. Lasch. 2000. Volume growth and survival graphs: a method for evaluating process-based forest growth models. Tree Physiology 20: 357–365.
Strandman, H., H. Väisänen and S. Kellomäki. 1993. A procedure for generating synthetic weather records in conjunction of climatic scenario for modelling ecological impacts of changing climate in boreal conditions. Ecological Modelling 70: 195–220.
Swartzman, G. L. and S. P. Kaluzny. 1987. Ecological Simulation Primer. MacMillan Publishing Company, New York, USA.
Väisänen, H., H. Strandman and S. Kellomäki. 1994. A model for simulating the effects of changing climate on the functioning and structure of the boreal forest ecosystem: an approach based on object-oriented design. Tree Physiology 14: 1081–1095.
Vanninen, P. and A. Mäkelä. 2005. Carbon budget for Scots pine trees: effects of size, competition and site fertility on growth allocation and production. Tree Physiology 25: 17–30.
Wang, K. -Y., S. Kellomäki and K. Laitinen. 1996. Acclimation of photosynthetic parameters in Scots pine after three years of exposure to elevated temperature and CO_2. Agricultural and Forest Meteorology 82: 195–217.
Wang, K. -Y. and S. Kellomäki. 2003. CLIMFOR: A Climate-Forest Model and its Applications. Sihuan Publishing House of Science and Technology, Chengdu, People's Republic of China.
Yli-Vakkuri, P. 1961a. Emergency and initial development of tree seedlings on burnt-over forest land. Seloste: Taimien syntymisestä ja ensikehityksestä kulotetuilla alueilla. Acta Forestalia Fennica 74: 1–51.
Yli-Vakkuri, P. 1961b. Kokeellisia tutkimuksia taimien syntymisestä ja ensikehityksestä kuusikoissa ja männiköissä. Summary: Experimental studies on the emergency and initial development of tree seedlings in spruce and pine stands. Acta Forestalia Fennica 74: 1–51.

PART V
Responses of Boreal Forest Ecosystem to Climate Change and Management

Impact of Climate Change on the Productivity of Boreal Forests

ABSTRACT

Atmospheric carbon is fixed in photosynthesis in gross primary production and lost in respiration processes in primary and secondary production. Under climate warming, gross and net primary production in boreal forests likely increase under moderate warming, especially in northern forests above N 63°. In southern boreal forests, the increase in primary production likely levels off or even reduces. This is especially the case for Norway spruce, whose growth may reduce due to increasing evaporation and the frequency of drought episodes. In peatland ecosystems, ground water is likely to lower under climate change. This is especially clear in drained peatlands, likely reducing CH_4 and increasing CO_2 effluxes from soil.

Keywords: gross primary production, net primary production, net ecosystem exchange, boreal forests, temperature sum

14.1 Carbon Uptake and Flow through Forest Ecosystem

Climate change is global, but its impact on forest ecosystems and forests is local. Processes controlling carbon uptake and carbon flow through the ecosystem link ecosystem dynamics to climatic change. At the scale of site and region, the impacts on carbon dynamics are dependent on: (i) how climate change modifies climatic and edaphic conditions; and (ii) how different tree species meet the changes in growing conditions. The sensitivity of processes controlling carbon uptake to climatic change indicates the future success of different tree species and needs to sustain the productivity of forest ecosystems.

In the forest ecosystem, Gross Primary Production (GPP) is mainly controlled by the interception of radiation in tree crowns and by the cycles of nutrients and water. In boreal forests, the scarce availability of Nitrogen (N) is among the main factors limiting gross primary production (Hyvönen et al. 2007). Uptake of CO_2 is further controlled by temperature, air humidity, atmospheric CO_2 and availability of soil water. Some of the carbon fixed in gross primary production is retained in the ecosystem for a while, but in the long term all the carbon fixed in gross primary production is lost in respiration and oxidative processes. In a geologic perspective, even the carbon accumulated in sediment returns to the atmosphere (Fig. 14.1; Table 14.1).

The difference between gross primary production and Autotrophic Respiration (RA) yields the Net Primary Production (NPP) (Fig. 14.1; Table 14.1). Net primary production represents the remaining share of GPP after carbon losses in autotrophic respiration of various organs: foliage, branches, stems, coarse roots, fine roots and mycorrhizae and chemicals for defense. On a global scale, the net primary production in boreal upland sites is 50–870 g C m^{-2} yr^{-1} (Gower et al. 2001), being greater in deciduous than in coniferous forests. The total and the above-ground net primary production are closely correlated, but in coniferous forests the share of below-ground production is greater (36% of the total primary production) than in deciduous forests (19%) (Gower et al. 2001).

The fall of foliage and fine root litter is the major contributor to litter on a seasonal basis, but the biomass of any organ will finally enter the detritus pool (Fig. 14.1; Table 14.1). The decomposition of litter and release of CO_2 (Heterotrophic Respiration, RH) and the growth of heterotrophic organisms, are dependent on substrate quality (e.g., the ratio between Carbon (C) and Nitrogen (N), (C/N ratio)) and the temperature and moisture conditions in the soil profile. Carbon potentially available for

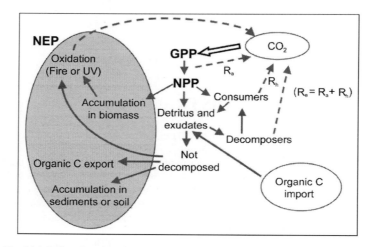

Fig. 14.1 Carbon flow through ecosystem (Lovett et al. 2006). Permission of Springer.

Table 14.1 Concepts used in analyzing the carbon flow through forest ecosystems on annual basis.

Concepts	Definition
Gross primary production, GPP [g m^{-2} yr^{-1}]	Gross carbon fixation into forest ecosystems.
Autotrophic respiration, RA [g m^{-2} yr^{-1}] • Maintenance, R_m • Growth, R_g	Loss rate of carbon through autotrophic respiration rate, $RA = R_m + R_g$
Net primary production, NPP [g m^{-2} yr^{-1}]	Gross production minus autotrophic respiration, $NPP = GPP - RA$
Heterotrophic respiration, RH [g m^{-2} yr^{-1}]	Loss rate of carbon through decay and respiration of secondary producers.
Net ecosystem production (*NEP*) or net ecosystem exchange, *NEE* [g m^{-2} yr^{-1}]	Gross primary production minus total respiration or net primary production minus heterotrophic respiration, $NEE = GPP - RA - RH = NPP - RH$
Net biome production, *NBP* [g m^{-2} yr^{-1}]	Net biome production $NBP = NEE$ plus carbon lost in disturbances, e.g., harvest, fires, etc.

longer storage is referred by Net Ecosystem Production (NEP), defined by the difference between NPP and RH or the Net Ecosystem Exchange (NEE). This indicates the difference between the GPP and carbon lost in ecosystem respiration summing autotrophic and heterotrophic respiration (Hamilton et al. 2002). Taking into account of the loss of carbon in disturbances (e.g., fire, harvest), Net Biome Production (NBP) is obtained, linking the uptake and loss of carbon with losses in disturbances (Table 14.1).

14.2 Response of Photosynthetic Production in Scots Pine across Finland

Calculations

In northern Europe, Scots pine is the main tree species, dominating the forest landscape throughout boreal forests (60°–70° N) with the potential to indicate the overall effects that the changing climate might have on the productivity of forests. The growth of Scots pine in the north (63°–70° N) is closely related to temperature, but in the south (60°–63° N) precipitation also plays a role. It might be expected that warming alone, or combined with increasing CO_2, would increase the productivity of Scots pine to a lesser degree in the south than in the north, as found by Kellomäki and Väisänen (1997). They used a process-based ecosystem model (FinnFor) to simulate climate change effects on the canopy net photosynthesis of Scots pine throughout Finland, related to the lengthening of the growing season. Based on the IS92a emission scenario (Houghton et al. 1995), changing climate involved: (i) an increase of 0.4°C in temperature per decade; (ii) a CO_2 increase of 3.5 µmol mol^{-1} per decade; and (iii) a combined increase in temperature and CO_2. The temperature rise in winter was 5–6°C and in summer 2°C by the simulation period. Precipitation increased/reduced following changes in cloudiness specific to the temperature conditions. In the simulations, the current CO_2 was 350 µmol mol^{-1}, increasing linearly from 350 to 700 ppm by the end of simulations. No management was used in simulations that extended over 100 years.

Changes in canopy photosynthesis

When assuming the current climate, the accumulated net canopy photosynthesis in the north was 1100–1800 Mg ha^{-1} over 100 years (Fig. 14.2). This was 55–80% of that in the south, where photosynthesis was 2,000–2,200 Mg ha^{-1} per 100 years. Rising CO_2 alone increased the net canopy photosynthesis by 15–30%, with a larger increase in the north. Similarly, increasing temperature alone increased the photosynthesis: 5% in the south and 25% in the north. The combined increase of both factors further increased the net canopy photosynthesis throughout the country, 25% in the south and 60% in the north. The linear fit of photosynthesis values against the length of the growing season showed a close correlation across the south-north temperature gradient (Fig. 14.2). This implied that in the south snow cover decreased from 100–160 days to 50–100 days, and in the north from 180–220 days to 150–180 days. Consequently, the length of the growing season (number of days with a mean T ≥ 5°C) increased 5% in the south and 20% in the north. At the same time, the moisture in surface soil frequently dropped below wilting point. Under the current climate, there were two to three dry days per year over a 100-year period in the south, but 15–20 days per year under a warming climate.

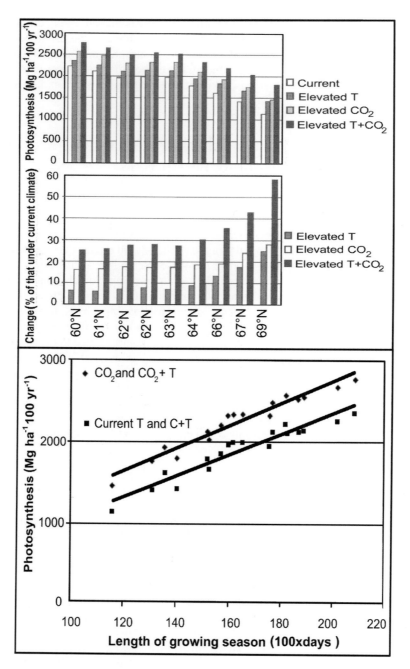

Fig. 14.2 Above: Total net canopy photosynthesis, and percentage change in the total net canopy photosynthesis over a period of 100 years in Scots pine stands across Finland under changing climate. Below: Net canopy photosynthesis in Scots pine stands across Finland over a period of 100 years as a function of the total length of growing season indicated by the sum of growing season days over 100 years. Upper line with filled squares standing on edge represent elevated CO_2 alone or combined with elevated temperature, and the lower line with filled squares standing on side represent the current or elevated temperature. Unpublished simulations (Seppo Kellomäki, University of Eastern Finland) based on a process-based forest ecosystem model (FinnFor) (Kellomäki and Väisänen 1997).

14.3 Response of photosynthetic production in Norway spruce across Finland

Calculations

The current distribution of Norway spruce is more southern than that of Scots pine, and it is most successful on mesic sites with high fertility. Climate change impacts on Norway spruce are therefore likely to be different than those on Scots pine, as shown by Ge et al. (2013). They used a process-based ecosystem model (FinnFor) to simulate the impact of climatic warming on the spatial pattern of primary production and carbon sequestration over the period 2000–2100 on sites dominated by Norway spruce. The country was divided into four climatic zones using the temperature sum: Zones I and II (> 1,000 d.d.) are located in the south and Zones III and IV (< 1,000 d.d.) in the north (Fig. 14.3). Only stands on the sites of medium (*Myrtillus* type, MT) and high (*Oxalis-Myrtillus* type, OMT) fertility selected from the database of the Finnish National Inventory (1999–2002) were used in the simulations, excluding management.

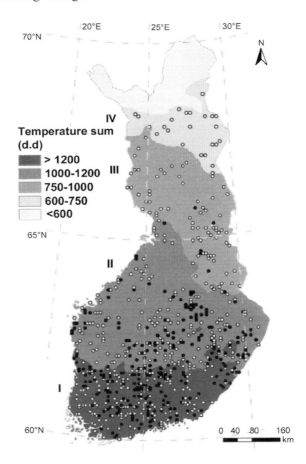

Fig. 14.3 Location of the sample plots dominated by Norway spruce in the temperature sum zones used in the analysis (Ge et al. 2013). Permission of Springer. The current climate was that from the period 1971–2000, with an atmospheric CO_2 of 352 ppm. The changing climate was based on the SRES A2 emission scenario, with a mean annual temperature increase of 4–6°C and an atmospheric CO_2 increase from 350 ppm to 840 ppm by 2100. The mean annual precipitation increased 10–15% in the south and 15–25% in the north. The temperature mainly increased in winter, and in summer remained nearly the same as currently (Carter et al. 2005).

Gross primary production

Table 14.2 shows that mean annual gross primary production (GPP) under climate change (CC) was 5–15% higher than under the current climate (CUR) in the first period (2000–2030). The increase was greater in the south (i.e., high current temperature sum) than in the north (i.e., low current temperature sum) and greater on fertile (OMT) than on medium fertile (MT) sites. This was the case across the south-north temperature gradient through the country. In the middle (2030–2060) and last (2060–2099) periods, climate change decreased gross primary production in southern and central Finland when the current temperature sum was greater than 1,000 d.d. The decline was greater in the southern than in the central part, following the differences in the prevailing temperature sum between the zones. In the north, gross primary production was higher under warming than under the current climate regardless of the simulation period. In the period 2060–2100, gross primary production was under warming climate 20–25% larger than under the current climate.

Net primary production

Regardless of the sites and climate, net primary production (NPP) was about 30% of gross primary production (Table 14.3). This was smaller than that (47%) found by Waring et al. (1998) for forests. The response of net primary production to climate change followed that of gross primary production, but the increase in net primary production in the period 2000–2030 tended to be slightly smaller than that in gross primary production. During the periods 2030–2060 and 2060–2099, the situation was opposite in the southern and central parts of boreal forests, where net primary production reduced more than gross primary production. At the same time, the increase in gross primary production in the north was greater than that of net primary production. The findings of Ge et al. (2010, 2013) about likely drought impacts on photosynthesis are in line with the measurements of Mäkinen et al. (2000, 2001). They demonstrated low precipitation in early summer, and high temperatures, to reduce the growth of Norway spruce in southern and central Finland on drought-sensitive sites.

Table 14.2 Mean annual gross primary production (GPP) of Norway spruce (periods 2000–2030, 2030–2060 and 2060–2099) under the current (CUR) and changing climate (CC) in the temperature sum (TS) zones for sites of varying fertility type (OMT and MT) (Ge et al. 2013). The values in parentheses are the percentage change (%) from that under the current climate.

Climatic zone (TS, d.d.)	Site type	Mean GPP (Mg C ha^{-1} yr^{-1}) per calculation period					
		2000–2030		2030–2060		2060–2099	
		CUR	CC	CUR	CC	CUR	CC
TS > 1,200	OMT	13.5	14.4 (6.8)	14.3	14.2 (−0.5)	12.8	11.8 (−7.5)
	MT	13.2	13.9 (4.8)	15.7	15.6 (−0.4)	13.3	12.1 (−9.4)
	Total	13.3	14.1 (5.8)	15.0	14.9 (−0.4)	13.1	11.9 (−8.5)
TS 1,000–1,200	OMT	12.3	13.3 (8.4)	13.3	13.4 (−0.1)	11.5	10.6 (−7.3)
	MT	12.4	13.3 (7.0)	14.2	14.1 (−0.1)	12.3	11.4 (−7.6)
	Total	12.4	13.3 (7.7)	13.7	13.7 (−0.1)	11.9	11.0 (−7.5)
TS 750–1,000	OMT	10.7	11.6 (9.1)	11.0	12.5 (13.0)	11.4	13.6 (19.4)
	MT	9.7	10.6 (9.0)	10.6	11.8 (11.5)	10.0	11.9 (19.2)
	Total	10.2	11.1 (9.0)	10.8	12.1 (12.3)	10.7	12.7 (19.3)
TS 600–750	OMT	–	–	–	–	–	–
	MT	3.7	4.2 (15.1)	3.9	4.6 (18.6)	4.5	5.7 (25.1)
	Total	3.7	4.2 (15.1)	3.9	4.6 (18.6)	4.5	5.7 (25.1)

Table 14.3 Mean annual net primary production (NPP) of Norway spruce (the periods 2000–2030, 2030–2060 and 2060–2099) under the current (CUR) and changing climate (CC) in the temperature sum (TS) zones for sites of varying fertility type (OMT and MT) (Ge et al. 2013). The values in parentheses are the percentage change (%) from that under the current climate.

Climatic zone (TS, d.d.)	Site type	Mean NPP (Mg C ha^{-1} yr^{-1}) per calculation period					
		2000–2030		2030–2060		2060–2099	
		CUR	CC	CUR	CC	CUR	CC
TS > 1,200	OMT	4.6	4.9 (5.2)	4.9	4.9 (−0.8)	4.4	3.9 (−12.7)
	MT	4.6	4.8 (4.4)	5.4	5.4 (−0.7)	4.6	4.0 (−13.9)
	Total	4.6	4.8 (4.8)	5.2	5.1 (−0.7)	4.5	4.0 (−13.3)
TS 1,000–1,200	OMT	4.2	4.5 (6.6)	4.6	4.6 (−0.2)	4.0	3.6 (−10.1)
	MT	4.3	4.6 (6.5)	4.9	4.9 (−0.1)	4.2	3.8 (−11.3)
	Total	4.3	4.5 (6.6)	4.7	4.7 (−.02)	4.1	3.7 (−10.7)
TS 750–1,000	OMT	3.7	4.0 (8.7)	3.8	4.2 (10.1)	3.9	4.6 (17.1)
	MT	3.4	3.6 (8.6)	3.6	4.0 (8.8)	3.4	4.0 (16.7)
	Total	3.5	3.8 (8.6)	3.7	4.1 (9.5)	3.7	4.3 (17.0)
TS 600–750	OMT	–	–	–	–	–	–
	MT	1.3	1.4 (12.6)	1.3	1.6 (18.0)	1.6	1.9 (23.8)
	Total	1.3	1.4 (12.6)	1.3	1.6 (18.0)	1.6	1.9 (23.8)

Field experiments also show that in the southern boreal zone irrigation increases the growth of Norway spruce, while there is no effect in the northern boreal zone (Bergh et al. 1999, Bergh et al. 2003).

Net ecosystem exchange

Ge et al. (2013) also found that in the period 2000–2030, the mean annual net ecosystem exchange (NEE) was 4–12% larger under climate change than under current climate regardless of climatic zone and site fertility (Table 14.4). The increase was the highest in the north and the smallest in the south. This indicates that climate warming increased tree growth more than ecosystem respiration. In the northern areas (TS < 1,000 d.d.), the net ecosystem exchange increased further in the periods 2030–2060 and 2060–2099, the increase being more than 20% in the north (TS < 750 d.d.). At the same time, the net ecosystem exchange reduced in the areas where the current temperature sum was more than 1,000 d.d.; i.e., in the south. The reduction was more than 10% in the period 2060–2099, and it was fairly similar on fertile (OMT) and medium fertile (MT) sites.

Figure 14.4 further demonstrates the spatial distribution of changes in NEE under a warming climate. In the south, the change in NEE was 2–10% lower and in the north 8–16% higher than under the current climate. In the south, the reduction was in line with the increase in temperature sum, and in the north the increase in the temperature sum enhanced NEE. Similarly, the change in GPP and NPP correlated negatively with the increase in temperature sum in the south but positively in the north. However, there was wide variation in the general tendencies. Especially in the south, the increase in temperature sum in many places even increased GPP and NPP and NEE. The regression slope of the functions indicates that the decrease (southern) in NEE was slightly higher, and the increase (northern) lower than those of GPP and NPP under the climate warming (Ge et al. 2013).

Table 14.4 Mean annual net ecosystem exchange (NEE) (the periods (2000–2030, 2030–2060 and 2060–2099) under the current (CUR) and changing climate (CC) in the temperature sum (TS) zones for sites of varying type (OMT and MT) (Ge et al. 2013). The values in parentheses are the percentage change (%) from that under the current climate.

Climatic zone (TS, d.d.)	Site type	2000–2030 CUR	2000–2030 CC	2030–2060 CUR	2030–2060 CC	2060–2099 CUR	2060–2099 CC
TS > 1,200	OMT	2.9	3.0 (5.1)	3.1	3.0 (–0.8)	2.7	2.4 (–13.2)
	MT	2.8	2.9 (3.9)	3.3	3.3 (–0.8)	2.8	2.5 (–13.8)
	Total	2.8	3.0 (4.5)	3.2	3.2 (–0.8)	2.8	2.4 (–13.5)
TS 1,000–1,200	OMT	2.6	2.8 (6.3)	2.8	2.8 (–0.4)	2.4	2.2 (–11.3)
	MT	2.6	2.8 (6.2)	3.0	3.0 (–0.3)	2.6	2.3 (–12.0)
	Total	2.6	2.8 (6.2)	2.9	2.9 (–0.4)	2.5	2.3 (–11.7)
TS 750–1,000	OMT	2.3	2.4 (7.5)	2.3	2.6 (9.3)	2.4	2.8 (16.2)
	MT	2.1	2.2 (7.3)	2.2	2.4 (8.6)	2.1	2.5 (16.1)
	Total	2.2	2.3 (7.4)	2.3	2.5 (9.0)	2.3	2.6 (16.2)
TS 600–750	OMT	–	–	–	–	–	–
	MT	0.8	0.9 (12.0)	0.8	1.0 (17.1)	1.0	1.2 (23.2)
	Total	0.8	0.9 (12.0)	0.8	1.0 (17.1)	1.0	1.2 (23.2)

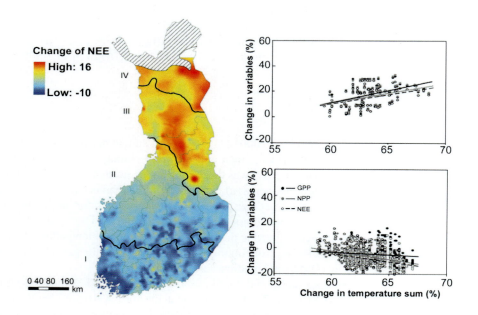

Fig. 14.4 Left: Change in the mean annual net ecosystem exchange (NEE) over 100 years under the warming climate compared to the values under the current climate. Right: Percentage change in mean annual gross primary production (GPP), net primary production (NPP) and net ecosystem exchange (NEE) as a function of the percentage change in temperature sum for southern (below) and northern Finland (upper) (Ge et al. 2013). Permission of Springer.

14.4 Response of Productivity to Climate Change on upland Sites

In growth and yield studies, productivity [m^3 ha^{-1} yr^{-1}] involves the potential stem wood growth in tree stands per unit area and unit time growing in natural or managed stands. Productivity varies according to the tree species, site fertility and structure of a stand (e.g., density), and realizes stem wood production at a given site with a given genotype (species) and specified management (Skovsgaard and Vanclay 2008). Productivity combines the effects of the physical and climate properties of sites on tree growth, including soil depth, texture, nutrient supply, precipitation, temperature, slope, elevation and aspect.

Productivity varies between regions as a function of the prevailing climate, which makes it sensitive to climate change, as demonstrated by Kellomäki and Kolström (1994). They calculated the probability distributions of the total stem wood growth of Scots pine, Norway spruce, Pendula birch and Pubescent birch for southern (61° N) and northern (66°) boreal sites of medium fertility (*Myrtillus* type, MT) using a gap-type growth and yield model. In the simulations, the density of the initial stands was 2,500 seedlings per hectare, regardless of tree species. Only temperature was allowed to increase at the rate of 0.04, 0.06 or 0.08°C yr^{-1} and precipitation remained as current. The simulations were extended over 160 years, and excluded management (Fig. 14.5).

In the south, the total stem wood growth of Scots pine was 970 m^3 ha^{-1} (probability > 0.95) under the current climate, and 1,020–1,030 m^3 ha^{-1} under warming (the increase 6%). In the north, the growth increased substantially more, from 400 m^3 ha^{-1} up to 800–900 m^3 ha^{-1}. Similarly, the growth of Norway spruce increased from 100 m^3 ha^{-1} up to 400–500 m^3 ha^{-1} in the north, and in the south remained as current or even reduced by 10% at the warming rate 0.08°C yr^{-1}. Regardless of the site, warming increased the total growth of birches, especially in the north, where growth increased from 20–30 m^3 ha^{-1} up to 200 m^3 ha^{-1} for Pendula birch, and from 60–70 m^3 ha^{-1} to 250 m^3 ha^{-1} for Pubescent birch. In the south, the growth of Pendula birch increased less, from 450 m^3 ha^{-1} to 550–570 whereas the growth of Pubescent birch was similar (250–300 m^3 ha^{-1}) regardless of climate. The temperature-induced changes in growth, especially in the north, are further demonstrated if mean annual stem wood growth is calculated (Table 14.5).

14.5 Response of Productivity to Climate Change on Peatlands

Impacts of climate change on peatland ecosystem

In Finland, peatlands cover 30% of the total land area (the total peatland area is about 10 million ha), and they contain 5,960 Tg of carbon in partly decayed organic matter or peat (Turunen 2008). Peatland is land covered by peat, with a thickness > 30 cm and a mineral content < 40%. In northern, and partly in central Finland, peatlands are mainly minerotrophic fens, whereas in the south ombrotrophic bogs dominate. In both cases, high ground water levels support ground vegetation dominated by sedges (*Carex* sp.) and *Sphagnum* mosses, which contribute the main part of organic matter accumulated in peat deposits. Ground water in fen sites is recharged by rain water and surface flow from the surrounding upland sites, whereas ground water in bog sites is mainly dependent on rain water (Fig. 14.6). Consequently, bogs represent low or intermediate site fertility, and they are mainly occupied by Scots pine if there is any tree growth. In fens, high ground water levels limit forest growth, even though site fertility in many cases may be high enough for Scots pine, Norway spruce and birch (mainly *Betula pubescens* Ehrh.). In many cases, only a lowering of ground water is necessary to enhance tree growth on peatlands. In Finland, nearly half the original peatlands are ditched for forestry, mainly in southern and central parts of the country.

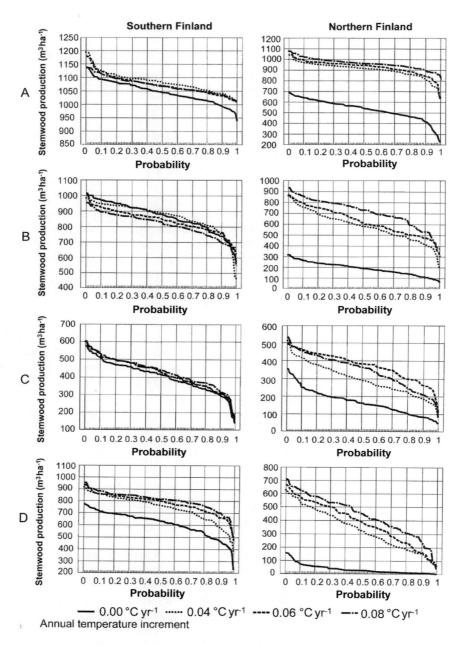

Fig. 14.5 Probability distributions of total stem wood growth of Scots pine (A, upper), Norway spruce (B, middle upper), Pubescent birch (C, middle lower) and Pendula birch (D, lower) in southern (left, 61° N) and northern (right 66° N) Finland over 160 years under the current climate and warming climate with varying warming rate at medium fertile sites (*Myrtillus* type, MT) (Kellomäki and Kolström 1994). Permission of Elsevier.

Table 14.5 Mean annual stem wood production in southern and northern Finland for Scots pine, Norway spruce and birch under the current climate and climate warming at varying rates (Kellomäki and Kolström 1994).

Tree species	Southern site, 61° N		Northern site, 66° N	
	m³ ha⁻¹ yr⁻¹	Change, %	m³ ha⁻¹ yr⁻¹	Change, %
Current temperature				
• Scots pine	6.5		2.4	
• Norway spruce	5.4		1.2	
• Pendula birch	3.8		0.2	
• Pubescent birch	2.5		1.0	
Warming, 0.04 °C yr⁻¹				
• Scots pine	6.7	3	5.7	138
• Norway spruce	5.4	0	3.6	200
• Pendula birch	4.6	22	2.1	100
• Pubescent birch	2.6	5	1.8	80
Warming, 0.06 °C yr⁻¹				
• Scots pine	6.7	3	5.8	138
• Norway spruce	5.1	−3	3.9	225
• Pendula birch	4.9	29	2.4	1,100
• Pubescent birch	2.6	5	2.3	130
Warming, 0.08 °C yr⁻¹				
• Scots pine	6.6	2	6.1	154
• Norway spruce	5.1	−6	4.4	267
• Pendula birch	5.0	32	2.7	1,250
• Pubescent birch	2.7	7	2.1	110

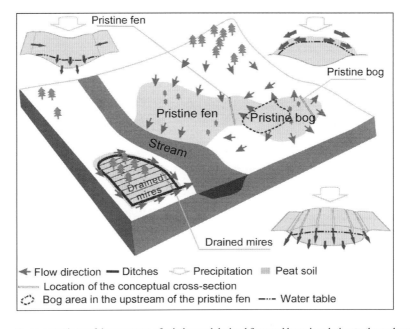

Fig. 14.6 Schematic presentations of the structure of pristine and drained fens and bogs in relation to the recharging of ground water from precipitation and surface water from surrounding sites, and the effects of drainage on ground water in mires (Gong et al. 2012). Permission of Elsevier.

Table 14.6 General responses of groundwater, soil temperature and carbon exchange in pristine peatland ecosystems to increases in atmospheric CO$_2$, temperature and precipitation on the scale of Finland. Downward arrows represent a decrease in parameter values, whereas upward arrows represent an increase. The increasing number of arrows indicates the increasing sensitivity of a parameter to the changes in a climate variable (Gong 2012).

Parameters	Increase in CO$_2$	Increase in temperature	Increase in precipitation
Ground water level	Not available	↓	↑
Temperature (10 cm depth)	Not available	↑↑↑	Change unclear
CH$_4$ source	↑	↓	↑
Net primary production	↑	↑	↑
Ecosystem respiration	↑	↑↑↑	↓
Net ecosystem exchange	↑	↓↓↓	↑

In peatland sites, climate change has major effects on vegetation through the changes in ground water level, depending on the water balance and the change in water storage:

$$P = Q + E + \Delta S \Rightarrow \Delta S = P - Q - E \tag{14.1}$$

where P is water in precipitation and in surface flow from surrounding sites, Q is runoff, E is evapotranspiration and ΔS is the change in water storage in soil. Following the conservation of mass, any water entering a site is lost either in evapotranspiration or surface runoff, or stored in the ground. In bogs, climate change affects the change of soil water storage through changes in precipitation and evapotranspiration, whereas in fen sites the water balance of surrounding sites may have an additional effect. In both cases, ditching enhances the runoff and lowers the ground water.

Based on model simulations, Gong (2012) found that an increasing temperature lowered the ground water level and increased soil temperature throughout Finland, if precipitation was same as currently (Table 14.6). This reduced the carbon lost in methane (CH$_4$) formed below the ground water level in oxygen-free conditions, whereas the increased precipitation under the current temperature made the ground water level higher and increased carbon lost in methane along with the decreasing soil temperature. Gong (2012) found further that net primary production increased along with rising precipitation and atmospheric CO$_2$, whereas the rising temperature tended to reduce net primary production due to increased autotrophic respiration. On fens, net primary production was more sensitive to changes in temperature than to precipitation, whereas in bogs the situation was opposite.

Change in ground water level and impacts on primary production

The impact of climate change on ground water varies substantially from site to site, depending on the type of peatland, in response to changes in the prevailing climate conditions and to seasonality. Gong (2012) found that climatic warming is likely to lower ground water mainly in spring (April–May), while the lowering is slight in summer and autumn (June–September). The simulations showed further that in the period 2000–2019 the ground water level mainly lowered in the southwestern part of the fen-dominated region and in the western parts of the bog-dominated region (i.e., the western coastal area) (Fig. 14.7). In the periods 2020–2059 and 2060–2099, the lowering of ground water increased, and the area of lowered ground water expanded further south and north. The ground water level lowered more in the fens than in the bogs (Gong 2012). In drained peatlands, the mean lowering of ground water was up to one cm over the whole simulation period throughout the country. The greatest lowering occurred in the western parts of country, where bogs are dominant (Gong et al. 2012).

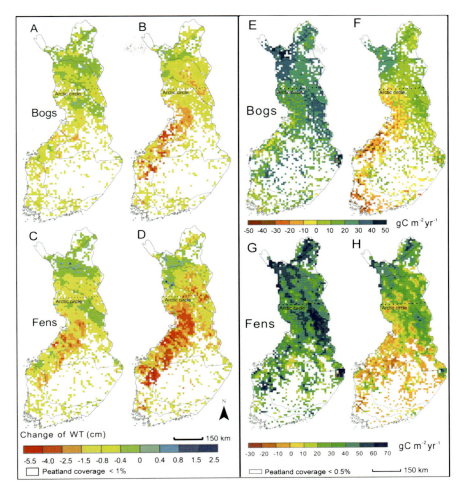

Fig. 14.7 Left: Spatial variation in current ground water in pristine bogs (left A, upper) and fens (left C, lower) and warming-induced change (2060–2099) of ground water in pristine bogs (left B, upper) and fens (left D, lower). Right: Spatial variation of current carbon accumulation in pristine bogs (right E, upper) and fens (right G, lower) and warming-induced change (2060–2099) in carbon accumulation in pristine bogs (right F, upper) and ferns (right H, lower) (Gong 2012). The simulations are based on the SRES A1B emission scenario, with the daily data for the current and changing climate in a 0.5° × 0.5° grid (Jylhä et al. 2009). Courtesy of Jinnan Gong, University of Eastern Finland, Finland.

Carbon sequestration in peatland ecosystems varies from site to site, depending on the type, in response to changes in climatic conditions. On a country scale, climate change tended to decrease the mean CO_2 sink by 21.5 ± 5.4 g C m^{-2} yr^{-1} and to increase the mean CH_4 emissions by 0.7 ± 0.3 g C m^{-2} yr^{-1} over the period 2000–2099 (Gong 2012, Fig. 14.7). The changes were the most pronounced in 2060–2099. In this period, the CH_4 emissions tended to decrease in the south along with the lowering ground water level, opposite to the north. Thus, the high number of fens in the north is likely to increase the overall CH_4 emissions on a country scale. The main area of pristine fens in the southern and western parts of country, and the pristine bogs on coastal areas are likely turn to CO_2 sources under climatic warming. Simulations also showed that carbon exchange in bogs is less sensitive to climate change than that in fens. On a country scale, the responses of CO_2/CH_4 fluxes to climatic warming are likely to decrease the total carbon sink strength of peatlands by 68% (Gong 2012, Gong et al. 2012).

14.6 Concluding Remarks

In Finland, climatic warming is likely to increase forest growth in the northern (above N 63°) boreal forests more than in the middle (N 62°–63°) and the southern (below N 62°) boreal forests, following the south-north temperature gradient across the country. However, the estimates are variable, and dependent on the model used in simulations. The largest growth increase for the north and smallest for the south are obtained when using growth and yield models in which the growth is driven by aggregated climate factors (e.g., temperature sum) on an annual basis. Using a gap-type model, Kellomäki and Kolström (1993, 1994) showed that growth would increase by 5% in the south but by 170% in the north, if the same temperature increase was used in simulations. Similarly, the regression between growth and temperature sum employed by Kauppi and Posch (1985) suggested that climatic warming could increase forest growth by 40% in southern and by 140% in northern Finland. When using eco-physiological models with a short time resolution, the growth increase/decrease due to climatic warming across the boreal zone is conservative (Kellomäki and Väisänen 1997, Bergh et al. 2003, Matala et al. 2003) in the range of 15–30%, regardless of tree species.

Under a warming climate, ground water in peatlands varies substantially from site to site in response to the changes in climate and seasonality. It is evident that ground water is likely to lower more in pristine fens than in bogs (Gong 2012). This is especially clear in drained peatlands. Nykänen et al. (1998) estimated that the lowering of the present ground water level by 10 cm would cause a 70% reduction in CH_4 emissions from fens and 45% from bogs. At the same time, the CO_2 effluxes from soil are likely to increase, and the carbon emissions in the enhanced growth of ground vegetation and trees may partly compensate for emissions to the atmosphere.

References

Bergh, J., S. Linder, T. Lundmark and B. Elfving. 1999. The effect of water and nutrient availability on the productivity of Norway spruce in northern and southern Sweden. Forest Ecology and Management 119: 51–62.

Bergh, J., M. Freeman, B. Sigurdsson, S. Kellomäki, K. Laitinen, S. Niinistö et al. 2003. Modelling the short-term effects of climate change on the productivity of selected tree species in Nordic countries. Forest Ecology and Management 183: 327–340.

Carter, T. R., K. Jylhä, A. Perrels, S. Fronzek and S. Kankaanpää. 2005. FINADAPT scenarios for the 21st century. Alternative futures for considering adaptation to climate change in Finland. Finnish Environmental Institute, FinAdapt Working Paper 2: 1–42.

Ge, Z. -M., X. Zhou, S. Kellomäki, K. -Y. Wang, H. Peltola, H. Väisänen et al. 2010. Effects of changing climate on water and nitrogen availability with implications on the productivity of Norway spruce in Southern Finland. Ecological Modelling 221: 1731–1743.

Ge, Z. -M., S. Kellomäki, H. Peltola, X. Zhou, H. Väisänen and H. Strandman. 2013. Impacts of climate change on primary production and carbon sequestration of boreal Norway spruce forests: Finland as a model. Climatic Change 118: 259–273.

Gong, J. 2012. Climatic sensitivity of hydrology and carbon exchanges in boreal peatland ecosystems, with implications on sustainable management of reed canary grass (*Phalaris arundinacea* L.) on cutaway peatlands. Ph. D. Thesis, University of Eastern Finland, Finland.

Gong, J., K. -Y. Wang, S. Kellomäki, C. Zhang and P. J. Martikainen. 2012. Modeling water table changes in boreal peatlands of Finland under changing climate conditions. Ecological Modelling 244: 65–78.

Gower, S. T., O. Krankina, R. J. Olson, M. Apps, S. Linder and C. Wang. 2001. Net primary production and carbon allocation patterns of boreal forest ecosystems. Ecological Applications 11(5): 1395–1411.

Hamilton, J. G., E. H. DeLucia, K. L. Naida, A. C. Finzi and W. H. Schlesinger. 2002. Forest carbon balance under elevated CO_2. Oecologia 131: 250–260.

Houghton, J. T., L. G. M. Filho, B. A. Callander, N. Harris, A. Kattenberg and K. Maskell. 1995. Climate Change 1995—the Science of Climate Change: Contribution of WGI to the Second Assessment Report of the Intergovernmental Panel on Climate Change, Intergovernmental Panel on Climate Change, Cambridge University Press, Cambridge, UK.

Hyvönen, R., G. I. Ågren, S. Linder, T. Persson, F. M. Cotrufo, A. Ekblad, M. Freeman et al. 2007. The likely impact of elevated [CO_2], nitrogen deposition, increased temperature and management on carbon sequestration in temperate and boreal forest ecosystems: a literature review. New Phytologist 173: 463–480.

Jylhä, K., K. Ruoteenoja, J. Räisänen, A. Venäläinen, H. Tuomenvirta, L. Ruokolainen et al. 2009. Arvioita Suomen muuttuvasta ilmastosta sopeutumistutkimuksia varten. ACCLIM-hankkeen raportti 2009. Raportteja 4: 1–102.

Kauppi, P. and M. Posch. 1985. Sensitivity of boreal forest to possible climatic warming. Climatic Change 7(1): 45–54.

Kellomäki, S. and M. Kolström. 1993. Computations on the yield of timber by Scots pine when subjected to varying levels of thinning under climate change in southern Finland. Forest Ecology and Management 59: 237–255.

Kellomäki, S. and M. Kolström. 1994. The influence of climatic change on the productivity Scots pine, Norway spruce and Pendula birch and Pubescent birch in southern and northern Finland. Forest Ecology and Management 65: 201–217.

Kellomäki, S. and H. Väisänen. 1997. Modelling the dynamics of the boreal forest ecosystems for climate change in the boreal conditions. Ecological Modelling 97(1/2): 121–140.

Lovett, G. M., J. J. Cole and M. L. Pace. 2006. Is net ecosystem production equal to ecosystem carbon accumulation? Ecosystems 9: 152–155.

Mäkinen, H., P. Nöjd and K. Mielikäinen. 2000. Climatic signal in annual growth variation in Norway spruce [*Picea abies* (L.) Karst.] along a transect from central Finland to the Artic timberline. Canadian Journal of Forest Research 30: 769–777.

Mäkinen, H., P. Nöjd and K. Mielikäinen. 2001. Climatic signal in annual growth variation in damaged and healthy stands of Norway spruce [*Picea abies* (L.) Karst.] in southern Finland. Trees 15: 177–185.

Matala, J., J. Hynynen, J. Miina, R. Ojansuu, H. Peltola, R. Sievänen et al. 2003. Comparison of a physiological model and a statistical model for prediction of growth and yield in boreal forests. Ecological Modelling 161: 95–116.

Nykänen, H., J. Alm, J. Silvola, K. Tolonen and P. Martikainen. 1998. Methane fluxes on boreal peatlands of different fertility and the effect of long-term experimental lowering of the ground water table on flux rates. Global Biogeochemical Cycles 12(1): 53–69.

Skovsgaard, J. P. and J. K. Vanclay. 2008. Forest site productivity: a review of the evolution of dendrometric concepts even-aged stands. Forestry 81(1): 13–31.

Turunen, J. 2008. Development of Finnish peatland area and carbon storage 1950–2000. Boreal Environment Research 13: 319–334.

Waring, R. H., J. J. Landsberg and M. Williams. 1998. Net primary production of forests: a constant fraction of gross primary production? Tree Physiology 18: 129–134.

15

Impact of Climate Change on the Growth and Development of Boreal Forests

ABSTRACT

In forestry, the productivity of the forest ecosystem refers to the rate of producing matter in a tree population/community for timber and biomass, which is dependent on the regeneration, growth and mortality of trees. Under boreal conditions, climate change enhances these processes compared to the situation under the current climate. This implies an increase in growth and the potential to produce timber and biomass, which may further be enhanced by selecting proper management to meet the potential provided by climate change.

Keywords: regeneration, growth, mortality, growth and development of forest stands, total growth, timber yield, thinning, impact of thinning on growth and timber yield

15.1 Growth and the Development of Tree Populations

Dynamics of mass in tree populations

In forestry, the productivity of forest ecosystems refers to the rate of producing matter [e.g., dry mass kg ha^{-1} yr^{-1}] in a tree population/community for timber and biomass. This equals the Net Primary Productivity (NPP); i.e., Gross Primary Production (GPP) minus photosynthates used for maintenance and growth processes. Over time, the mass in a tree population/community occupying a given site develops as a function of the birth (regeneration), growth and death of trees:

$$M(t) = M(t-1) + G(t) + S(t) - D(t) \tag{15.1}$$

where M(t) is the mass of trees on the site at time t, G(t) is growth of trees, S(t) is the mass of new seedlings and D(t) is the mass of dying trees. This is further:

$$M(t) = h(t) \times \bar{s} + n(t) \times \bar{m}(t-1) + n(t) \times \bar{g}(t) - k(t) \times (\bar{m}(t) + \bar{g}(t)) \tag{15.2}$$

Table 15.1 Selected concepts used in forestry for assessing growth and yield in a population/community of trees. Volume generally refers to stem wood but it can be converted to mass if necessary using wood density.

Concept	Explanation
Growth rate, m³ ha⁻¹ yr⁻¹	Annual rate of net primary growth of surviving trees plus ingrowth and the subsequent growth of ingrowth in a given inventory cycle.
Total growth, m³ ha⁻¹	Accumulation of growth rate over time.
Mortality rate, m³ ha⁻¹yr⁻¹	Difference between total growth of trees in inventories in a given measuring cycle and the initial total tree growth.
Total mortality, m³ ha⁻¹	Accumulation of trees that died over time.
Net growth rate m³ ha⁻¹ yr⁻¹	Difference between growth and mortality rates of trees in inventories in a given measuring cycle.
Total net growth, m³ ha⁻¹	Accumulation of net growth of trees over time.
Stocking, m³ ha⁻¹	Amount of accumulated net growth at a given point of time.

where h(t) is the density of new seedlings, n(t) is the density of trees, and k(t) is the density of trees dying while $\overline{s(t)}$ is the mean mass of seedlings, $\overline{m(t)}$ is the mean mass of trees, and $\overline{g(t)}$ is the mean growth of trees. The stock of trees at a given moment is the balance based on the mass in the previous moment plus the growth of trees and the mass of new trees and their growth minus the mass of dying trees and their growth (Table 15.1).

Interaction between regeneration, growth and mortality in forest dynamics

Kilkki (1984) demonstrated the dynamics of a tree population by applying the graphic model in Fig. 15.1. The left-hand boxes represent the naturally (A) growing and developing population/community of trees, and the right-hand boxes the same with thinning (B). In both cases, the initial stocking is divided into that surviving to the end of the period (e) and that dying (a) naturally or removed in cutting. The growth during a given period is divided into the growth of trees surviving throughout the period (Z_E), the growth of trees removed naturally or in cutting during the period (Z_A) and the growth of establishing seedlings (ingrowth) dying during the period (S_K). Consequently: (i) stocking at the beginning of the period is $B = e + a$; (ii) stocking at the end of the period is $E = e + Z_E + K$; (iii) total growth during the period is $Z_{br} = Z_E + Z_A + K + S_K$; (iv) total removal during the period is $A = a + Z_A + S_K$ and; and (v) net growth during the period is $Z_N = E - B = Z_{br} - A = Z_E - a + K$.

15.2 Regeneration under Climate Change

Regeneration under the climate warming in boreal forest

Under boreal conditions, the quantity and quality of seed crops are most affected by temperature (Henttonen et al. 1986, Pukkala 1987, Pukkala et al. 2010) with impacts on flowering and forming of seeds. Under this assumption, a process-based ecosystem model (FinnFor) was used to simulate seed crop, germination of seeds and establishment of seedlings. Kellomäki and Väisänen (1995) found that climatic warming is likely to increase the frequency of bumper crops in Scots pine as expected on the basis of empirical measurements (Koski and Tallqvist 1978), particularly in the north (Fig. 15.2). Even there, the annual seed crop was large enough to yield established seedlings every year, unlike under the current climate. The fraction of germinated seeds increased both in the south and north with the increasing temperature, but in the south this was less than that possibly due to limiting soil moisture (Kellomäki and Väisänen 1995).

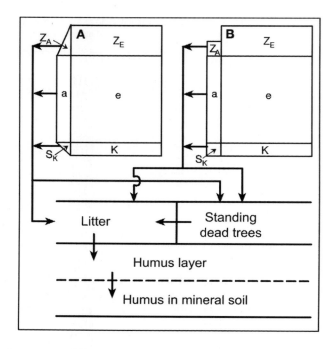

Fig. 15.1 Relationships between growth, removal and stocking in natural developing (A) and managed (B) forests over a given period, based on Kilkki (1984). Legend: e is Initial stocking surviving to the end of period; a is Initial stocking removed naturally or in cutting during the period; K is Growth surviving to the end of period for the trees exceeding the measuring accuracy at the beginning of the period; S_K is Share of growth removed during the period for the trees exceeding the measuring accuracy at the beginning of the period; Z_E is Share of growth for initial stocking surviving to the end of the period; and Z_A is Growth of initial trees removed naturally (died) during the period.

Figure 15.3 shows further that the mean number of established seedlings varied substantially from year to year following temperature fluctuation. Under the current climate, the number of seedlings was particularly abundant every 4–6 years, with a shorter interval in the south than in the north (Koski and Tallqvist 1978, Pukkala 1987), where no seedlings were established in three out of 20 years. The same pattern of fluctuation in the establishment was repeated under the warming climate, but even in the north there were no years without establishment, unlike under the current climate. In the latter case, the probability of seedlings dying was highest just after establishment, and under the current climate only some of the original seedlings survived over the 20 years used in assessing regeneration success. Under the climate warming, the survival of seedlings was enhanced, with the consequent increase in the final density of seedling stands (Kellomäki and Väisänen 1995).

After the regenerative cut (parent tree method), most of seedlings had a diameter < 1 cm, regardless of the temperature conditions (Fig. 15.3). The share of seedlings with a diameter < 1 cm increased under the climate warming, while the share of tallest seedlings reduced, particularly in the south. However, the absolute number of tall seedlings increased substantially due to enhanced growth under the elevated temperature, even in the north. The height of oldest and tallest seedlings was more than two meters in the south, if the current climate was assumed. In the north, the height of tallest seedlings was about 1.4 m, which is about 60–70% of that in the south. Under elevated temperature, the height of tallest seedlings was about 2.7 m in the south, and about 2 m in the north; i.e., 70–80% of that in the south. Height growth was thus enhanced more in the north than in the south (Kellomäki and Väisänen 1995).

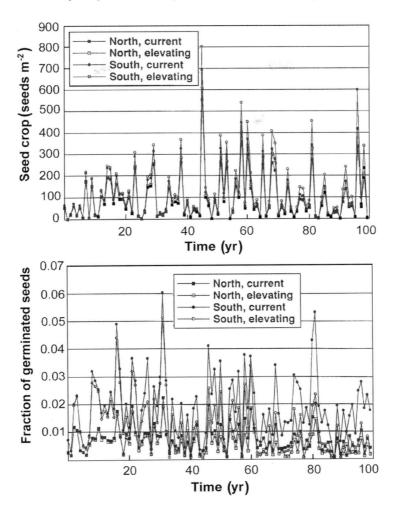

Fig. 15.2 Above: Seed crop, and Below: Fraction of germinated seeds of Scots pine as a function of temperature conditions (Kellomäki and Väisänen 1995). Permission of Canadian Science Publishing. Simulations were run for sites of medium fertility (*Myrtillus* type, MT) for southern (61° N) and northern (66° N) boreal forests. The moisture content in the top soil layer was 21 m³ m⁻³ at the wilting point and 54 m³ m⁻³ at field capacity. The current mean annual temperature represented the period 1960–1980, while under warming the mean annual temperature was 5°C higher than the current one. Regardless of the site and climate, the mean annual precipitation was 600 mm throughout the simulations.

Regeneration under the climate warming at timber line

At high latitudes, full forest cover reduces gradually through the timber and tree lines to Artic heaths and tundra. Globally, the Artic tree line seems to be a thermal boundary preventing the survival, growth and reproduction of trees at high latitudes. In northern Europe, the Artic tree line in the north is in the altitudinal range of 300–600 m, higher under continental than under oceanic conditions (Körner 1998, Körner and Paulsen 2004). In northernmost Finland (69° N and above), Scots pine and mountain birch (*Betula pubescent var. tortuosa*) form the timberline, which is the ecotone from the boreal forests to Arctic tundra. In these conditions, the length of the growing season varies from year to year from 350 d.d. to 950 d.d., with a long-term mean 650 d.d. (Juntunen and Neuvonen 2006). According to Grace et al. (2002), low temperature limits regeneration and growth much more than photosynthesis in the temperature range 5–20°C implying that climatic warming at the current

222 Managing Boreal Forests in the Context of Climate Change

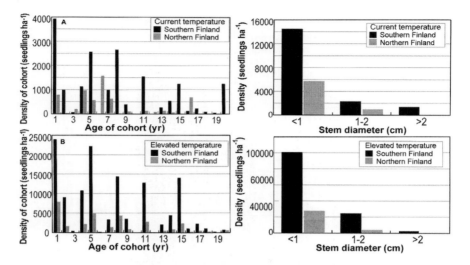

Fig. 15.3 Left: Density of a 20-year-old Scots pine seedling stand as a function of the age of the seedling cohorts under current (upper) and elevated temperature (lower). Right: Diameter distribution (diameter 1.3 m above ground level) of Scots pine seedlings in 20-year-old stand under the current (upper) and elevated (lower) temperature (Kellomäki and Väisänen 1995). Permission of Canadian Science Publishing. For simulations, see Fig. 15.2.

timber line may trigger a rapid advance of the timber line northwards if not curbed by the prevailing land use, such as intensive grazing in timber line forests.

Kellomäki et al. (1997b) used a process-based model (FinnFor) to determine whether a warming climate would enhance the regeneration of Scots pines at the Artic timber line (N 69°) in Finland. The simulations showed that temperature elevation increased the length of the growing season (the number of days with a mean daily air temperature $\geq +5°C$) from the current 110–120 days to 150–160 days. This implied that in summer (June–August), the temperature in the uppermost soil layer increased from 8–10°C up to 13–15°C. The increasing soil temperature in summer was associated with a small reduction (10%) in soil water content (Kellomäki et al. 1997b).

The simulations showed further that the total seed crops increased by 5% over the simulation period under warming alone or combined with elevating CO_2 (Kellomäki et al. 1997b). The increase was especially great when the temperature increased by 1–1.5°C from the current one, as also found by Juntunen and Neuvonen (2006). Thereafter, the occurrence of zero crops became rare. The potential success of regeneration was further increased by the increasing fraction of mature seeds under elevated temperature (Fig. 15.4). Under the current climate or elevating CO_2, the fraction of mature seeds was larger than zero only in three years out of 100, whereas under the warming climate the fraction of mature seeds was more than zero 67 times out of 100 (Kellomäki et al. 1997b). At the end of simulation period, the stand density was 600–700 seedlings per hectare for the current climate and 7,000 seedlings per hectare for the warming climate. Regardless of climate, the diameter distribution represented the dominance of small seedlings, but the CO_2 and temperature elevation alone or combined clearly increased diameter and height growth. Under CO_2 elevation alone the total volume of seedlings remained small, mainly due to the low stand density unlike under the combined elevation of CO_2 and temperature (Kellomäki et al. 1997b).

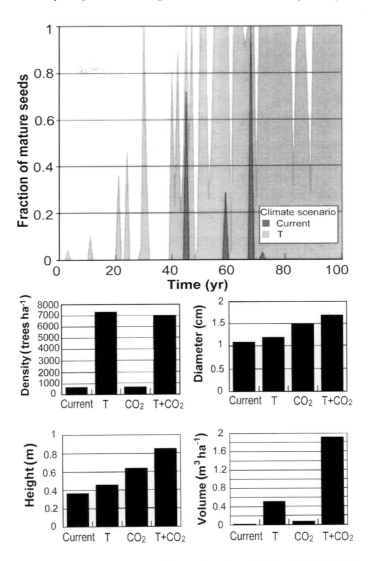

Fig. 15.4 Above: Fraction of mature seeds in seed crop of Scots pine at the artic timberline in Finland under the current and warming (T) climate. Below: Density, mean diameter and height and volume in stand at the end of 100-year simulation period under gradual warming (Kellomäki et al. 1997b). Permission of Springer. The simulations were done for a site of medium fertility (*Myrtillus* site, MT) at the timber line in northernmost Finland (N 69°). The density of parent Scots pines was 10 trees ha^{-1}, with a height and diameter of 15 m and 20 cm. The moisture in surface soil was 0.21 m^3 m^{-3} at the wilting point and 0.54 m^3 m^{-3} at the field capacity. Under the warming climate, temperature increased from the current –2°C to +2°C in 100 years. Precipitation remained at the current level of 300–400 mm for both climate conditions, and atmospheric CO_2 increased from 350 ppm to 700 pp under the changing climate.

15.3 Growth and Timber Yield under Climate Change at a Stand Scale

Management/climate change interaction

In general, the productivity of boreal forests is likely to increase in response to elevated CO_2 and temperature due to longer growing seasons and enhanced mineralization of nitrogen. To avoid increasing self-thinning and to utilize the increasing growth and yield, earlier and/or more intensive thinning and shortened rotation period may be necessary, as found by Kellomäki et al. (1997a). They used a process-based ecosystem model (FinnFor) to study how the increasing CO_2, temperature and precipitation would affect the timber yield from stands of boreal Scots pine in the south (61° N). The timing of thinning and the length of rotation were related to the dynamics of tree stands in compliance with the thinning rules applied in practical forestry.

In general, growth was larger in unthinned than thinned stand, because the stocking through the rotation was lower in the latter case. In both cases, trees matured faster under the changing climate than under the current one (Fig. 15.5), particularly when temperature, precipitation and CO_2 increases were combined. More rapid growth and development under the changing climate

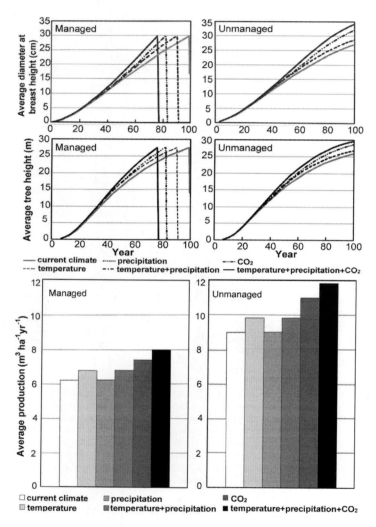

Fig. 15.5 Unthinned and thinned Scots pine stands under varying climate in southern boreal conditions (61° N) (Kellomäki et al. 1997a). Permission of Elsevier. Above: Development of diameter and height, and Below: Mean annual growth. Simulations used the initial stand density of 2,500 seedlings per hectare established on a site of medium fertility (*Myrtillus* type, MT), with a water content of 0.21 m³ m⁻³ at the wilting point and 0.53 m³ m⁻³ at field capacity. Under the current climate, the mean annual temperature was +3.6°C, with the precipitation of 610 mm and the CO_2 concentration of 350 ppm. The warming climate included: (i) a temperature increase of 0.4°C per decade; (ii) a precipitation increase of 9 mm per decade; (iii) a CO_2 increase of 33 ppm per decade; and (iv) a combined increase of temperature, precipitation and CO_2 based on the IS92a emission scenario (Carter et al. 1995).

reduced the rotation length and made thinnings earlier than under the current conditions. The average total stem wood production over the rotation was 6.2 m³ ha⁻¹yr⁻¹ under the current CO_2 and temperature and increasing precipitation alone. When temperature alone was increased or along with precipitation, the mean annual growth was 6.8 m³ ha⁻¹ yr⁻¹ (+9%). Similarly, the annual growth was 7.4 m³ ha⁻¹yr⁻¹ (+19%) for elevating CO_2 alone and 8.0 m³ ha⁻¹ yr⁻¹ (+28%) for the combined elevation of temperature, precipitation and CO_2.

The changes in climatic conditions were introduced gradually, and thus the differences in growth between climatic scenarios were small in the early rotation. The first thinning was made at almost the same time in all cases, and only the timing of the last two thinnings and the final cut differed substantially. Under the current climate, the rotation length was 99 years (Fig. 15.6). A rise in precipitation had no effect on rotation length, but temperature increase alone shortened rotation by nine years, as did a combined increase in temperature and precipitation. Increasing CO_2 had the most pronounced effect: the CO_2 increase alone shortened the rotation by 17 years and the combined effect of increased CO_2, temperature and precipitation shortened the rotation length by 26 years. The mean timber yield per year increased by 10% due to the temperature increase or combined with the increase in precipitation. Similarly, elevating CO_2 alone increased the timber yield by 20%. Under a combined increase of all the factors, the timber yield increased by 30% (Kellomäki et al. 1997a).

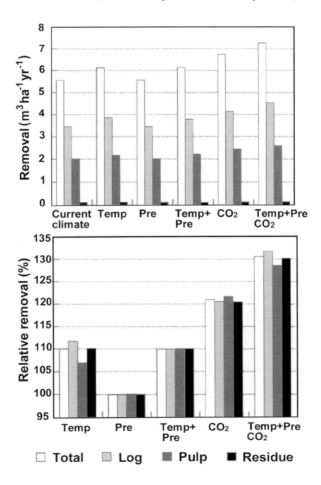

Fig. 15.6 Above: Mean timber yield under the current and changing climate. Below: Percentage of timber yield under the changing climate compared to that under the current climate in southern boreal conditions (Kellomäki et al. 1997a). Permission of Elsevier. For simulations, see Fig. 15.5.

> **Box 15.1 Simulations used to analyze how thinning affects growth and timber yield in southern and northern boreal forests**
>
> The current climate was that for the period 1961–1990, with a constant CO_2 of 350 ppm. The current mean annual temperature was 3.5°C in the south and 1.1°C in the north, and precipitation was 480 and 460 mm. The changing climate included the HadCM2 and ECHAM4 scenarios based on greenhouse emission scenario IS92a (Houghton et al. 1995). Under the climate warming, CO_2 increased linearly from 350 to 653 ppm by the end of the simulation. At the end of simulations using the changing climate, the mean annual temperature was 7–9°C in the south and 5–6°C in the north, while precipitation was 560–590 mm and 590–610 mm, respectively. Thinning included no thinning (UT) and basic thinning (BT (0,0)). Furthermore, two regimes with an increase of 15% or 30% in the triggering thinning and in the removal of remaining basal area ((BT (15, 15), BT (30, 30)) with higher stocking were used. Regardless of tree species and site, clear cut was made when the mean diameter of trees exceeded 30 cm (Briceño-Elizondo et al. 2006a,b). For thinning regimes, see Fig. 13.17.

Response of growth and timber to climate and thinning regime

Briceño-Elizondo et al. (2006a,b) used a process-based model (FinnFor) to study how total growth and timber yield under varying thinning regimes responded to climate change in southern (62° N) and northern (66° N) boreal forests (Box 15.1; Tables 15.2–15.4). In the south, the total growth of Scots pine was 700 m³ ha⁻¹ without thinning, of which 45% of total growth was lost in mortality, and 320 m³ ha⁻¹ of timber was obtained in the final cut. Under basic thinning (BT (0, 0)), the total growth was about 500 m³ ha⁻¹, but the mortality was marginal. The same pattern held for Norway spruce; i.e., without thinning the total growth was 1,080 m³ ha⁻¹ of which 700 m³ ha⁻¹ was obtained in final cut. The share of mortality was 35% of total growth. Under basic thinning (BT (0, 0)), the total growth of Norway spruce was 760 m³ ha⁻¹ in the south, nearly all (750 m³ ha⁻¹) being harvested for timber. Similarly, the total growth of birch was 540 m³ ha⁻¹ without thinning, with the timber yield being 290 m³ ha⁻¹. About 45% of the total growth was lost in the mortality. When basic thinning (BT (0,0)) was used, the total growth of birch was 340 m³ ha⁻¹ with a timber yield of 330 m³ ha⁻¹. In the north, the total growth and timber yield were up to two thirds of those in the south regardless of tree species. In general, high stocking in thinned stands increased the total timber yield regardless of species and site.

Changing climate increased the total growth and timber yield regardless of the site conditions and thinning regime (Briceño-Elizondo et al. 2006a,b). Over the management regimes, the total growth of Scots pine increased 28% in the south and 54% in the north (Tables 15.2–15.4), whereas the growth of Norway spruce increased 24% in the south and 41% in the north. The growth increase in birch remained slightly smaller than in conifers: 23% in the south and 35% in the north. The timber yield increased in the same way as the growth, regardless of tree species and site conditions. Under thinning, Scots pine timber increased by 26% in the south and by 50% in the north, whereas the increase in Norway spruce was slightly smaller, by 23% in the south and 40% in the north. In birch, climate change increased timber less than in conifers, by 20% in the south and 33% in the north. Under climate change, the differences between the south and north in total growth and timber yield were similar to those under the current climate, even though climate change clearly increased the absolute values. Under no thinning, this was especially evident regardless of site. In general, high stocking in thinned stands increased the total timber yield, regardless of species and site.

Table 15.2 Growth and timber yield [m³ ha⁻¹ per rotation] of Scots pine as a function of the thinning regime and site under the current (CUR) and changing climate (CC) in southern (62° N) and northern (66° N) boreal conditions (Briceño-Elizondo et al. 2006a,b). Both sites were of medium fertility (*Myrtillus* type, MT) suitable for Scots pine, with an initial density of 2,500 seedlings per hectare. For thinning regimes, see Fig. 13.17.

Thinning regime	Total growth, m³ha⁻¹ CUR	CC	Mortality, m³ha⁻¹ CUR	CC	Timber, m³ha⁻¹ CUR	CC	Change in timber, % of BT(0,0) CUR	CC
South (62° N)								
• BT (0,0)	504	643	1.3	4	503	638	---	---
• BT (15,0)	524	669	1.9	2	522	667	+4	+5
• BT (15,15)	548	701	1.3	2	547	699	+9	+9
• BT (30,0)	540	704	1.3	3	538	702	+7	+10
• BT (30,30)	592	757	1.7	2	591	755	+17	+18
• UT (0,0)	700	873	318	440	323	433	−24	−32
North (66° N)								
• BT (0,0)	295	466	1.8	17	293	449	---	---
• BT (15,0)	305	460	0.9	1	304	459	+4	+2
• BT (15,15)	321	482	1.2	1	319	481	+9	+7
• BT (30,0)	320	477	0.9	1	319	475	+9	+6
• BT (30,30)	345	522	1.4	2	343	521	+17	+16
• UT (0,0)	425	695	158	312	267	383	−9	−15

BT (0,0) indicates current thinning rules (basic thinning), and, e.g., BT (15,0) indicates a thinning regime with a 15% increase in the upper limit of stocking and 0% change in the remaining stocking compared to that under basic thinning (BT (0,0)), while UT (0,0) indicates no thinning before the terminal cut (Fig. 13.17).

Table 15.3 Growth and timber yield [m³ ha⁻¹ per rotation] of Norway spruce as a function of thinning regime and site under the current (CUR) and changing climate (CC) in southern (62° N) and northern (66° N) boreal conditions (Briceño-Elizondo et al. 2006a,b). Both sites were of medium fertility (*Myrtillus* type, MT) and suitable for Norway spruce, with an initial density of 2,500 seedlings per hectare. For thinning regimes, see Fig. 13.17.

Thinning regime	Total growth, m³ha⁻¹ CUR	CC	Mortality, m³ha⁻¹ CUR	CC	Timber, m³ha⁻¹ CUR	CC	Change in timber, % of BT(0,0) CUR	CC
South (62° N)								
• BT (0,0)	755	942	2.5	12	752	930	----	---
• BT (15,0)	776	929	0.3	9	776	920	+3	−1
• BT (15,15)	811	1012	3.1	6	808	1005	+8	+8
• BT (30,0)	807	998	0.3	7	807	991	+7	+7
• BT (30,30)	859	1080	0.3	7	858	1073	+14	+15
• UT(0,0)	1076	1328	379	516	697	866	−7	−7
North (66° N)								
• BT(0,0)	519	732	0.3	7	519	726	----	---
• BT(15,0)	553	756	1.6	1	552	755	+6	+4
• BT(15,15)	574	784	2.0	0	573	784	+10	+8
• BT(30,0)	567	789	0.3	0	566	789	+9	+9
• BT(30,30)	613	841	3.6	1	610	840	+18	+16
• UT (0,0)	740	1110	228	393	512	717	−1	−1

BT (0,0) indicates current thinning rules (basic thinning), and, e.g., BT (15,0) indicates a thinning regime with a 15% increase in the upper limit of stocking and 0% change in the remaining stocking compared to that under basic thinning (BT (0,0)), while UT (0,0) indicates no thinning before the terminal cut (Fig. 13.17).

Table 15.4 Growth and timber yield [m³ ha⁻¹ per rotation] of birch as a function of the thinning regime and site under the current (CUR) and changing climate (CC) in southern (62° N) and northern (66° N) boreal conditions (Briceño-Elizondo et al. 2006a,b). Both sites were of medium fertility (*Myrtillus* type, MT), suitable for birch, with an initial density of 2,500 seedlings per hectare. For thinning regimes, see Fig. 13.17.

Thinning regime	Total growth, m³ha⁻¹ CUR	CC	Mortality, m³ha⁻¹ CUR	CC	Timber, m³ha⁻¹ CUR	CC	Change in timber, % of BT(0,0) CUR	CC
South (62° N)								
• BT (0,0)	338	416	6.6	7	332	409	----	---
• BT (15,0)	356	430	5.5	5	350	425	+6	+4
• BT (15,15)	374	454	5.8	6	367	448	+11	+10
• BT (30,0)	363	457	5.3	6	357	451	+8	+10
• BT (30,30)	405	494	6.1	6	399	488	+20	+19
• UT(0,0)	542	640	247	315	294	315	−11	−23
North (66° N)								
• BT (0,0)	250	323	8.9	5	240	318	----	---
• BT (15,0)	258	339	5.9	6	252	333	+5	+5
• BT (15,15)	266	353	5.8	6	260	348	+8	+9
• BT (30,0)	266	362	6.0	6	259	356	+8	+12
• BT (30,30)	291	384	7.3	7	284	377	+18	+19
• UT(0,0)	353	508	148	244	205	264	−15	−17

BT (0,0) indicates current thinning rules (basic thinning), and, e.g., BT (15,0) indicates a thinning regime with a 15% increase in the upper limit of stocking and 0% change in the remaining stocking compared to that under basic thinning (BT (0,0)), while UT (0,0) indicates no thinning before the terminal cut (Fig. 13.17).

15.4 Growth and Timber Yield under Climate Change at a Regional Scale

Impacts on growth

Garcia-Gonzalo et al. (2007a,b) also used a process-based (FinnFor) model to analyze the effects of climate change and management on forest production over an area representing a middle boreal forest area (63° N) managed regularly for timber (Fig. 15.7). Norway spruce covered 64% (933 ha) of the area, Scots pine 28% (412 ha) and birch 7% (106 ha). Norway spruce and birch dominated the fertile (*Oxalis-Myrtillus* type, OMT) sites, whereas Scots pine was the main species on the poor (*Vaccinium* type, VT) sites. Norway spruce also dominated medium fertile sites (*Myrtillus* type, MT) mixed with birch and Scots pine. Regardless of tree species, the tree stands were mainly in the seedling and thinning phases (60–70% of the area), whereas the share of mature stands was small.

Over the management unit, the total growth of stem wood under no thinning (UT) was 864 m³ ha⁻¹ per 100 years; but one third of the total growth was lost through mortality (34%, 293 m³ ha⁻¹) (Table 15.5). The total growth was 37% higher than under the basic thinning (BT (0,0)) (630 m³ ha⁻¹), but in the latter case the amount of dead wood was small (5%, 34 m³ ha⁻¹) as was also the case for other thinning regimes (1–6%). An increase of 15 and 30% in the upper limit triggering thinning (BT(15,0), BT(30,0)) increased total growth by 2 and 5% compared to that under basic thinning (BT(0,0)). This tendency was further enhanced if the basal area remaining after thinning was increased: the increase was 6% for BT (15,15) and 11% for BT (30,30). The increased total

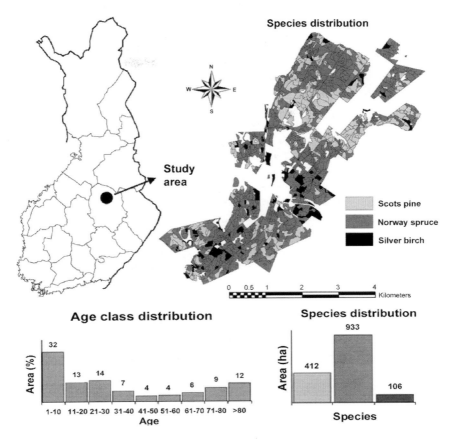

Fig. 15.7 Location and distribution of forest in separate compartments in the study area, with the main statistics of tree stands (Garcia-Gonzalo et al. 2007b). Permission of Elsevier. The current climate was that for the period 1961–1990, while the mean annual temperature was 3.1°C, and the mean annual precipitation 478 mm. Using warming climate under the HadCM2 and ECHAM4 scenarios, the mean annual temperature increased by 4.2 and 5.5°C and the mean annual precipitation by 85 and 113 mm. The current CO_2 concentration was 350 ppm, which increased to 653 ppm under both warming scenario. Thinning regimes were the same as used by Briceño-Elizondo et al. (2006a,b) at the stand scale (Box 15.1), using the rotation of 100 years regardless of tree species.

growth was related to the increased stocking throughout the rotation. The effects of management regimes on the total growth followed the same pattern regardless of tree species.

Under the HadCM2 warming scenario, the total growth for the whole management unit increased by 20–22% depending on the thinning regime (excluding unthinned); and under the ECHAM4 scenario the increase was 24–27%. Regardless of tree species and warming scenario, the increase in growth was slightly higher without thinning than under thinning. The increased growth also implied an increase in mortality: without thinning, the mortality for the whole management unit was 36% of the total growth under both warming scenarios. When thinning was done, the share of dead wood for the whole management unit increased from 5% (34 $m^3\,ha^{-1}$) under the current climate to 6% under warming scenarios (45 $m^3\,ha^{-1}$ for HadCM2 and 48 $m^3\,ha^{-1}$ for ECHAM4).

Impacts on timber yield

Under the current climate, the average timber yield without thinning (UT) was 471 $m^3\,ha^{-1}$ over the rotation (100 years) for the whole management unit (Garcia-Gonzalo et al. 2007a,b)

(Table 15.5). This was 24% less than under basic thinning (BT(0,0)) (619 m³ ha⁻¹). The timber yield tended to increase if thinning was done later than under basic thinning (BT (0,0)). Over the whole management unit, the increase in timber yield was 3 and 5% under the thinning regimes BT(15,0) and BT(30,0), respectively. This tendency was further enhanced if the basal area remaining after thinning was also kept higher than under basic thinning (BT(0,0)): the increases were 6 and 12% for BT(15,15) and BT(30,30), respectively.

Under the current climate, Norway spruce produced 718 m³ ha⁻¹ and Scots pine 474 m³ ha⁻¹ of timber if basic thinning (BT(0,0)) was used (Table 15.5). Both conifers showed a reduction of 24% in timber yield if no thinning (UT) was used. Similarly, birch produced 313 m³ ha⁻¹ of timber in BT (0,0), which was reduced by 11% without thinning. Regardless of the species, the largest timber yield was obtained under the management scenario BT (30,30): the percentage of saw logs increased along with the increasing thinning threshold, especially if the remaining basal area was also increased. Over the whole management unit, the percentage of saw logs ranged from 64% for the BT (0,0) to 69% for BT (30,30). However, the percentage of saw logs was clearly larger if

Table 15.5 Effect of varying thinning regimes on the total timber yield [m³ ha⁻¹ per 100 years] for different tree species in the management unit under the changing climate, the change of timber yield (%) in relation to the timber yield under basic thinning (BT (0,0)) and the share (%) of saw logs in the total timber yield (Garcia-Gonzalo et al. 2007a,b). For thinning regimes, see Fig. 13.17.

Thinning regime	Current climate			Changing climate, ECHAM4			Changing climate, HadCM2		
	Timber, m³ ha⁻¹	Δtimber, %	Logs, %	Timber, m³ ha⁻¹	Δtimber, %	Logs, %	Timber, m³ ha⁻¹	Δtimber, %	Logs, %
Scots pine									
• BT (0,0)	474	-	69	573	-	72	557	-	71
• BT (15,0)	494	4	71	601	5	74	582	5	73
• BT (30,0)	512	8	72	629	10	75	609	9	74
• BT(15,15)	513	8	71	623	9	74	603	8	73
• BT(30,30)	548	15	73	669	17	76	646	16	75
• UT(0,0)	358	−24	85	420	−27	89	408	−27	75
Norway spruce									
• BT (0,0)	718	-	62	776	-	60	770	-	60
• BT (15,0)	735	2	64	795	2	62	793	3	61
• BT (30,0)	748	4	65	829	7	63	824	7	63
• BT(15,15)	754	5	65	820	6	62	818	6	63
• BT(30,30)	793	11	68	870	12	64	867	13	65
• UT(0,0)	543	−24	86	633	−19	87	626	−19	87
Birch									
• BT (0,0)	313	-	67	368	-	69	362	-	69
• BT (15,0)	322	3	68	384	4	71	377	4	71
• BT (30,0)	334	7	68	389	6	71	383	6	70
• BT(15,15)	339	8	69	399	9	71	391	8	71
• BT(30,30)	364	17	70	427	16	73	419	16	72
• UT (0,0)	278	−11	82	312	−15	86	308	−15	85
Management unit (mean)									
• BT (0,0)	619	-	64	689	-	63	679	-	63
• BT (15,0)	636	3	66	710	3	65	703	3	64
• BT (30,0)	651	5	67	740	7	66	731	8	66
• BT(15,15)	655	6	67	733	6	66	725	7	65
• BT(30,30)	692	12	69	781	13	68	771	14	68
• UT(0,0)	471	−24	85	549	−20	88	541	−20	87

BT (0,0) indicates current thinning rules (basic thinning), and, e.g., BT (15,0) indicates a thinning regime with a 15% increase in the upper limit of stocking and 0% change in the remaining stocking compared to that under basic thinning (BT (0,0)), while UT (0,0) indicates no thinning before the terminal cut (Fig. 13.17).

no thinning was applied: 85% for the conifers and 82% for silver birch. This implies that small-dimensioned trees suitable for pulp wood were not harvested during the rotation, and were lost to mortality.

Over the thinning regimes, climate change increased the mean timber yield 11–12% relative to that under the current climate (Table 15.5). All tree species followed this pattern, but the response was species-specific. Regardless of the thinning regime, the mean timber increase for Scots pine was 18–22%, for birch 16–18% and for Norway spruce 9%. For Norway spruce only, the timber yield over the management unit increased more without thinning (UT) than under any thinning. Climate change also slightly decreased the share of saw logs in Norway spruce (–2%), but not in Scots pine and birch if basic thinning (BT (0,0)) was used (Table 15.5). Both climate change scenarios tended to trigger thinning earlier and thus increase the number of thinning, regardless of the species. However, the thinning regime affected the timber yield in the same way as under the current climate. The timber yield was therefore highest (771 m^3 ha^{-1} for the HadCM2 and 781 m^3 ha^{-1} for the ECHAM4 warming scenario) for the thinning BT (30,30): an increase of 14% (HadCM2) and 13% (ECHAM4) relative to basic thinning BT (0,0). Under warming, the reduction in timber yield without thinning was slightly higher for Scots pine and birch compared to that under the current climate. For Norway spruce the loss was reduced to 19% under both climate change scenarios.

15.5 Concluding Remarks

Process-based simulations indicated that temperature elevation alone could increase the seed crop throughout the boreal forests, thus enhancing the potential for natural regeneration. Even at the northern timber line, seed crops with fully maturated seeds in Scots pine may be large enough for the establishment of closed stands on a reasonable time scale. Simulations showed further that under a warming climate alone or combined with the increase in atmospheric CO_2 enhanced the establishment of seedlings, with a rapid increase in stand density (Juntunen and Neuvonen 2006). This was partly due to the increased soil temperature, which enhanced the germination of seeds if they were fully mature. This is in line with the findings from reforestation experiments, where disturbing the humus layer and thereby increasing soil temperature substantially improves germination and the establishment of seedlings (Pohtila 1977). This is especially the case in northern boreal forests, where climatic warming increases the soil temperature, which otherwise remains below the values for optimal germination.

Simulations also showed that an increase in temperature and precipitation, with a concurrent increase in CO_2, may enhance stand-scale growth by 24–53%, the increase being smaller in the south than in the north. This implies that the total timber yield would increase by 20–40% compared to that under the current climate if current management rules are used. The increase was larger in the north than in the south, where the absolute values of timber yield were still larger than in the north regardless of tree species. However, the thinning regime had a clear effect on the total growth and timber yield (Kellomäki et al. 1988): an increase in the thinning threshold with a consequent increase in stocking increased the total growth and timber yield. This tendency was further enhanced if the remaining basal area in thinning was increased concurrently with the increase in the threshold for thinning. The current thinning guidelines might therefore not be adequate for management under altered climatic conditions (Briceño-Elizondo et al. 2006a,b).

References

Briceño-Elizondo, E., J. Garcia-Gonzalo, H. Peltola and S. Kellomäki. 2006a. Carbon stocks and timber yield in two boreal forest ecosystems under current and changing climatic conditions subjected to varying management regimes. Environmental Science & Policy 9: 237–252.

Briceño-Elizondo, E., J. Garcia-Gonzalo, H. Peltola, J. Matala and S. Kellomäki. 2006b. Sensitivity of growth of Scots pine, Norway spruce and silver birch to climate change and forest management in boreal conditions. Forest Ecology and Management 232: 152–167.

Carter, T., M. Posch and H. Tuomenvirta. 1995. SILMUSCEN and CLIGEN User's Guide. Publication of Academy of Finland 5/95: 1–62.

Garcia-Gonzalo, J., H. Peltola, E. Briceño-Elizondo and S. Kellomäki. 2007a. Changed thinning regimes may increase carbon stock under climate change: a case study from a Finnish boreal forest. Climatic Change 81: 431–454.

Garcia-Gonzalo, J., H. Peltola, E. Briceño-Elizondo and S. Kellomäki. 2007b. Effect of climate change and management on timber yield in boreal forests, with economic implications: a case study. Ecological Modelling 209: 220–234.

Grace, J., F. Berninger and L. Nagy. 2002. Impacts of climate change on the tree line. Annals of Botany 90: 537–544.

Henttonen, H., M. Kanninen, M. Nygren and R. Ojansuu. 1986. The maturation of Scots pine seeds in relation to temperature climate in northern Finland. Scandinavian Journal of Forest Research 1: 234–249.

Houghton, J. T., L. g. M. Filho, B. A. Callander, N. Harris, A. Kattenberg and K. Maskell. 1995. Climate Change 1995—the Science of Climate Change: Contribution of WGI to the Second Assessment Report of the Intergovernmental Panel on Climate Change, Intergovernmental Panel on Climate Change. Cambridge University Press, Cambridge, UK.

Juntunen, V. and S. Neuvonen. 2006. Natural regeneration of Scots pine and Norway spruce close to the timberline in Northern Finland. Silva Fennica 40(3): 443–458.

Kellomäki, S. and H. Väisänen. 1995. Model computations on the impact of changing climate on natural regeneration of Scots pine in Finland. Canadian Journal of Forest Research 25: 929–942.

Kellomäki, S., T. Karjalainen and H. Väisänen. 1997a. More timber from boreal forests under changing climate. Forest Ecology and Management 94: 195–208.

Kellomäki, S., H. Väisänen and T. Kolström. 1997b. Model computations on the effect of elevating temperature and atmospheric CO_2 on the regeneration of Scots pine at timber line in Finland. Climatic Change 37: 683–708.

Kilkki, P. 1984. Metsänmittausoppi. Joensuun yliopisto, metsätieteellinen tiedekunta. Silva Carelica 3: 1–222.

Körner, C. 1998. A re-assessment of high elevation treeline positions and their explanations. Oecologia 115: 445–459.

Körner, C. and J. Paulsen. 2004. A world-wide study of high altitude treeline temperatures. Journal of Biogeography 31: 713–732.

Koski, V. and R. Tallqvist. 1978. Tuloksia monivuotisista kukinnan ja siemensadon määrän mittauksista metsäpuilla. Summary: Results of long-time measurements of the quality of flowering and seed crop of forest trees. Folia Forestalia 364: 1–60.

Pohtila, E. 1977. Reforestation of ploughed sites in Finnish Lapland. Communicationes Instituti Forestalis Fenniae 91(4): 1–98.

Pukkala, T. 1987. Simulation model for natural regeneration of *Pinus sylvestris, Picea abies, Betula pendula* and *Betula pubescens*. Silva Fennica 21: 37–53.

Pukkala, T., T. Hokkanen and T. Nikkanen. 2010. Prediction models for the annual seed crop of Norway spruce and Scots pine in Finland. Silva Fennica 44(4): 629–642.

16

Management of Boreal Forests for Timber and Biomass under Climate Change

ABSTRACT

At a stand scale, management is widely based on recommendations or predetermined instructions. Rules specify the length of rotation and timing and the intensity of thinning which is likely to sustain the yield. At a regional scale, the sustainable yield depends on the overall composition of forests in the region. The identification of stand-level management is based on the potential set of management regimes for optimizing the management schedules for the whole region. Both rule-based and optimized management is applied in order to identify how climate change may affect the potential to produce sustainable yields of timber and energy biomass.

Keywords: sustainable management, rule-based management, optimized management, timber biomass, climate change

16.1 Management in Controlling Genotype/Environment Interaction

Management operations in controlling genotype/environment interaction

Management controls the long-term functional and structural development of forest ecosystems (succession) so that ecosystems produce goods and services aimed at management. Management regimes (combinations of management measures) and management measures are tools to direct the succession towards the necessary structures and functions of ecosystem set by the management objectives.

Management of the properties of tree populations and sites, or both, can be used to maintain or increase production. Figure 16.1 demonstrates how selected management measures control the interaction between the genotype and environment in producing, for example, timber (saw

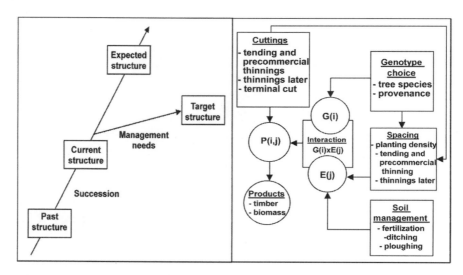

Fig. 16.1 Left: Schematic presentations of how the future functioning and structure of the forest ecosystem is affected by the past and current structures. If the future structure deviates from that expected, there is a need to direct the ecosystem dynamics (succession) through management which produces a more optimal ecosystem structure and functioning as regards management objects (Kellomäki et al. 2009). Right: Schematic presentations of how selected management measures affect the interaction between the genotype and environment and how they affect the production of timber and biomass alone or combined.

logs, pulpwood) and biomass alone or combined. The choice of proper genotype (tree species or provenance of selected tree species) for a given site is fundamental in management, and the period from planting to the last thinning before the terminal cut controls the competition for resources. Spacing makes resources available for trees, as is the case in soil management. In management, the choice of genotype, spacing and soil management should be combined in such a way that the site conditions are optimized (tailored) for a given genotype over the whole production cycle (rotation).

Rule-based and/or optimized management

Forest management may be considered at the stand scale and/or at the scale of the forest area (region, landscape). At the stand scale, management is widely based on recommendations or predetermined instructions: management is rule-based. The rules specify, for example, the length of rotation and the timing and intensity of thinning, which is likely to sustain management and maximize the yield. Management rules may be based on long-term practical experiences and/or long-term experiments in optimizing production in relation to the management objectives and constraints set on the production and growth of tree stands. At a regional scale, the maximum sustainable yield depends, however, on the overall composition of forests in the region. If the region is dominated by mature stands, the immediate cutting of all mature stands for regeneration will substantially decrease the maximum sustainable cuttings in the future. Similarly, if the region is dominated by young stands, the maximum sustainable yield is increased in the future (Nuutinen et al. 2006). This implies that optimized management is region-specific, as defined by the properties of sites and tree stands throughout the region.

16.2 Growth and Development of Boreal Forests under Rule-based Management

Growth, stocking and cutting drain

Kellomäki et al. (2005, 2008) analyzed the growth and development of managed boreal forests throughout Finland under climate change by applying the current management rules for planting, tending in pre-commercial thinning, commercial thinning and rotation length (Box 16.1). Figure 16.2 shows that the mean annual growth over all tree species currently varies from less than 1 m^3 ha^{-1} yr^{-1} in the north to 6 m^3 ha^{-1} yr^{-1} in the south. Under a warming climate, growth in the north increased up to 90% at the maximum, while the increase in the south was up to 20%.

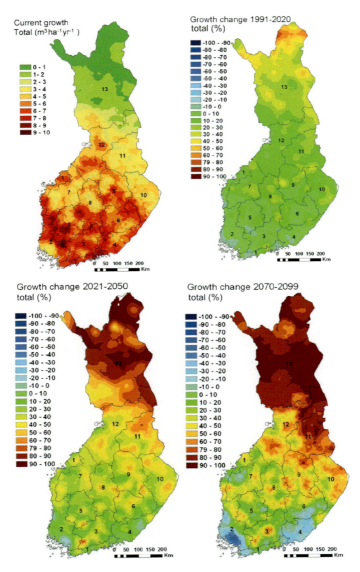

Fig. 16.2 Effect of climate change on the integrated growth of Scots pine, Norway spruce and birch simulated with the Sima model described in Box 16.1 (Kellomäki et al. 2005). Numbers in Figures refer to the administrative regions.

Box 16.1 Simulation of the growth and development of boreal forests under a set of management operations and climate change

Kellomäki et al. (2005, 2008) used a gap-type growth and yield model (Sima) to investigate the growth and development of managed boreal forest. Growth was made responsive to climatic and edaphic factors, and to a set of management operations (Fig. 16.3).

Dynamics of a tree stand/community is controlled by the regeneration, growth and mortality of trees. Growth is based on diameter growth ΔD [cm yr^{-1}] = $\Delta D_o \times M_1 \times,\ldots,\times M_n$, where; ΔD_o is diameter growth [cm yr^{-1}] in optimal conditions; and $M_1,\ldots,M_n$ are multipliers representing the temperature sum (TS; +5°C threshold), prevailing light conditions, soil moisture and nitrogen supply. Growth in optimal conditions is related to no shading and no limitation of soil moisture and nitrogen supply, the maturity of trees (diameter of tree, D cm) and atmospheric CO_2 [ppm]:

$$\Delta D_o = \exp\left(a + \frac{b}{0.01 \times CO_2}\right) \times D \times e^{DGRO \times D} \qquad (16.1)$$

where a, b and DGRO are the parameters. The parameter estimation was based on the data generated with a process-based model (FinnFor) using the climate combining the CO_2 elevation from 350 to 700 ppm for the medium fertile site (*Myrtillus* type, MT) in the middle boreal zone at the latitude 62° N. Based on the generated data, the growth for single Scots pines of varying diameter was calculated under no shading and ample water supply and constant foliage nitrogen (Fig. 16.4).

Diameter is further used to calculate the mass of stems, foliage, branches and roots (Mass (i, j), kg tree^{-1}):

$$Mass(i, j) = \exp\left[a(i, j) + b(i, j) \times \frac{D}{c(i, j) + D}\right] \qquad (16.2)$$

where a (i, j), b (i, j) and c (i, j) are parameters specific to species i and mass component j. Litter from any organ and the mortality of whole trees transfers carbon and nitrogen into the soil, where litter and humus (soil organic matter) decay, with the consequent nitrogen release for the reuse in growth.

In the simulations, the permanent sample plots of the National Forest Inventory established by the Finnish Forest Institute in 1985 were used in a grid of 16 km x 16 km in the south and a grid of 16 km x 32 km in the north. In the simulations, a 10 km x 10 km grid for current (1991–2000), and a 50 km x 50 km grid for warming climate were used for the periods 1991–2020, 2021–2050 and 2070–2099 based on the SRES A2 emission scenario (Ruosteenoja et al. 2005). Atmospheric CO_2 increased from 350 ppm in 1990 to 840 ppm in 2100. The mean annual temperatures increased 4°C in summer and 6°C in the winter throughout the country.

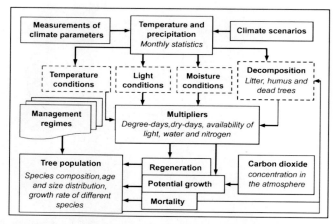

Fig. 16.3 Outline of the model (Sima) used for simulating the growth and development of managed boreal forests over Finland under climate change (Kellomäki et al. 2005, 2008).

Box 16.1 contd....

Box 16.1 contd....

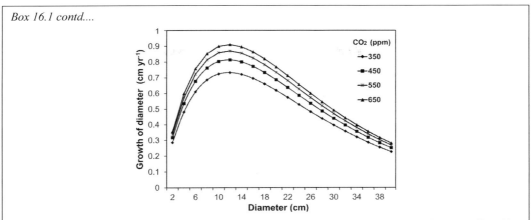

Fig. 16.4 Potential diameter growth of boreal Scots pine as a function of diameter and atmospheric CO_2 (Kellomäki et al. 2005, 2008).

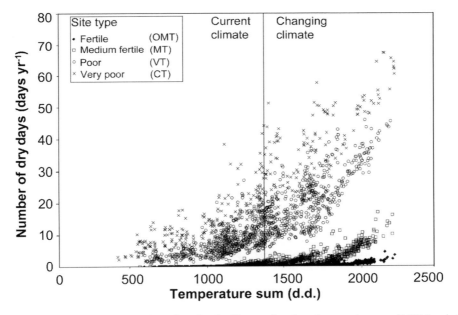

Fig. 16.5 Number of dry days in sites on sites of varying fertility as a function of temperature sum (+5°C threshold, d.d.) under current and warming climate (Kellomäki et al. 2005, 2008). Legend: OMT (*Oxalis-Myrtillus* type) refers to fertile sites, MT (*Myrtillus* type) to medium fertile sites, VT (*Vaccinium* type) to poor sites and CT to extremely poor sites (*Calluna* type).

In many places in the south, the growth increase in Norway spruce was small, or growth even reduced towards the end of the century due to the clear increase in the frequency and length of drought episodes (Fig. 16.5). Norway spruce was competitive on fertile sites (*Oxalis-Myrtillus* type, OMT) with birch and other deciduous species, even though the dominance of birch increased. At the same time, the dominance of Scots pine increased on the medium fertile sites (*Myrtillus* type, MT) currently occupied by Norway spruce. The growth of Scots pine increased throughout southern and middle boreal forests, except in the south-western and south-eastern corners of the country. These

Table 16.1 Mean current growth, stocking and potential cutting drain of forests at different time slices under climate change divided between southern and northern Finland (Kellomäki et al. 2005).

Region	Growth, $m^3 ha^{-1} yr^{-1}$ (in parenthesis % change)			
	Current	1990–2020	2021–2050	2050–2099
Southern Finland	5.5	5.9 (7)	6.3 (11)	6.8 (12)
Northern Finland	2.2	2.6 (18)	3.7 (68)	4.6 (109)
Total	4.1	4.5 (10)	5.3 (29)	5.9 (44)

	Total stocking, $m^3 ha^{-1}$ (in parenthesis % change)			
	Current	1990–2020	2021–2050	2050–2099
Southern Finland	144	146 (1)	173 (20)	144 (0)
Northern Finland	78	82 (5)	125 (59)	145 (86)
Total	117	120 (2)	154 (31)	144 (23)

	Potential cutting drain, $m^3 ha^{-1} yr^{-1}$ (in parenthesis % change)			
	Current	1990–2020	2021–2050	2050–2099
Southern Finland	3.2	3.3 (3)	4.2 (31)	5.0 (56)
Northern Finland	1.1	1.2 (9)	2.2 (100)	3.0 (168)
Total	2.3	2.4 (4)	3.3 (52)	4.2 (82)

Northern Finland includes the forests above 63° N and southern Finland forests below 63° N. The percentage change in relation to the current climate is given in parenthesis. Note that growth represents only upland sites on mineral soils, excluding peatlands.

changes in growth reduced total growth locally, but country-scale total growth increased, especially in northern Finland.

Table 16.1 summarizes the mean annual growth under the current and warming climate, making a clear distinction between southern and northern boreal regions. In the south, the growth in the regional scale may increase up to 12% in this century. This is substantially less than that in the north, where the growth may be doubled compared to that under the current climate. Over the whole country, an increase of 44% was obtained, mostly affected by the large increase in the northern part of the country.

Stocking is the total stem wood volume of trees per hectare at a given point in time. Currently, mean stocking in the south is up to 150 $m^3 ha^{-1}$, and in the north up to 80 $m^3 ha^{-1}$ (Table 16.1). Under climate change, the mean stocking in the south increased in the 50-year perspective but reduced towards the end of this century to a similar level to that currently. In the north stocking is likely doubled compared to the current situation by the end of this century following rapid growth increase.

The potential cutting drain is the commercial round wood removal plus logging residue. The figures in Table 16.1 represent the maximum sustainable removals when applying current management recommendations. In southern Finland, the potential cutting drain may increase up to 56% by the end of this century. In northern Finland, the increase is much greater (up to 170%), but the absolute value (3 $m^3 ha^{-1} yr^{-1}$) is still less than two thirds of that in the south (5 $m^3 ha^{-1} yr^{-1}$).

Tree species composition

Tree species composition is the percentage share of stem wood volume of different tree species in total stem wood volume (stocking). Under changing climate, the share of Scots pine in the south may increase through the whole simulation period, up to 60% of the total volume, later in this century (Table 16.2). At the same time, the share of Norway spruce may reduce to 10%. Most of the reduction might occur later in this century, when birch seems to replace Norway spruce on fertile

Table 16.2 Tree species composition in percentage of the mean stocking of forests currently and at different time slices under climate change, divided between southern and northern Finland (Kellomäki et al. 2005).

Region and species	Current	1991–2020	2021–2050	2070–2099
Southern Finland				
• Scots pine, %	42	44	54	62
• Norway spruce, %	49	45	33	8
• Birch, %	9	11	13	30
Northern Finland				
• Scots pine, %	62	63	68	77
• Norway spruce, %	27	26	22	14
• Birch, %	11	11	10	8
Over Finland				
• Scots pine, %	47	49	59	68
• Norway spruce, %	43	39	29	12
• Birch, %	10	12	12	20

Northern Finland includes the forests above 63° N and southern Finland includes forests below 63° N. The percentage change in relation to the current climate is given in parenthesis. Note that the tree species composition represents only upland sites on mineral soils, excluding peatlands.

sites in many places. In the north, too, the share of Norway spruce may reduce. There, Norway spruce is replaced by Scots pine, whose share exceeds 70% of the total stem wood volume. In the north, the share of birch seems to remain similar to current values, or even slightly reduced. Over the whole country, the total volume comprised 68% Scots pine, 12% Norway spruce and 20% birch at the end of the simulation period.

Carbon sequestration

Carbon in the forest ecosystems (trees and soil) currently follows the same pattern as that of growth and stocking of trees. In the south, the total amount of carbon is 100 Mg C ha^{-1} over larger areas, whereas in the north the amount of total carbon remains 50 Mg ha^{-1}. Under climate change, the amount of carbon increased more in the north than in the south, as one may expect on the basis of the changes in growth (Table 16.3) (Alam et al. 2008, 2010). However, the mean total amount of carbon remains somewhat smaller in the north than in the south, where the total amount of carbon may reduce in some areas due to the reduction in growth and stocking of Norway spruce. Over the whole country, the total amount of ecosystem carbon was still close to 30% more than currently.

16.3 Growth and Development of Boreal Forests under Optimized Management

Simulations

Kärkkäinen et al. (2008) used the MELA management model (Siitonen et al. 1996) to optimize the management of timber production and the consequent production of logging residues for energy over Finland. The MELA model is based on integrated simulation and optimization, in which the management of stands is solved endogenously, based on forest-level (e.g., forest region) goals (Nuutinen et al. 2006). The identification of stand-level management is based on the potential set of management regimes that is used as input in optimizing management schedules for each stand (Fig. 16.6). Optional management schedules control the dynamics of tree stands: regeneration,

Table 16.3 Mean amount of carbon currently in the forest ecosystem (trees, soil) and at different time slices under climate change divided between southern and northern Finland (Kellomäki et al. 2005).

Region	Carbon, Mg C ha^{-1} (in parenthesis % change)			
	Current	1991–2020	2021–2050	2070–2099
Trees				
• Southern Finland	49	52 (5.9)	57 (17.1)	53 (8.3)
• Northern Finland	25	28 (11.9)	36 (45.8)	40 (60.8)
• Total mean	39	42 (8.3)	50 (28.8)	51 (29.7)
Soil				
• Southern Finland	40	40 (0.1)	43 (7.2)	50 (23.9)
• Northern Finland	32	32 (0)	34 (6.1)	41 (28.4)
• Total mean	37	37 (0)	39 (6.8)	46 (25.7)
Trees and soil				
• Southern Finland	89	90 (1.5)	100 (12.6)	102 (14.7)
• Northern Finland	57	58 (2.2)	70 (23.0)	82 (45.3)
• Total mean	76	77 (1.8)	89 (16.8)	96 (27.1)

Northern Finland includes forests above 63° N and southern Finland forests below 63° N. The percentage change in relation to the current climate is in parenthesis. Note that the carbon only represents upland sites on mineral soils, excluding peatlands.

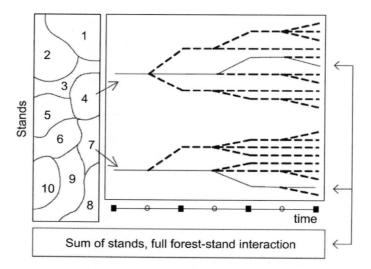

Fig. 16.6 Outline of integrated stand and forest-level optimization. A decision tree consists of simulated management schedules over time for each stand in the forest area. Optimal stand management (i.e., management schedule) is derived for each stand on the basis of forest-level objectives and constraints (Nuutinen et al. 2006). Permission of Springer.

growth and mortality. Management schedules include planting with a selection of tree species, clearing regeneration areas, soil preparation, tending young stands, cuttings, ditching, fertilization and pruning of Scots pine if appropriate. The linear programming is used in selecting an optimal combination of management schedules for individual management units. The forest-level optimization task is given as an objective function and a set of utility constraints (Nuutinen et al. 2006).

The MELA model uses an empirical growth and yield model (Motti), which was made sensitive to the elevation of CO_2 and temperature (Box 16.2). The simulations covered the whole of Finland, including 23 million ha of land used in forestry. The simulations were run for the current climate and

Box 16.2 Making a statistical growth and yield model responsive to climate change

Matala et al. (2005) made a statistical growth and yield model (Motti) responsive to climate change, utilizing a process-based physiological model (FinnFor). In the latter case, biomass growth is related to the photosynthetic production responsive to climate change, allowing a transfer of the climate change effect into the Motti model. The transfer variable indicates the relative climate change effect (RSEv(t)) on the volume growth:

$$RSEv(t) = \frac{\Delta V(t)_{scen} - \Delta V(t)_{current}}{\Delta V(t)_{current}} \tag{16.3}$$

where $\Delta V(t)_{scen}$ refers to the growth under a given climate change scenario and $\Delta V(t)_{current}$ to growth under the current climate. Consequently: $\Delta V(t)_{scen} = \Delta V(t)_{current} + RSEv(t) \times \Delta V(t)_{current}$. Using a process-based model, the values of RSEv(t) were obtained for correcting the growth in the Motti model to include climate-induced changes in growth. The values of $\Delta V(t)_{scen}$ represent the relative scenario effect, which is dependent on stand density (RDF), the competition status of trees (RDFL), site fertility (ST) and the location of the site, indicated by temperature sum (TS) (Fig. 16.7): $RSEv(t) = f(CO_2, T, RDF, RDFL, ST, TS)$.

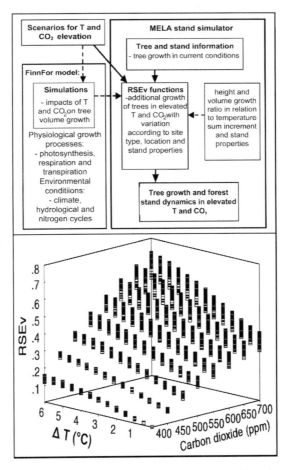

Fig. 16.7 Above: Outline of how climate change impacts on growth were introduced into the growth and yield model used in the MELA forest planning system. Below: Performance of transfer variable RSEv in Equation (16.3) as a function of the elevating CO_2 and temperature (Matala 2005). Courtesy of the Finnish Society of Forest Sciences and the Finnish Forest Research Institute, Permission of Elsevier.

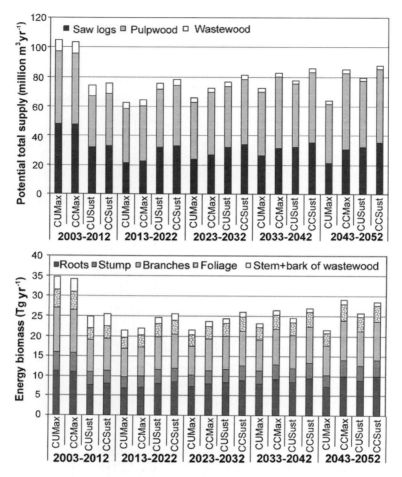

Fig. 16.8 Mean annual removal of timber (upper) and energy biomass (lower) for 10-year periods under the current (CU) climate and changing climate (CC), when applying the maximum cutting (Max) and the sustainable cutting (Sust) scenarios in management (Kärkkäinen et al. 2008). Permission of Elsevier.

a warming climate. In the latter case, the mean annual temperature increased gradually by 6°C along with the concurrent doubling of atmospheric CO_2 from 350 to 683 ppm by 2099.

Management objectives

Two optional management objectives were used in the simulations. First, management was aimed at "maximizing the net present value for timber production using a 5% interest rate (max cutting scenario)" (Kärkkäinen et al. 2008) (Fig. 16.8). Second, management was aimed at "maximizing net present value from timber production using a 4% interest rate with non-decreasing flow of wood, saw logs and net income over a given period and net present value after the 50-year period greater than or equal to the beginning (Sust cutting scenario)". The simulations were applied across the whole of Finland based on National Forest Inventory data using a five-year time step. Wood prices and the costs of management were based on the deflated (1999) average realized prices and costs during the 10-year period 1990–1999. Simulations were run for five 10-year periods: 2003–2012, 2013–2022, 2023–2032, 2033–2042, 2043–2052 applying the management rules recommended for practice.

Yields of timber and energy biomass and carbon sequestration

The production of timber and energy biomass is closely linked with each other, as they represent different parts of the total tree biomass. In the first phase, the cut-to-length harvest system produces saw logs, pulpwood and logging residue (foliage, living and dead branches, stem tops too small for timber, trees of low quality and/or too small even for pulpwood), which may be used for energy biomass. The potential energy biomass may further include stumps and the coarse roots that remained attached to stumps, extracted for energy biomass. The close link between timber and energy biomass implies that the long-term supply of energy biomass is greatly affected by the cutting regime, which optimizes timber production under given objectives and constraints placed on production.

Kärkkäinen et al. (2008) found that the share of energy biomass was 44% of the total removal, including timber and energy biomass regardless of cutting (management) and climate scenarios. Using the maximum cutting scenario, the supply of timber in the period 2003–2013 varied from 103 to 105 million m^3 yr^{-1} (in both cases about 4.5 m^3 ha^{-1} yr^{-1}) depending on the climate scenario (Fig. 16.8). At the same time, the obtainable amount of energy biomass was 35 Tg yr^{-1} (79 million m^3 yr^{-1}, 3.4 m^3 ha^{-1} yr^{-1}). During the period 2043–2053 under the current climate, the potential timber supply was 64 million m^3 yr^{-1}, (2.8 m^3 ha^{-1} yr^{-1}) and energy biomass 22 Tg yr^{-1} (49 million m^3 yr^{-1}, 2.1 m^3 ha^{-1} yr^{-1}), whereas under climate change the timber supply was 85 million m^3 yr^{-1} (3.7 m^3 ha^{-1} yr^{-1}) and energy biomass 29 Tg yr^{-1} (66 million m^3 yr^{-1}, 2.9 m^3 ha^{-1} yr^{-1}), respectively.

Using the sustainable cutting scenario, the timber supply in the period 2003–2013 was 74 and 76 million m^3 yr^{-1} (about 3.2 m^3 ha^{-1} yr^{-1}) depending on climate scenario, whereas the potential supply of energy biomass was 25 Tg yr^{-1} (57 million m^3 yr^{-1} (2.5 m^3 ha^{-1} yr^{-1})) for both climate scenarios. During the period 2043–2053, under the current climate, the potential supply of timber was 80 million m^3 yr^{-1} (3.5 m^3 ha^{-1} yr^{-1}) and energy biomass 26 Tg yr^{-1} (59 million m^3 yr^{-1}, 2.6 m^3 ha^{-1} yr^{-1}). Under climate change, the supply of timber and energy biomass was 88 million m^3 yr^{-1}, (3.8 m^3 ha^{-1} yr^{-1}) and 29 Tg yr^{-1} (65 million m^3 yr^{-1}, 2.8 m^3 ha^{-1} yr^{-1}).

Kärkkäinen et al. (2008) further found that in the maximum cutting scenario the amount of potential timber supply decreased by the end of the simulation period (2053). This was because final cuttings in the first period changed the structure of forests towards the reduced dominance of mature Norway spruce forests in final cuts, thus reducing future cutting potential. In the sustainable cutting scenario, the number of cuttings was more stable due to the constraints set on cutting in optimization. Regardless of the cutting and climate scenarios, the proportion of saw logs in cuttings is likely to decrease in the coming 50 years, whereas the amount of pulpwood is likely to increase. This will affect the yields of energy biomass provided by cutting residue. The number of stands in the thinning phase is likely to increase, with the consequence that the proportion of residues will be greater compared to that of stems harvestable for timber.

Based on the simulations of Kärkkäinen et al. (2008) and Matala et al. (2009), the carbon sequestration in the growing stock was analyzed further under different cutting and climate scenarios for the period 2003–2053. At the beginning of the simulations, the amount of carbon in the growing stock was 765 Mton (2,802 Tg CO_2). By the end of the simulations, the carbon stock increased to 894 Mton (3,275 Tg CO_2) when assuming a sustainable cutting scenario under the current climate (CUSust). Similarly, the carbon stock increased to 906 Mton (3,321 Tg CO_2) under the current climate and management aimed at maximizing net present value (CUMax). Under warming, the carbon stock increased further to 1,060 Mton (3,885 Tg CO_2) when assuming a sustainable cutting scenario (CCSust), and to 1,026 Mton (3,758 Tg CO_2) under management maximizing the net present value (CCMax).

16.4 Concluding Remarks

The application of optimized management in simulations showed that forest growth is likely to increase under climate change but the share of saw logs may reduce and the share of pulp wood increase in the next 50 years, affected by the current structure of forests. This further implies a reduction of logging residues harvestable for energy biomass related to the reduction of mature Norway spruce stands. This is likely to be compensated for by the increase of growth under climate change. This is possible during the next 50 years, when the increase of evapotranspiration is still moderate in relation to the increase in precipitation. Rule-based management showed this to be the case especially in northern Finland, where the low temperature is currently greatly limiting forest growth. In this time perspective, the potential cutting drain is likely to increase even in southern Finland, where the occurrence and duration of drought episodes may substantially increase later in this century.

References

Alam, A., A. Kilpeläinen and S. Kellomäki. 2008. Impacts of thinning on growth, timber production and carbon stocks in Finland under changing climate. Scandinavian Journal of Forest Research 23: 501–512.

Alam, A., A. Kilpeläinen and S. Kellomäki. 2010. Potential energy wood production with implications to timber recovery and carbon stocks under varying thinning and climate scenario in Finland. Bioenergy Research 3: 362–372.

Kärkkäinen, L., J. Matala, K. Härkönen, S. Kellomäki and T. Nuutinen. 2008. Potential recovery of industrial wood and energy wood raw material in different cutting and climate scenarios for Finland. Biomass and Bioenergy 32: 934–943.

Kellomäki, S., H. Strandman, T. Nuutinen, H. Peltola, K. T. Korhonen and H. Väisänen. 2005. Adaptation of forest ecosystems, forests and forestry to climate change. Finnish Environment Institute, FinAdapt Working Paper 4: 1–50.

Kellomäki, S., H. Peltola, T. Nuutinen, K. T. Korhonen and H. Strandman. 2008. Sensitivity of managed boreal forests in Finland to climate change, with implications for adaptive management. Philosophical Transactions of the Royal Society B363: 2341–2351.

Kellomäki, S., V. Koski, P. Niemelä, H. Peltola and P. Pulkkinen. 2009. Management of forest ecosystems. pp. 252–373. *In*: S. Kellomäki (ed.). Forest Resources and Sustainable Management. Gummerus Oy, Jyväskylä, Finland. Second edition.

Matala, J. 2005. Impacts of climate change on forest growth: a modelling approach with applications to management. Ph. D. Thesis. University of Eastern Finland, Joensuu, Finland.

Matala, J., R. Ojansuu, H. Peltola, R. Sievänen and S. Kellomäki. 2005. Introducing effects of temperature and CO_2 elevation on tree growth into a statistical growth and yield model. Ecological Modelling 181: 173–190.

Matala, J., L. Kärkkäinen, K. Härkönen, S. Kellomäki and T. Nuutinen. 2009. Carbon sequestration in the growing stock of trees in Finland under different cutting and climate scenarios. European Journal of Forest Research 128: 493–504.

Nuutinen, T., J. Matala, H. Hirvelä, K. Härkönen, R. Ojansuu, H. Peltola, H. Väisänen et al. 2006. Regionally optimized forest management under changing climate. Climatic Change 79(3–4): 315–333.

Ruosteenoja, K., K. Jylhä and H. Tuomenvirta. 2005. Climate scenarios for FINADAPT studies of climate change adaptation. Finnish Environment Institute, Finadapt Working Paper 15: 1–15.

Siitonen, M., K. Härkönen, H. Hirvellä, J. Jämsä, H. Kilpeläinen, O. Salminen et al. 1996. Mela Handbook. The Finnish Forest Research Institute, Helsinki, Finland.

Disturbances and Damage Affecting Boreal Forests under Climate Change

ABSTRACT

Disturbance in forest ecosystems involves damage due to excessive variability in environmental conditions, and lack of balance between organisms occupying forest sites. Disturbances are due to abiotic factors (frost, wind, snow, fire, drought) and biotic factors (insect and fungal pests, herbivores, invasion of exotic species), which alter the forest communities regardless of management. Abiotic and biotic factors damaging trees and forests are modified by changes in temperature and precipitation, making boreal forests prone to climate change. This concerns both currently existing and alien insects, fungi and vertebrates that are likely to invade under a warming climate.

Keywords: boreal forest disturbance, abiotic damage, biotic damage, climate change

17.1 Risk of Abiotic and Biotic Damage and Climate Change

Disturbance or damage in forest ecosystems involves pronounced changes in the ecosystem structure and subsequent changes in ecosystem functioning. Damage is related to excessive variability in abiotic conditions and uncontrolled balance between organisms occupying the site. Abiotic and biotic damages may be realized at given probabilities, referring to the risk of damage or losses.

Abiotic damage is related to strong wind, excessive snowfall and wild fire, which destroy single trees and populations/communities of trees, from a few hectares to thousands of hectares. Wind and snow damage are related to mechanical forces due to the wind and gravity, and fire kills and damages trees due to excessive heat load and high temperature. High summer temperatures may also damage trees, including foliage, stems, and especially roots if high temperature is combined with drought. Under boreal conditions, trees may also be damaged by low temperatures, especially in spring, summer and/or autumn, whenever low temperatures exceed the frost resistance of tissue in foliage, branches and shoots, stems and roots. Frost damage may retard the growth of trees, but seldom causes the immediate death of domestic trees except small seedlings.

Following abiotic damage, trees are highly susceptible to consequential biotic damage; e.g., broken and uprooted trees can lead to detrimental insect attacks on other trees, due to the increased

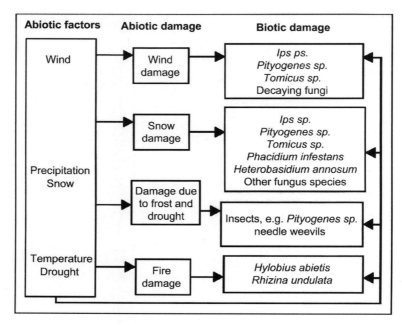

Fig. 17.1 Outline of how weather and climate affect outbreaks of populations of damaging insects and fungi related to abiotic damage.

availability of breeding material. High summer temperatures with increased drought episodes further increase the growth of insect populations, which attack trees with reduced growth. Abiotic and biotic disturbances interact with one another, or with changes in temperature and precipitation, thus increasing the death of trees. This process may feed forward and enhance the mortality of trees. Thus, outbreaks of many damaging insects and fungi are affected by the occurrence of dead and dying trees related to abiotic damage (Fig. 17.1).

17.2 Risk of Frost Damage under Climate Warming

Phenology of trees and risk of frost damage

Frost damage is damage due to below-zero temperatures, whenever the frost resistance of living tissues is low. Damage may occur at any time of year, and is related to the phenology of trees. The risk of frost damage is greatest in early summer and early autumn, when night frost may coincide with low frost resistance/hardiness during the transition from winter dormancy to active growth in early spring and summer or during the transition from active growth to winter dormancy in early autumn. The adaptation to warming climate includes trade-offs between utilizing the full growing season and avoiding frost damage through the proper timing of dehardening in spring and hardening in autumn (Saxe et al. 2000).

Bud burst is preceded by low chilling temperatures during winter. However, even under elevating temperature chilling of boreal trees is likely to be fulfilled, and earlier bud burst may be expected in response to higher spring temperatures (Myking and Heide 1995, Häkkinen et al. 1998). Since the middle of the 1800s, the flowering and bud burst of several local deciduous species in southern and northern boreal forests have advanced 3.3 to 11.0 days per century, implying an increase of 1.8°C in ambient temperature (Linkosalo et al. 2009). Throughout Europe, climatic warming has advanced bud burst from two to four days per 1°C increase of winter temperature as was found by

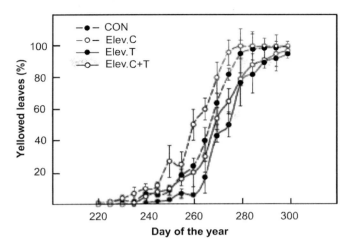

Fig. 17.2 Yellowing of leaves in young birches as a percentage of the total leaves in seedlings grown under ambient conditions (CON), elevated CO_2 (Elev. C) and temperature (Elev. T) and combined both (Elev. C + T) in closed chambers (Kellomäki and Wang 2001). Permission of Oxford University Press. Each point is based on four replicates per treatment. Whole leaf was yellow if more than 50% of the leaf area was yellow.

Kramer (1995). The greatest advance has occurred in Norway spruce and birch. Similarly, Pudas et al. (2008a,b) found that bud burst and flowering of selected deciduous trees and dwarf shrubs in northern boreal region (> 66° N) has advanced 1–2 days per year in the 10-year period 1997–2006. The advance was largest in the north, and it was related to changes in the mean May temperature and the snow melt.

Warming may also prolong the growing season, e.g., for lime [*Tilia cordata* L.] and beech [*Fagus sylvatica* L.] (Kramer 1995). Similarly, Kellomäki and Wang (2001) found that elevated temperature in closed chambers affected the ontogenetic development of Pendula birch (*Betula pendula* Roth) seedlings. This implied that the elongation of growing period increased the retention of leaf areas (Fig. 17.2). Yellowing was delayed two to three weeks, but the delay had only marginal effect on the annual photosynthetic production. Under the combined elevation of CO_2 and temperature, yellowing was delayed in a similar way to that under elevated temperature alone, whereas elevated CO_2 alone turned the leaves yellow earlier than in ambient conditions. On the contrary, Pudas et al. (2008a) found no systematic change in the timing of autumn phenophases of Pubescent birch (*Betula pubescens* Ehrh.).

Timing of height growth of Scots pine under elevated CO_2 and temperature

Using closed chambers, Kilpeläinen et al. (2006) studied how the elevation of CO_2 and temperature, alone or combined, may affect the course of height growth in Scots pine. The onset of height growth (Fig. 17.3) took place on the days 102 and 98 from the beginning of the year under elevated temperature alone or combined with elevated CO_2, while the onset took place on the days 112 and 110 under ambient conditions and elevated CO_2 alone. On the other hand, the cessation of height growth took place on the day 182 under ambient conditions and under elevated CO_2 alone, but on the days 162 and 159 under elevated temperature alone or combined with elevated CO_2. The duration of height growth varied between 61 and 73 days depending on the treatment: 71 days under ambient conditions, 73 days for elevated CO_2 alone, 61 days for elevated temperature alone and 62 days for the combined elevation of CO_2 and temperature. Slaney et al. (2007) also found that the initiation and cession of shoot growth in Norway shoots were earlier under elevated temperature than under ambient conditions.

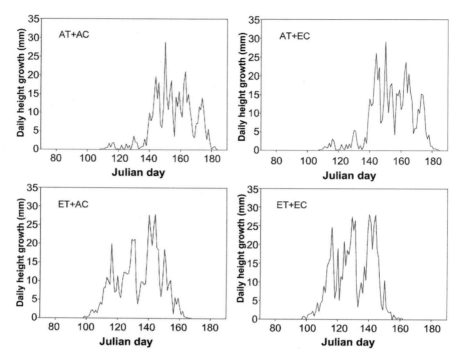

Fig. 17.3 Daily growth rate of Scots pine grown under ambient conditions (AT + AC), elevated temperature (ET + AC) and CO_2 (AT + EC) alone, and combined both (ET + EC) in a chamber experiment (Kilpeläinen et al. 2006). Temperature was elevated by 6°C in winter (December to February), 4°C in spring (March to May) and autumn (September to November) and 2°C in summer (June to August) (Kellomäki et al. 2000). Permission of Elsevier.

Table 17.1 Onset, cessation and duration of height growth of boreal Scots pine in calendar days and temperature sum (+ 5°C threshold) grown under ambient conditions and elevated CO_2 and temperature (T) alone or the combined elevation of CO_2 and temperature. Calendar day is from the beginning of the year (Kilpeläinen et al. 2006).

Treatment	Onset, calendar day	Cessation, calendar day	Length of growing period, days	Temp. sum at onset, d.d.	Temp. sum at cessation, d.d.	Length of growing season, d.d.
Ambient	112	182	71	25	468	443
CO_2	110	182	73	20	479	459
T	102	162	61	76	530	454
CO_2 + T	98	159	62	68	518	450

Kilpeläinen et al. (2006) related the onset, cessation and duration of height growth to the annual temperature sum (Table 17.1). Under ambient conditions and elevated CO_2, the onset took place at the temperature sum (threshold temperature of +5°C) 25 d.d. and 20 d.d., while the cessation occurred at the sum 468 d.d. and 479 d.d. At the same time, the onset and cessation under elevated temperature alone or combined with elevated CO_2 took place at substantially higher temperature sum: the onset at 76 d.d. and 68 d.d. and the cession at 530 d.d. and 518 d.d. Consequently, the duration of growth in terms of temperature sum was similar from 440 d.d. to 450 d.d. regardless of climatic treatment, but in terms of calendar days the growth period under elevated temperature alone or combined with elevated CO_2 was up to two weeks shorter than that under the current temperature or under elevated CO_2 alone. Similarly, Slaney et al. (2007) found a high correlation between temperature sum (day degree ≥ 0°C) and shoot elongation in Norway spruce but no precise timing of bud burst in relation

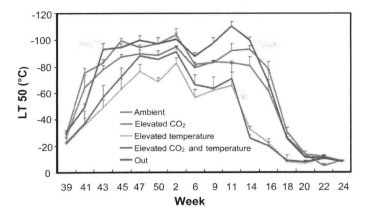

Fig. 17.4 Effects of climate change on the annual course of frost hardiness of Scots pine in a chamber experiment (Seppo Kellomäki, University of Eastern Finland, unpublished). LT50–values indicate the temperature at which half the needles subjected to low temperatures are damaged if grown in ambient conditions or under elevated CO_2 or temperature or combined both in the Mekrijärvi chamber experiment.

to the temperature sum was found. Rousi and Heinonen (2007) also found that under the middle boreal conditions the bud burst of birch (*Betula pendula* Roth) was related to the temperature sum (+ 5°C threshold), varying in the range from 34 d.d. to 71 d.d. They further found that in the period 1926–2005 bud burst had become earlier, at a rate of 1.2 days per decade.

Frost hardiness of Scots pine under elevated CO_2 and temperature

Based on open-top chambers, Repo et al. (1996) found that the temperature alone or combined with elevated CO_2 hastened bud burst of Scots pine up to two months and advanced the springtime loss of frost hardiness by 29 days on average. This experiment used an extreme elevation of winter temperature (i.e., 5–20°C higher than outside the chambers). However, even a smaller elevation of temperature outside the growing season may cause earlier spring loss of frost hardiness, as demonstrated in Fig. 17.4. The elevation of winter temperature was up to 6°C compared to ambient conditions (Seppo Kellomäki, University of Eastern Finland, unpublished). These measurements showed that the loss of spring frost hardiness may be advanced by 14 days compared to ambient temperature. Frost hardiness under elevated temperature alone or combined with elevated CO_2 was from –70 to –80°C in deep winter, and under the other treatments frost hardiness was from –80 to –110°C, respectively. During the bud burst, the frost hardiness was more than –20°C under the elevated temperature.

The sensitivity of frost hardiness/resistance to climate change may further be elaborated using the model for frost hardiness developed by Leinonen et al. (1995). The model is based on: (i) response of frost hardiness to the change in environment depending on the phase of annual development; (ii) response to photoperiod during hardening; and (iii) response of frost hardiness to short-term temperature variability. The daily change in frost hardiness is dependent on the difference between the stationary (target) level of frost hardiness ($\hat{R}(t)$) and the actual frost hardiness (R (t)):

$$\frac{dR(t)}{dt} = \frac{1}{\tau} \times \left[\hat{R}(t) - R(t) \right] \qquad (17.1)$$

where τ is a time constant. The stationary level of frost hardiness is modeled as a function of the additive effect of night length and temperature, and the effect of annual development of hardening competence (C_R), indicating the capacity of environmental factors (short photoperiod, low temperatures) to induce or retain frost hardiness (Leinonen et al. 1995, Saxe et al. 2000):

$$\hat{R}(t) = \hat{R}_{\min} + C_R \times \left[\Delta \hat{R}_T(t) + \Delta \hat{R}_P(t) \right] \quad (17.2)$$

where $\hat{R}_{\min}$ is the minimum level of frost hardiness, $\Delta \hat{R}_T(t)$ is the increase in the stationary level of frost hardiness induced by temperature and $\Delta \hat{R}_P(t)$ is the increase in the stationary level of frost hardiness induced by night length.

Frost may damage trees whenever their frost resistance is less than the temperature, and may occur at any time of the year. The basic question is how climate change affects the phenological cycle of trees and consequent frost hardiness in relation to the annual temperature cycle and variability. Leinonen (1997) explored frost damage for boreal Scots pine with model calculations that utilized parallel simulations of daily frost hardiness and daily minimum temperature under the current climate and under climate change. Leinonen (1997) found that under the current climate less than 5% of the needle area was damaged in a 100-year period and only once up 25% of the needle area was affected. The elevation of temperature by 6°C in the simulation increased loss of needle areas up to 10%. However, needle loss only slightly affected total light interception and photosynthesis because the reduced within-crown shading likely compensated for the effects of reduced needle area on total photosynthesis.

17.3 Risk of Snow Damage

Climatic conditions affecting the occurrence of snow damage

Snow damage is related to the properties of climate and weather, and the properties of trees to resist snow loading alone or combined with loading wind. Snow damage refers to the breakage of crown or stem, bending of stem, or uprooting of a whole tree under a loading of snow (Fig. 17.5). Damage takes place, if the bending moment on a tree under snow loading alone, or in combination with wind loading, exceeds the maximum resistive moment of crown, stems or roots (Solantie 1994, Peltola et al. 1997). The risk of excessive snow load on the crown is the greatest if wet snow falls during low wind (velocity < 9 m s^{-1}) associated with an air temperature around zero. A large snow load may especially be expected if snow accumulates rapidly and evenly on crowns at temperatures in the range of +0.6 to –3°C (Solantie 1994). Low or moderate risk is evident if snowfall < 40 mm (in water), and snowfall > 40 mm implies a moderate to high risk (Peltola et al. 1997, 1999a,b). In Finland, the mean return period of severe snow damage is 5–15 years, in such a way that the return is shorter in the south than in other parts of the country.

Primarily, snow loading depends on the quality (wet, dry) and quantity of snow, which affect the amount of snow intercepted on the crown. A low wind speed (< 9 m s^{-1}) allows snow loading, and a high wind speed (> 9 m s^{-1}) may dislodge attached snow. Rain and fog droplets may further be intercepted in the crown at subzero temperatures in the form of ice and rime. The risk of damaging snow loads is, however, quite local, following variability in precipitation, temperature and wind velocity in relation to topography. The effect of altitude is particularly pronounced, higher altitudes presenting greater risk of high snow loading than lower altitudes. In northern Europe, sites at altitudes greater than 150–200 m above sea level seem to be most susceptible to snow damage due to high snow fall and the long duration of snow load (Valinger and Lundqvist 1994).

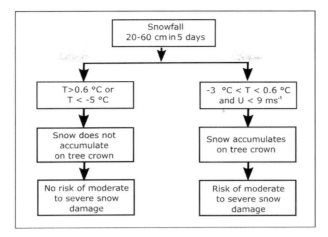

Fig. 17.5 Effect of temperature (T) and wind speed (U) on snow accumulation on tree crown and the risk of severe snow damage (Nykänen et al. 1998). The limit is set at 40 mm for low or moderate risk and at over 60 mm for severe risk (Solantie 1994). Courtesy of the Finnish Society of Forest Science and the Finnish Forest Research Institute.

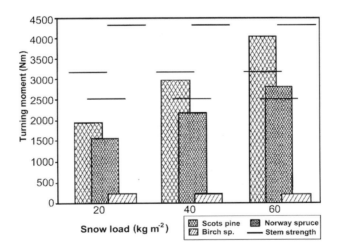

Fig. 17.6 Snow loading of 60 kg m^{-2} may cause breakage in Scots pine, but it is not sufficient to break birch (Peltola et al. 1997, 1999a,b). Permission of Canadian Science Publishing. Columns indicate combined loading of snow and wind, when assuming a constant wind velocity of 8 m s^{-1} above the canopy. The lines indicate critical values of stem strength. The trees were 12 m tall, with a taper of (diameter/height) 1:120.

Structural properties of trees and forests affecting the occurrence of snow damage

In general, coniferous trees are turned down if snow loads exceed 50 kg m^{-2} (Peltola et al. 1997, 1999a,b). The critical limits of damaging forces may vary substantially, however, depending on the properties and size of trees and the distribution of foliage along the stem. In Scots pine, for example, snow load increases the total bending moment most in the upper stem, breaking stem at the level of the crown bottom. In Norway spruce, there is no clear pattern, but the stem may break in variable sections within the crown area. Peltola et al. (1999a,b) found that slender (slightly tapering, e.g., diameter/height 1:120) Scots pines and Norway spruces with a height of 12 m may be most likely damaged by snow, and especially if the snow load exceeds 60 kg m^{-2} (Fig. 17.6). In these computations, the resistance to stem breakage was assumed to be related to the third power of the breast height diameter.

Risk of snow damage under climate change

Under changing climate, both the quantity and quality of snow, and the growth and development of trees affect the occurrence of snow damage. The increasing fall of wet snow, with the increased frequency of around-zero temperatures may increase risks while reduced snow duration may reduce risks. Especially, young stands (e.g., mature seedling stands and stands mature for first thinning) are vulnerable to snow induced damage as pointed by Kilpeläinen et al. (2010a). They studied the risk of snow-induced damage under climate change by simulating the number of days per year when the snow accumulation damages trees. The snow accumulation in five days ≥ 20 kg m^{-2} is likely to damage trees.

The simulations were run for the whole of Finland, assuming that the forests were managed in the current way for timber production. In modeling, snow accumulation was assumed to be a continuous process based on cumulative snowfall on crowns and the loss of snow from crowns related to air temperature and wind speed (Box 17.1; Fig. 17.7). Snow accumulation was linked to a process-based forest ecosystem (Sima), which was used to study the interaction of snow accumulation and structure of forests under the changing climate, with the impact on the risk of damage. The simulations used the inventory plots of the Finnish National Forest Inventory allowing generalization of the findings over 23 million hectares of forest land across the boreal zone (Kilpeläinen et al. 2010a).

Kilpeläinen et al. (2010a) found that under the current climate (1961–1990) there was a clear risk of snow damage (20 days per year) in several parts of northern boreal forests (Fig. 17.8), but in the western and southern parts of middle and southern boreal forests the risk was low. Under the climate warming, the risk of snow damage reduced throughout the country by a few days in 1991–2020. The risk decreased further in 2021–2050, especially in northwestern part of boreal forests, and in 2070–2099 the risk of snow damage was evident in 0–6 days per year in the south, and to 6–12 days in the north. Over the whole country, the frequency of heavy snowfall (i.e., ≥ 20 kg m^{-2} snow fall over a 5-day period) reduced from 18 to 8 days, with a 56% reduction in the number of risk days. Kilpeläinen et al. (2010a) also found that the share of forest land at risk of snow-induced damage was < 3% in 1991–2020 over the whole country. In the period 2021–2050, the situation was similar, excluding the north-eastern part of country. By the end of the century (2070–2099), the share of the risk area had reduced to less than 2% of the total forest area.

Box 17.1 Calculating snow accumulation on tree crowns

Snow accumulation (Snowacc$_n$, kg m^{-2}) includes snow fall and snow loss (Snowloss, %) related to air temperature (T, °C) and wind speed (U, m s^{-1}) (Gregow et al. 2008):

$$Snowloss(T) = 11.502 \times T^{2.6361} \quad \text{if} \quad T \succ 0\,°C$$
$$Snowloss(T) = 0 \quad \text{if} \quad T \prec 0\,°C \qquad (17.3)$$

$$Snowloss(U) = 0.0038 \times U^3 - 0.2176 \times U^2 + 0.8605 \times U \qquad (17.4)$$

Based on the three-hour time step indicated by n, the snow accumulation is (Gregow et al. 2008):

$$Snowacc_n = Prec_n + Snowacc_{n-1} \times (1 - (Snowloss_n(T) + Snowloss_n(U))/100) \qquad (17.5)$$

where Prec$_n$ is the precipitation in snow, in a three-hour period. On any day, there is a risk of snow damage if the snow accumulation is in a three-hour period ≥ 20 kg m^{-2} (Fig. 17.7).

In simulations, the current (1961–1990) and changing (2000–2099, SRES A2 scenario) temperature and precipitation values were interpolated to a 50 × 50 km grid (Ruosteenoja et al. 2005). During climate warming, CO$_2$ elevated from 350 ppm to 840 ppm by 2099, with the increase in the mean temperature 4°C in summer and 6°C in winter. Precipitation increased by 20% in winter but remained unchanged in summer. Wind velocity was the same as under the current and changing climate.

Box 17.1 contd....

Box 17.1 contd....

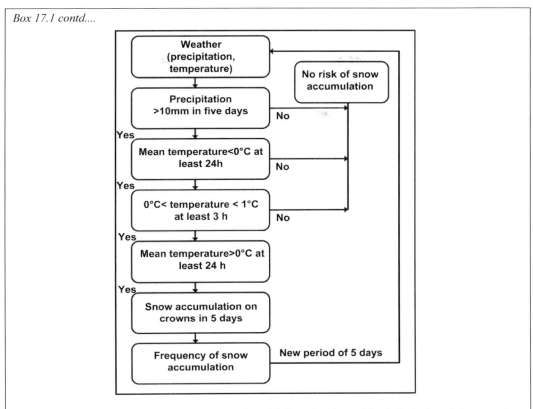

Fig. 17.7 Accumulation of snow on tree crown as a function of the intensity of snow fall, wind velocity and air temperature, based on Kilpeläinen et al. (2010a).

Kilpeläinen et al. (2010a) applied the model findings to estimate the amount of stocking at risk of being destroyed by excess snow fall. They found that young trees and stands prior to first commercial thinning (the dominant height 12–15 m) were particularly prone. On the other hand, the structure of forests changed gradually throughout the simulation period due to management and timber harvest. Climatic warming changed growth, thus changing the share of young trees and stands from sub-period to sub-period. Kilpeläinen et al. (2010a) found that the mean annual stocking per hectare at risk increased substantially from the period 1991–2020 to the period 2021–2050 but reduced again substantially in the period 2070–2099 (Fig. 17.8). In the period 2021–2050, the amount of stocking at risk was greatest in the north-eastern and north-western parts of the country, where the share of prone trees and stands was especially large.

17.4 Risk of Wind Damage

Factors affecting risk of wind damage

The recent storm events in Europe demonstrate how devastating an extreme wind episode may be to forests. For example, in two storms in France, up to 170 million m^3 of wood were destroyed in winter 1999. In southern Sweden, 80 million m^3 was destroyed in a single storm episode in winter 2005 (Fitzgerald and Lindner 2013). In Finland, strong winds blew down more than 11 million m^3 of timber between 1975 and 1985 (Laiho 1987). More recently, storms blew down 7.3 million m^3

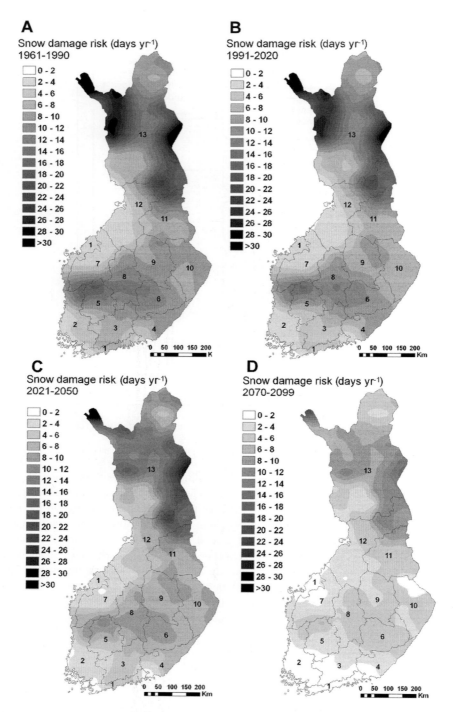

Fig. 17.8 Number of days with the risk of snow damage per year. A: baseline period (1961–1990), B: near-term (1991–2020), C: mid-term (2021–2050) and D: long-term (2070–2099) future. The numbers in the maps refer to the administrative regions (forest center) used in forestry (Kilpeläinen et al. 2010a). Permission of Springer. For calculations, see Box 17.1.

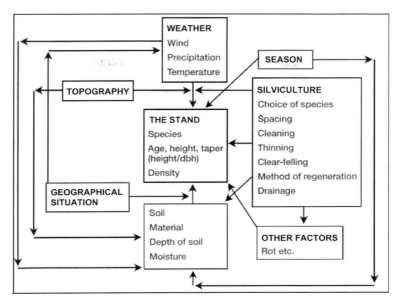

Fig. 17.9 Main factors affecting the risk of wind damage based on Peltola et al. (2009a,b).

of timber in November 2001, and in July 2002 one million m³ across southern and western Finland. These events have involved high wind speeds (> 19 m s⁻¹) and/or snow fall, and such damage cannot be avoided or even reduced through careful forest management. In northern Europe wind-induced damage frequently also occurs on a smaller scale, resulting in continuous losses of timber, with a close correlation to forest structure modified by management and timber harvest.

The occurrence of wind damage is closely linked to the occurrence of high wind speeds, but the susceptibility of trees and forests to wind damage is related to the properties of trees and stands, including tree species, height and diameter, crown area, rooting depth and spacing (Coutts 1986) (Fig. 17.9). Risks of wind damage are also site-specific, following variability in soil conditions, topography and local climate. A mean regional wind speed > 14 m s⁻¹ with wind speeds up to 30 m s⁻¹ in gusts seems to be enough for wind damage. The risk is highest due to sudden changes in the exposure of trees not acclimatized to strong winds, as is the case in stands adjacent to recent clear-cut areas or in stands recently thinned intensively (Peltola et al. 1999a,b). In such cases, wind penetrates deeper into the canopy, with a subsequent increase in the wind load imposed on the trees. The probability of damage decreases with the time elapsed since thinning; i.e., the growth of remaining trees increases the strength of stem and roots to resist dynamic wind loading.

Mechanism of wind damage

Wind damage involves stem breakage or turning-down of trees due to wind force exceeding the resistance of the stem or root system. Wind may damage single trees and/or whole stands (several trees damaged at the same time). The susceptibility of trees and stands to wind damage is controlled by the properties of the wind (wind velocity, wind gustiness) and the structure of trees and forests, as defined by tree species, distribution of tree height and diameter, crown area, rooting depth and width, stand density and stocking, soil type and topography. The threshold wind speeds (critical wind speeds) required to break or uproot trees are affected by properties of the crown, stem and roots, which anchor tree to soil (Box 17.2).

Box 17.2 Mechanisms of wind damage

A tree may fall down or break if the total bending moment exceeds the support provided by the root-soil plate (Coutts 1986), or if the breaking stress exceeds the stem strength (Petty and Worrel 1981). Wind-induced force at height z [m] on the stem is (Fig. 17.10):

$$F_1(z) = 0.5 \times \rho \times C_d \times A(z) \times U(z)^2 \quad (17.6)$$

where $F_1(z)$ is wind force [N], $U(z)$ is wind speed [m s^{-1}], $A(z)$ is the projected area of the tree against wind [m^2], C_d is the drag coefficient (dimensionless) and ρ is air density [kg m^{-3}]. Wind profile along the crown is: $U(z) = (U_o/k) \times \log((z-d)/Z_o)$ where Z_o is the roughness length [m], d is the zero plane displacement [m], k is von Karman's constant (dimensionless) and U_o is the friction velocity [m s^{-1}].

Once any large bending of a tree occurs, there is an additional force at the height z: $F_2(z) = M(z) \times g$ where $F_2(z)$ is the force due to gravity [N], $M(z)$ is the green mass of stem and crown [kg] and g is the gravity constant [m s^{-2}]. The bending moment determined for each segment by wind drag and horizontal deflection of segment weight is: $T(z) = F_1(z) \times z + F_2(z) \times x(z)$ where $T(z)$ is the bending moment [Nm], z is the height along the stem [m], and x(z) is the horizontal displacement of the stem from the upright position [m].

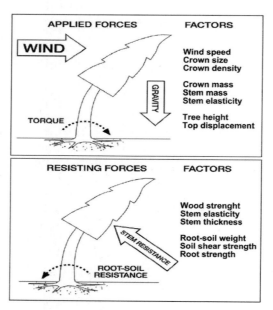

Fig. 17.10 Forces affecting a tree in wind (Peltola and Kellomäki 1993).

Critical wind speed damaging trees

Wind speed uprooting or breaking trees decreases along with tree size (Fig. 17.11); i.e., the resistance increases more slowly with tree size than does the bending moment. The crown to stem weight ratio affects the damaging wind speed. Tall and slender trees with a large crown are more susceptible than otherwise similar trees with smaller crowns. For example, a tree with a stem taper (ratio stem diameter to height) of 1:100 is broken at a wind speed of 28–26 m s^{-1} if the tree height of 12–20 m is assumed. Similarly, the critical wind speed for trees of the same height but with a taper of 1:70 is 43–41 m s^{-1}. The stem taper also affects the critical wind speed required to uproot a tree but the critical wind speed for uprooting is usually much lower than that for breaking regardless of stem taper. For example, a taper of 1:100 requires wind speeds of 15–12 m s^{-1} to overturn trees of 12–20 m in height, and for a taper of 1:70 the critical wind speeds are 25–21 m s^{-1}.

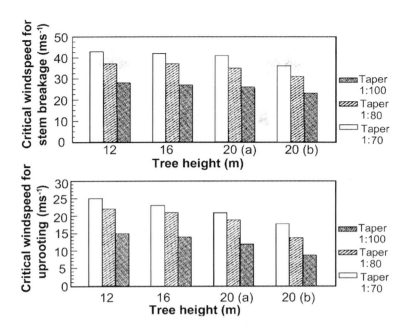

Fig. 17.11 Critical wind speeds for stem breakage (upper) and for uprooting (lower) of Scots pine as a function of tree height, stem taper and crown weight/stem weight ratio a = 0.3 and b = 0.5 (Peltola and Kellomäki 1993). Courtesy of the Finnish Society of Forest Science and the Finnish Forest Research Institute.

Impact of climate change on the risk of wind damage

So far, there is no evidence that the changing climate has increased the mean wind velocity over Finland (Gregow 2013). On the other hand, the warming climate will reduce the duration of soil frost, and thus, strong winds in autumn and spring will occur more frequently during periods with no soil frost anchoring trees. Based on model simulations, Peltola et al. (1999) found that under the warming climate the frequency of damaging winds during the period of non-frozen soil may increase from 55 to 80% in the south and from 40 to 50% in the north. A concurrent increase in soil moisture may further increase the risk of wind damage due to the reduction of anchorage (Kellomäki et al. 2010). Furthermore, climate warming is likely to enhance height growth, thus increasing the wind drag on the crown. At the same time, forest management effects on the properties of tree stands, like spacing, species composition, distribution of height, diameter and stocking, and thus future risks of wind-induced damage (Fig. 17.12).

Currently, the mean values of damaging wind speeds are lowest (i.e., the risk of damage is high) in southern Finland (Fig. 17.12), where mature Norway spruce forests are common. In northern Finland, damaging wind speeds are currently much higher (i.e., the risk is low), where the main part of forests is dominated by Scots pine at the intermediate developmental phase (Kellomäki et al. 2008). Under climatic warming, the differences in wind-induced risks between the south and north are likely to change, controlled by tree species composition and age class distribution. In the south, the more wind-resistant Scots pine and birches are likely become dominant at the expense of Norway spruce, thus reducing wind-induced damage. At the same time, the mean values of damaging wind speeds may substantially decrease in the north due to enhanced growth and maturing of trees. However, there may locally be large differences in the risk of wind-induced damage due to differences in the structure and properties of tree stands.

258 Managing Boreal Forests in the Context of Climate Change

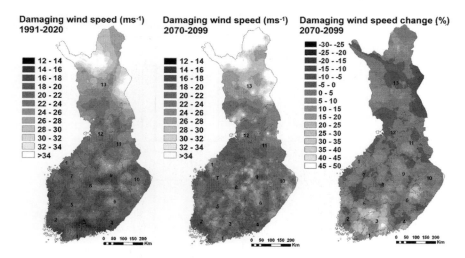

Fig. 17.12 Distribution of mean values of damaging wind speeds across Finland currently (left, 1991–2020) and under climate change (middle, 2070–2099), and the changes (right, %) due to climate change (Kellomäki et al. 2009). The changes are related to the changes induced by climate change in the structure of forests (e.g., tree species composition, mean height and diameter, stocking) when applying regular management based on current rules.

17.5 Risk of Fire Damage

Factors affecting fire risk and fire damage

Fire damage is any damage to trees, including the death of whole trees, due to fire. Fire risk refers to the occurrence or return of damaging fire. Return or the return period indicates the length of time period in years, in which a given site may be burnt again. In purely natural conditions, lightning is a common cause of forest fire, with varying return periods depending on site conditions. In northern Europe, the return period of natural fires is 30–300 years, the lower values representing xeric sites and higher values mesic sites (Päätalo 1998, Päätalo et al. 1999). Currently, most fires are caused by humans, but weather affects the ignition of fire and the moisture of fuel with the subsequent area and intensity of burning (Fig. 17.13).

Under boreal conditions, the area burnt annually is climate limited, and it is strongly dependent on the pattern and properties of the forest landscape, e.g., the mosaic structure of peatlands and lakes modify the landscape. This is also the case for Finland, where the percentage of forest land burnt annually is < 0.005% of the total area of forest land, with the mean size of the burnt area being < one hectare (Forest Fire Statistics 1993). The spreading rate of fire is related to the topography, but also wind velocity and the availability of fuel affect the spreading rate. All these factors have an effect on the intensity of burning and the height of flames, which damage the crown, foliage and stem cambium. Small trees with a low crown and thin stem bark (e.g., Norway spruce) die even in fires with low intensity and flame height, and tall trees with thick stem bark (e.g., Scots pine) may survive intensive fires with high flames.

Long- and short-term variability in precipitation, temperature and air humidity in the main growing season affect the moisture of fuel and thus the risk of fire and intensity of burning. One to two weeks without rain is needed to significantly increase the fire risk under current climate (van Wagner 1983), but even a few millimeters of rainfall are enough to saturate litter surface and temporarily lower the fire risk. Climate also determines the length of the fire season in the interaction with fuel accumulation and soil conditions. Wind also has a very strong effect on fire

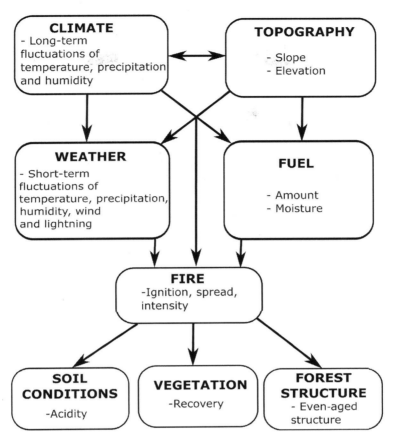

Fig. 17.13 Main factors affecting fire risk, consequent fire occurrence and the effects of fire on the forest ecosystem based on Päätalo (1998) and Päätalo et al. (1999). Courtesy of the Finnish Society of Forest Science and the Finnish Forest Research Institute.

damage in making vegetation more flammable, and increasing the spread of fire once ignited. In general, the rate of spread is doubled for each 4 m s^{-1} increase in wind speed (van Wagner 1983).

In Finland, the mean annual precipitation varies in the range of 300–700 mm in such a way that precipitation is fairly evenly distributed over the whole year. Currently, low precipitation in growing season is enough to maintain the high moisture of litter and humus due to low evaporation, and thus reduce the risk of fire in general. Under climate change, longer summers and the consequent increase of dry spells may increase fire risk. This claim is supported by findings suggesting that fires are most likely when the summer temperature is above the long-term average (Zackrisson 1977). This seems to be especially true if the July temperature is above the long term mean. On the other hand, changes in the amount and seasonal distribution of precipitation may change the risk of damaging fires. For example, early snow melt combined with only a small increase in precipitation in the spring and early summer are likely to increase fire risk even in humid boreal conditions. A temperature increase of 3–5°C in June–August may increase the potential fire area by 15 to 50 times in western Europe, if there is no increase in precipitation (Suffling 1992).

> **Box 17.3 Calculations of forest fire potential under climate change over Finland**
>
> Based on Venäläinen and Heikinheimo (2003), water loss from the surface layer for calculating fire potential is estimated using the drying/wetting curve; i.e., the drier the soil the less water is lost from the surface layer in potential evaporation. The curve was also used to calculate the wetting effect of any rain event. Thereafter, the volumetric moisture content of soil surface layer is calculated by adding the water in any rain event to the previous value of water content and by subtracting water lost in evaporation. Finally, the moisture content is scaled to forest fire index values (F_w, 1–6): $F_W = 30.71 \times W^2 + 30.88 \times W - 8.76$, where W is the volumetric water content [$m^3\ m^{-3}$] in the surface soil. If the value of the fire index exceeds 4, a forest fire alert is in force (Venäläinen and Heikinheimo 2003). The forest fire index is also used to estimate the likely frequency of forest fires per year based, using the surface moisture for the period 1961–2000 from several monitoring sites throughout Finland. The data is converted to annual forest fire potential, with the annual number of days when the fire index value is ≥ 4.

Fire risk under climate change

Kilpeläinen et al. (2010b) studied forest fire potential and frequency under climate change based on simulations over the whole territory of Finland (60°–70° N). Forest fire potential involves the number of days per year when forest fire alerts are declared by firefighting authorities, and fire frequency is the probable number of fire episodes per year. Fire potential is indicated by the forest fire index based on the weather observations used to calculate the moisture of surface soil (Venäläinen and Heikinheimo 2003). Moisture in surface soil is given as the volumetric water content [$m^3\ m^{-3}$] of the soil surface layer down to 60 mm. The value is calculated based on daily precipitation and potential daily evaporation, using the Penman-Monteith equation (Box 17.3).

Kilpeläinen et al. (2010b) found that forest fire potential in the period 1961–1990 was highest in the coastal and southern areas of Finland, varying from 60 to 100 days per year (Fig. 17.14). In the central and eastern parts, the fire potential varied from 35 to 60 days per year and in the north from 15 to 20 days per year. Under the warming climate, the fire potential in the first period (1990–2020) increased in southern and central Finland, whereas in northern and eastern Finland the fire potential may decrease by 10%. In the second period (2021–2050), the fire potential may increase over the whole country, albeit less than 20%. In the third period (2070–2099), the increase in forest fire potential is likely highest in the south.

In the period 1970–1997, the mean annual number of forest fires over the whole country was 626, and calculated values were 744 fires (Kilpeläinen et al. 2010b). Under a warming climate, the number of fires in the first periods (1990–2020 and 2021–2050) increased up to 776, and in the third period (2070–2099) up to 894. The relative increase in the number of forest fires by the end of the century was 20% compared to that under the current climate. The largest increase in the number of forest fires per 1,000 km^2 was in the coastal areas and in the southernmost part of the country. In the period 2070–2099, the number of fires in these areas was 6–9 fires per 1,000 km^2, implying an increase of 24–29% compared to current values. In these areas, the fire frequency was closely related to the enhanced evaporation and only a small increase in annual precipitation. An increase in the annual precipitation more than 20% is needed to compensate for enhanced evaporation, and reduce the increased fire frequency due to the increasing temperature.

Disturbances and Damage Affecting Boreal Forests under Climate Change 261

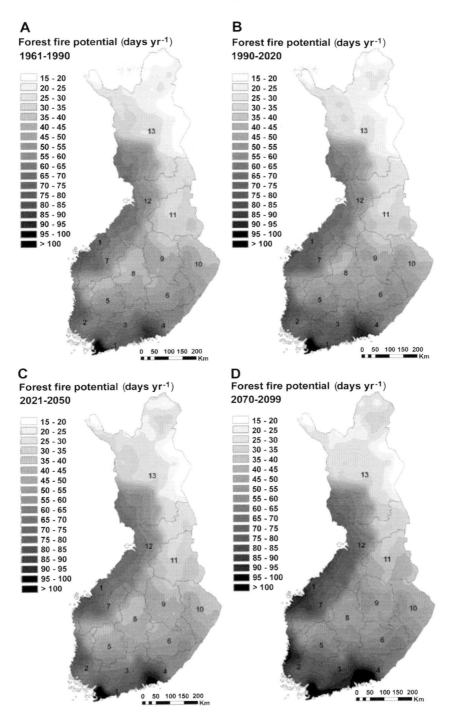

Fig. 17.14 Mean annual forest fire potential in Finland for the current climate (A), and for the near-term (1990–2020), mid-term (2021–2050) and long-term (2070–2099) future (B–D) (Kilpeläinen et al. 2010b). Permission of Springer. The current climate is represented by the period 1961–1990, and the warming climate was based on the SRES A2 emission scenario. In this case, the current atmospheric CO_2 of 350 ppm increased up to 840 ppm by 2100. The mean annual temperatures increased 4°C in summer and 6°C in winter. Precipitation increased more than 20% during winter by 2100, but remained nearly unchanged in summer (Ruosteenoja et al. 2005). Numbers on maps refer to forest centers.

17.6 Risk of Outbreaks of Existing and Invasive Pests under Climate Change

Pest/host interaction under climate change

Biotic damage may be divided into those related to the currently existing insects, pathogens and herbivores and to those related to alien ones, which may invade under a warming climate. Outbreaks of damaging insects and pathogens are likely to increase in frequency and intensity. This is particularly the case in the margins of the host tree species, where the northward expansion of several pests is likely to most change the current pest/host interaction (Harrington et al. 2001, Battisti 2004). The distribution of insects and pathogens is ultimately determined by climatic factors such as day length, temperature, precipitation and length of growing season. Changes in winter temperatures are especially important in boreal conditions, since higher minimum temperatures are shifting pest distribution further north (Niemelä et al. 2001).

Profound changes in the success/loss of many insects and pathogens are dependent on (Evans et al. 2002, Battisti 2004): (i) the survival and reproduction of species (life cycle); (ii) the natural enemies (parasites, predators) of species; (iii) the availability and properties (e.g., nutrients, fiber content, secondary compounds) of biomass in host trees; (iv) the vigor and defense capabilities of host trees; and (v) the phenological synchrony of damaging species and host trees. On the other hand, the short life cycle (month, years) of damaging insects and fungi facilitates a rapid adaptation to warming climate, while the adaptation of tree populations is extended over decades/centuries. This may increase the vulnerability of trees, and make it difficult to predict damages under climate warming. Insects, especially those occurring in periodic outbreaks, have a great potential for genetic adaptation to new environments, and to migrate. At the same time, the effects of defoliators, wood borers and bark beetles could become more damaging due to a lengthening of the growing season and expanding populations (Virtanen et al. 1996, Niemelä et al. 2001, Bale et al. 2002).

Pests likely to increase in a damaging capacity

Table 17.2 lists some existing and alien insects and fungi, which are likely damaging trees under changing climate in Europe. The list is based on expert assessments rather than systematic experiments or monitoring of how damaging insects and fungi are responding to the elevation in temperature and CO_2 and changes in precipitation (Perry 2000).

Outbreaks of many damaging insects and fungi are closely related to the occurrence of dying and dead trees, which may lead to detrimental attacks on the remaining trees. Many will have greater success, such as the European spruce bark beetle (*Ips typographus*) (Fischlin et al. 2009). The flight period of the bark beetle is initiated when the daily mean temperature exceeds 18°C in spring. Two generations of bark beetles may be born in warmer summers, thus increasing the risks of major damage to Norway spruce. Currently, the risk is highest in southern boreal forests (< 62° N) but in the future risk may be high even in the southern parts of northern boreal forests (< 64° N). Similarly, the populations of European pine sawfly (*Neodiprion sertifer*), common pine sawfly (*Diprion pini*), and pine beauty (*Panolis flammea*) may grow if the summer temperature increased by 2–3°C (Virtanen et al. 1996). On the other hand, cold winters have controlled outbreaks of many existing pests in the boreal zone. For example, eggs of European sawfly cannot survive in temperatures below –36°C (Niemelä and Veteli 2006) (Fig. 17.15). Under climate warming, such low winter temperatures are likely reducing, thus increasing the risk of saw fly outbreaks even in northern boreal forests (Virtanen et al. 1996, Niemelä et al. 2001).

Climatic warming combined with increasing precipitation is likely to increase also the occurrence of several existing diseases (fungi, bacteria, viruses and mycoplasma-like organisms) (Fischlin

Table 17.2 Selected existing and alien insects and fungi, which may potentially damage trees and forests under climate change in Finland (Perry 2000, Fischlin et al. 2009).

Existing insects likely to increase damage	Alien insects likely to cause damage
• *Aradus cinnamomeus* • *Tomicus* spp. • *Ips typographus* • *Neodiprion sertifer* • *Diprion pini* • *Bupalus piniarius* • *Panolis flammea* • *Operophtera brumata*	• *Scolytus scolotys, Scolotys multistriatus* • *Totrix viridana* • *Lymantria dispar* • *Lymantria monacha* • *Bursaphalencus xylophilus*
Existing fungi likely to increase damage	Alien fungi likely to cause damage
• *Heterobasidion annosum* • *Armillaria* spp. • *Lachnellula pini* • *Lophodermella sulcigena* • *Phacidium infestans* • *Lophodermium seditiosum* • *Coleosporium tussilaginis* • *Ascocalyx abietina*	• *Ceratocystis ulmi*

et al. 2009, Müller et al. 2012) (Table 17.2). This is the case for root rot (*Heterobasidion annosum*) common in the southern and middle boreal forests. Root rot is likely to expand further to the north, where climate warming reduces soil frost and increases the soil temperature closer to the optimum (22–28°C) of root rot. Müller et al. (2012) estimate that climatic warming (SRES A2 scenario, with the 5°C increase by 2100) may increase the activity of root rot by 50% in southern and 90% in northern boreal forests. Furthermore, root rot is likely acclimating rapidly to changing environment, which increases the risk of root rot throughout the boreal zone. Higher winter temperatures and high humidity may increase further epidemics of damaging fungi such as Scleroderris canker (*Gremmenniella abietina*) and pine needle cast (*Lophodermella sulcigena*). However, high summer temperatures, along with drought, may dampen epidemics of some fungi, but they may flourish in cool rainy summers even under the changing climate (Straw 1995, Perry 2000).

Invasive species damaging trees

In general, European forests seem to have been buffered but not excluded against damage caused by new species invading from outside Europe (Mattson et al. 2007). A good example of new organisms with highly damaging potential is the pine wood nematode (*Bursaphalencus xylophilus*) originating from North America. This nematode is transported in fresh timber, but its success is closely related to temperature conditions. Until now, low summer temperatures and short growing seasons are effectively limiting the success of this species outside northern Europe, even though it has frequently been found in imported timber. Climate change also seems to remove limitations for the ambrosia beetle (*Gnathotrichus materiarius*), originating in North America. It was introduced into Europe in the 1930s, and since then, it has become established in central Europe. In the late 1990s, the ambrosia beetle was first recorded in the southern parts of northern Europe in warm summers. Until now, low summer temperatures and short growing seasons have limited the spread of this species towards the north, but climate warming may change the situation (Fischlin et al. 2009).

Higher temperatures may also expand the habitat of the nun moth (*Lymantria monacha*) far above 60° N in northern Europe. Currently, this insect mainly causes damage to Norway spruce in central and southern Europe (Bejer 1988), where the mean temperature in July exceeds 16°C and the mean temperature in September exceeds 10.5°C. High summer temperatures and associated drought episodes allow the nun moth to expand further north, as noted by Vanhanen et al. (2007).

They used the CLIMEX model to estimate the potential changes in its distribution in Europe (Fig. 17.16). The CLIMEX model simulates the values of the Ecoclimatic Index (EI) on a regional basis, for assessing the suitability of sites for a given species. Under the current climate, the nun moth is present in southern boreal forests, but an annual mean temperature 5.8°C higher than the current one may result in a northward shift of the species potentially beyond the Arctic Circle (N 66°).

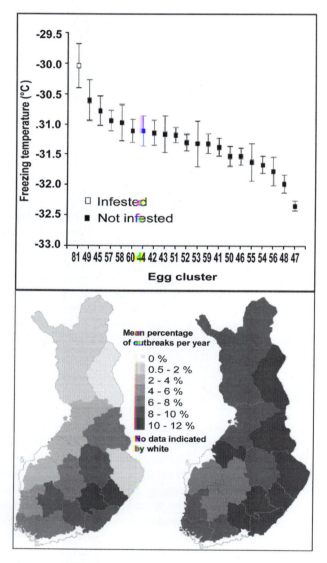

Fig. 17.15 Above: Distribution of temperatures, below which eggs of European saw fly (*Neodiprion sertifer*) cannot survive (freezing temperature). Each point indicates the egg cluster of one female, with eggs infested (open squares) or not infested (full squares) by parasites (Veteli et al. 2005, Niemelä and Veteli 2006). Courtesy of Wiley. Below: Mean percentage of outbreaks of pine sawfly per year in Finland in 1961–1990 (left side of panel) and expected mean percentages of outbreaks in 2050 (right side of panel) (Virtanen et al. 1996). Courtesy of the Finnish Society of Forest Science and the Finnish Forest Research Institute.

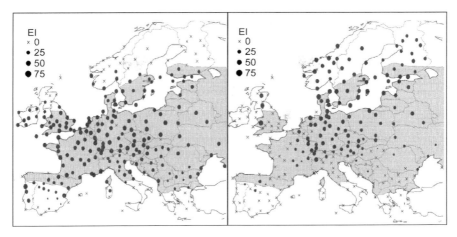

Fig. 17.16 Distribution of the nun moth (*Lymantria monacha*). Left: Under the current climate, and Right: Under temperature increase by 5.8°C (Vanhanen et al. 2007). The shaded area indicates the current distribution and dots the distribution simulated with the CLIMEX model as indicated by the ecoclimatic index (EI) (Box 17.4). Courtesy of the Finnish Society of Forest Science and the Finnish Forest Research Institute.

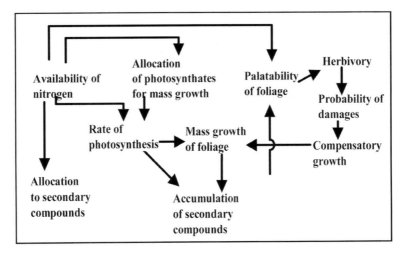

Fig. 17.17 Schematic presentation of how the availability of nitrogen may affect growth, secondary compounds, and herbivory, with compensatory growth of foliage. Compensatory growth is the growth increase initiated by herbivory.

Secondary compounds and herbivory

Temperature is the main climatic factor controlling the development, survival, range and abundance of damaging insects. This makes the risk of insect damage directly sensitive to climate warming and to variability in temperature patterns (i.e., seasonal, annual). The effects of increasing CO_2 are indirect rather than direct, mediated through the changes in the properties of host plants. For example, elevating CO_2 may increase the content of defensive chemicals in plant tissues and reduce the nitrogen content in the foliage of coniferous and deciduous species (Veteli et al. 2002, 2007, Battisti 2004). An increase in the Carbon/Nitrogen (C/N) balance of tissues reduces the quality of food for many defoliating insects. Furthermore, the increase of fibers in foliage material reduces the palatability and nutritional value of forage. Consequently, larvae of damaging insects may grow and develop more slowly with a higher risk of being attacked by natural enemies (Fig. 17.17).

> **Box 17.4 Ecoclimatic index for assessing success of invasive species**
>
> Ecoclimatic index (EI, 0–100) may be used to indicate the potential distribution and the relative abundance of a given species (Suthers and Maywald 1985):
>
> $$EI = \left[\frac{100}{52}\sum_{w=1}^{52}(TI_w \times MI_w \times DI_w)\right] \times \left[\left(1-\frac{CS}{100}\right)\times\left(1-\frac{HS}{100}\right)\times\left(1-\frac{DS}{100}\right)\times\left(1-\frac{WS}{100}\right)\right] \quad (17.11)$$
>
> where TI_w is the weekly temperature index, MI_w is the weekly moisture index and DI_w is the weekly diapause index. CS is the Cold Stress, HS is the Heat Stress, DS is the Dryness Stress and WS the Wetness Stress, calculated on an annual basis to indicate the climatic requirements of the species. The values of EI range from 0 to 100 indicating climatic suitability of the location for a species. At an EI value of 0, a species cannot establish a viable population at the location, whereas values > 20 indicate a favorable climate for the species (Suthers and Maywald 2005). Vanhanen et al. (2007) used a mean annual temperature increase of 5.8°C from the current in calculations, based on SRES A1 and B2 emissions scenarios.

Many secondary compounds (e.g., terpenoids, phenolics, tannins, alkaloids) make plant material bitter or toxic, thus providing a chemical defense against phytophagous insects and mammals. Regarding selected boreal deciduous tree species, Veteli et al. (2007) found that elevated CO_2 (750 ppm) generally stimulated increased carbon partitioning to various phenolic compounds, and an increase in temperature (2°C) had an opposite effect. The combined effects of both elevated CO_2 and temperature were additive, thus cancelling the separate effects. Woody plants can also respond to herbivory by releasing a wide variety of volatile compounds from resin ducts. On the other hand, volatile compounds emitted from insect-damaged plant organs may attract parasitic and predatory insects that are natural enemies of feeding insects. This may further induce defense responses in neighboring plants (Paré and Tumlinson 1999). Consequently, the larvae of damaging insects may grow and develop slowly, with a higher risk of being attacked by natural enemies. This mechanism may alter the dynamics of the whole food web system, but the long-term effects of these changes on the dynamics of forest ecosystems are still poorly known.

Damage due to mammal herbivores

Currently, mammal herbivores such as hare (*Lepus timidus*), moose (*Alces alces*) and white-tailed deer (*Odocoileus virginianus*) are among the most important species grazing in young forests in northern Europe, including Finland (Kuokkanen et al. 2004). The direct impact of climate on populations, and the current and future distribution of mammal herbivores will be affected by high temperature in the summer and the depth and seasonal distribution of snow. Moose, for example, become stressed by temperatures above –5°C in winter and above 14°C in summer. Consequently, winter weather is likely to impact on the migration, use of habitats and foraging patterns of moose. Warming with the expected changes in tree species composition, with the accompanying changes in the seasonal patterns of the quality and quantity of food, will impact on the distribution and population densities of mammal herbivores.

The elevating CO_2 may further affect the success of large herbivores, as claimed by Mattson et al. (2004). They found that the elevation of CO_2 may decrease the palatability of birch (*Betula pendula* Roth) with respect to the hare (*Lepus timidus*). They grew one-year-old seedlings in closed-top chambers for one summer and autumn in pots containing an unfertilized commercial peat with three different soil Nitrogen (N) levels (low = 0 kg N ha^{-1}, medium = 150 kg N ha^{-1}, high = 500 kg N ha^{-1}) at two temperatures (T) (ambient and ambient + 3°C) under atmospheric CO_2 at either 362

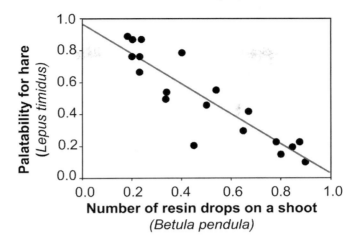

Fig. 17.18 Palatability of birch (*Betula pendula* Roth) shoots for hare as a function of number of resin drops on a shoot (Niemelä and Veteli 2006). Courtesy of the Finnish Forest Research Institute.

or 700 ppm. Phytomass was used in feeding trials with caged hares. Mattson et al. (2004) found that elevated CO_2 reduced feeding by 48%, and temperature elevation had no effect on feeding. The palatability of birch shoots to hare seemed to be inversely related to the density of resin drops on shoots (Fig. 17.18). The density of resin drops was especially enhanced under elevated CO_2, thus reducing the forage value of birch shoots for hare (Niemelä and Veteli 2006).

17.7 Concluding Remarks

Climate warming is likely to increase the diversity of pests (insects and fungi) and to increase their feeding and predation. Changes in temperature and precipitation control further the occurrence, timing, frequency, duration, extent and intensity of abiotic disturbances (e.g., wild fire, drought, strong wind) damaging trees thus enhancing biotic damages in forest (Dale et al. 2001). This is evident in British Columbia, where more than 14 million ha of lodgepole pines (*Pinus contorta* Douglas) have been destroyed since the early 1990s. In this case, outbreaks of mountain pine beetle (*Dendroctonus ponderosae*) and *Dothistroma* needle blight (*Dothistroma septosporum*) were related to climate warming, with disturbances in host/pest interactions (Woods et al. 2010).

The multiple effects of climate change on the growth of trees and on herbivory/host interactions may alter the successional path of the ecosystem from that expected based only on the physiological responses of trees. In the boreal forests, mammals, especially moose, are among the major agents destroying Scots pines and birches in seedling phases in winter grazing sites (Niemelä et al. 2001). The effects of warming winters are the opposite for voles, which may locally miss the sheltering snow cover. Consequently, the typical cycling of vole populations and subsequent damage to trees may be damped down, and the density of vole populations is likely to reduce. The herbivore-mediated responses to climatic warming will probably vary regionally, but climatically induced changes in phytophagous animal populations may make it important to modify the management of boreal forests in producing timber, biomass and other ecosystem services in sustainable way (Veteli et al. 2007).

References

Bale, J. S., J. Gregory, J. G. Masters, I. D. Hodkinson, C. Awmack, T. M. Bezemer et al. 2002. Herbivory in global climate change research: direct effects of rising temperature on insect herbivores. Global Change Biology 8: 1–16.

Battisti, A. 2004. Forests and climate change–lessons from insects. Forest 1: 17–24.

Bejer, A. 1988. The nun moth in European spruce forests. pp. 211–231. *In*: A. Berryman (ed.). Dynamics of Forest Insect Populations: Patterns, Causes, Implications. Plenum Publishing Corporation, New York, USA.

Coutts, M. P. 1986. Components of tree stability in Sitka spruce on peaty soil. Forestry 59(2): 173–197.

Dale, V. H., L. A. Joyce, S. McNulty, R. P. Neilson, M. P. Ayres, M. D. Flannigan et al. 2001. Climate change and forest disturbances. BioScience 51(9): 723–734.

Evans, H., N. Straw and A. Watt. 2002. Climate change implication for insect pests. pp. 99–118. *In*: M. Broadmedow (ed.). Climate Change and UK Forests. Forestry Commission, Edinburgh, UK.

Fischlin, A., M. Ayres, D. Karnosky, S. Kellomäki, S. Louman, C. Ong et al. 2009. Future environmental impacts and vulnerabilities. *In*: R. Seppälä, A. Buck and P. Katila (eds.). Adaptation of Forests and People to Climate Change–A Global Assessment Report. IUFRO World Series Vol. 22.

Fitzgerald, J. and M. Lindner (eds.). 2013. Adapting to Climate Change in European Forests—Results of the MOTIVE Project. Pensoft Publishers, Sofia, Bulgaria.

Forest Fire Statistics 1991–1992. 1993. ECE/TIM/70.FAO. UNECE. United Nations, New York, USA.

Gregow, H., U. Puranen, H. Peltola, S. Kellomäki and D. Schulz. 2008. Temporal and spatial occurrence of strong winds and large snow load amounts in Finland during 1961–2000. Silva Fennica 42(4): 515–534.

Gregow, H. 2013. Impact of strong winds, heavy show loads and soil frost conditions on the risk to forests in northern Europe. Ph.D. Thesis, University of Eastern Finland, Joensuu, Finland.

Häkkinen, R., T. Linkosalo and P. Hari. 1998. Effects of dormancy and environmental factors on timing of bud burst in *Betula pendula*. Tree Physiology 18: 707–712.

Harrington, R., R. A. Fleming and P. I. Woiwod. 2001. Climate change impacts on insect management and conservation in temperate regions: can they be predicted? Agricultural and Forestry Entomology 3(4): 233–240.

Kellomäki, S., K. -Y. Wang and M. Lemettinen. 2000. Controlled environment chambers for investigating tree response to elevated CO_2 and temperature under boreal conditions. Photosynthetica 38: 69–81.

Kellomäki, S. and K. -Y. Wang. 2001. Growth and resource use of birch seedlings under elevated carbon dioxide and temperature. Annals of Botany 87: 669–682.

Kellomäki, S., H. Peltola, T. Nuutinen, K. T. Korhonen and H. Strandman. 2008. Sensitivity of managed boreal forests in Finland to climate change, with implications for adaptive management. Philosophical Transactions of the Royal Society B363: 2341–2351.

Kellomäki, S., V. Koski, P. Niemelä, H. Peltola and P. Pulkkinen. 2009. Management of forest ecosystems. pp. 252–373. *In*: S. Kellomäki (ed.). Forest Resources and Sustainable Management. Gummerus Oy, Jyväskylä, Finland. Second Edition.

Kellomäki, S., M. Maajärvi, H. Strandman, A. Kilpeläinen and H. Peltola. 2010. Model computations on the climate change effects on snow cover, soil moisture and soil frost in the boreal conditions over Finland. Silva Fennica 44(2): 213–233.

Kilpeläinen, A., H. Peltola, I. Rouvinen and S. Kellomäki. 2006. Dynamics of daily height growth in Scots pine trees at elevated temperature and CO_2. Trees 20: 16–27.

Kilpeläinen, A., H. Gregow, H. Strandman, S. Kellomäki, A. Venäläinen and H. Peltola. 2010a. Impacts of climate change on the risk of snow-induced forest damage in Finland. Climatic Change 99(1-2): 193–209.

Kilpeläinen, A., S. Kellomäki, H. Strandman and A. Venäläinen. 2010b. Climate change impacts on forest fire potential in boreal conditions in Finland. Climatic Change 103: 383–398.

Kramer, K. 1995. Phenotypic plasticity of the phenology of seven European tree species in relation to climatic warming. Plant, Cell and Environment 18: 93–104.

Kuokkanen, K., P. Niemelä, J. Matala, R. Julkunen-Tiitto, J. Heinonen, M. Rousi et al. 2004. The effects of elevated CO_2 and temperature on the resistance of winter-dormant birch seedlings (*Betula pendula*) to hares and voles. Global Change Biology 10: 1504–1512.

Laiho, O. 1987. Metsiköiden alttius tuulituhoille Etelä-Suomessa. Summary: susceptibility of forest stands to wind throw in Southern Finland. Folia Forestalia 706: 1–24.

Leinonen, I., T. Repo, H. Hänninen and K. E. Burr. 1995. A second-order dynamic model for the frost hardiness of trees. Annals of Botany 76: 89–95.

Leinonen, I. 1997. Frost hardiness and annual development of forest trees under changing climate. Ph. D. Thesis, University of Eastern Finland, Joensuu, Finland.

Linkosalo, T., R. Häkkinen, J. Terhivuo, H. Tuomenvirta and P. Hari. 2009. The time series of flowering and leaf bud burst of boreal trees (1846–2005) support the direct temperature observations of climatic warming. Agricultural and Forestry Meteorology 149: 453–461.

Mattson, W. J., K. Kuokkanen, P. Niemelä, R. Julkunen-Tiitto, S. Kellomäki and J. Tahvanainen. 2004. Elevated CO_2 alters birch resistance to Lagomorpha herbivores. Global Change Biology 10: 1402–1413.

Mattson, W. J., H. Vanhanen, T. O. Veteli, S. Sivonen and P. Niemelä. 2007. Few immigrant phytophagous insects on woody plants in Europe: legacy of the European crucible? Biological Invasion 9(8): 957–974.

Müller, M. M., T. Piri and J. Hantula. 2012. Ilmaston lämpeneminen haastaa nykyistä tehokkaampaan juurikäävän torjuntaan. Metsätieteellinen Aikakauskirja 4/2012: 312–315.
Myking, T. and O. M. Heide. 1995. Dormancy release and chilling requirements of buds of latitudinal ecotypes of *Betula pendula* and *B. pubescens*. Tree Physiology 15: 697–704.
Niemelä, P., F. S. Chapin III, K. Danell and J. P. Bryant. 2001. Herbivory-mediated responses of selected boreal forests to climatic change. Climatic Change 48: 427–440.
Niemelä, P. and T. Veteli. 2006. Ilmastonmuutoksen vaikutukset metsätuhoihin ja -tauteihin boreaalisessa vyöhykkeessä. *In*: J. Riikonen and E. Vapaavuori (eds.). Ilmasto muuttuu - mukautuvatko metsät. Metsäntutkimuslaitoksen tiedonantoja 944: 92–98.
Nykänen, M. -L., H. Peltola, C. P. Quine, S. Kellomäki and M. Broadgate. 1997. Factors affecting snow damage of trees with particular reference to European conditions. Silva Fennica 31: 193–213.
Päätalo, M. -L. 1998. Factors influencing occurrence and impacts of fires in northern European forests. Silva Fennica 32(2): 185–202.
Päätalo, M. -L., H. Peltola and S. Kellomäki. 1999. Modelling the risk of snow damage to forests under short-term snow loading. Forest Ecology and management 116: 51–70.
Paré, P. M. and J. H. Tumlinson. 1999. Plant volatiles as a defense against insect herbivores. Plant Physiology 121: 325–331.
Peltola, H. and S. Kellomäki. 1993. A mechanistic model for calculating wind throw and stem breakage of Scots pine at stand edge. Silva Fennica 27(2): 99–111.
Peltola, H., M. -L. Nykänen and S. Kellomäki. 1997. Model computations on the critical combination of snow loading and windspeed for snow damage of Scots pine, Norway spruce and birch sp. at stand edge. Forest Ecology and Management 95: 229–241.
Peltola, H., S. Kellomäki and H. Väisänen. 1999a. Model computations on the impacts of climatic change on soil frost with implications for wind throw risk of trees. Climatic Change 41: 17–36.
Peltola, H., S. Kellomäki, H. Väisänen and V. -P. Ikonen. 1999b. A mechanistic model for assessing the risk of wind and snow damage to single trees and stands of Scots pine, Norway spruce and birch sp. Canadian Journal of Forest Research 29: 647–661.
Perry, M. L. (ed.). 2000. Assessment of potential effects and adaptation for climate change in Europe: The European ACACIA Project. Report of concerted action of the environment programme of Research Directorate General of the Commission of the European Communities 1465–458X 1465–458X, Jackson Environment Institute, University of East Anglia, Norwich, UK.
Petty, J. A. and R. Worrell. 1981. Stability of coniferous tree steams in relation to damage by snow. Forestry 54(2): 115–128.
Pudas, E., M. Leppälä, A. Tolvanen, J. Poikolainen, A. Venäläinen and E. Kubin. 2008a. Trends in phenology of *Betula pubescens* across the boreal zone in Finland. International Journal of Biometeorology 52: 251–259.
Pudas, E., A. Tolvanen, J. Poikolainen, T. Sukuvaara and E. Kubin. 2008b. Timing of plant phenophases in Finnish Lapland 1997–2006. Boreal Environment Research 13: 31–43.
Repo, T., H. Hänninen and S. Kellomäki. 1996. The effect of long-term elevation of air temperature and CO_2 on frost hardiness of Scots pine. Plant, Cell and Environment 19: 209–216.
Rousi, M. and J. Heinonen. 2007. Temperature sum accumulation effects on within-population variation and long-term trends in date if bud burst of European white birch (*Betula pendula*). Tree Physiology 27: 1019–1025.
Ruosteenoja, K., K. Jylhä and H. Tuomenvirta. 2005. Climate scenarios for FINADAPT studies of climate change adaptation. Finnish Environment Institute, FinAdapt Working Paper 15: 1–15.
Saxe, H., M. G. R. Cannell, Ø. Johnsen, M. G. Ryan and G. Vourlitis. 2000. Tree and forest functioning in response to global warming. Tansley review 123. New Phytologist 149: 369–400.
Slaney, M., G. Wallin, J. Medhurst and S. Linder. 2007. Impact of elevated carbon dioxide concentration and temperature on bud burst and shoot growth of boreal Norway spruce. Tree Physiology 27: 301–312.
Solantie, R. 1994. Effect of weather and climatological background on snow damage of forest in southern Finland. Silva Fennica 28(3): 203–211.
Straw, N. A. 1995. Climate change and the impact of green spruce aphid, *Elatobium abictinum* (Walker), in the UK. Scottish Forestry 49: 134–145.
Suffling, R. 1992. Climate change and boreal forest fires in Fennoscandia and Central Canada. Greenhouse impact on cold-climate ecosystems and landscapes. Catena Supplement 22: 111–132.
Suthers, R. W. and G. F. Maywald. 1985. A computerised system for matching climates in ecology. Agriculture, Ecosystems & Environment 13: 281–299.
Valinger, E. and L. Lundqvist. 1994. Reducing wind and snow induced damage in forestry. Department of Silviculture. Swedish University of Agricultural Sciences, Umeå. Reports 37: 1–11.
Vanhanen, H., T. Veteli, S. Päivinen, S. Kellomäki and P. Niemelä. 2007. Climate change and range shifts in two insect defoliators: gypsy moth and nun month–a model study. Silva Fennica 41(4): 621–638.
van Wagner, C. E. 1983. Fire behavior in northern conifer forests and shrublands. pp. 65–80. *In*: R. W. Wein and D. A. MacLean (eds.). Role of Northern Circumpolar Ecosystems. John Wiley and Sons, New York, USA.
Venäläinen, A. and M. Heikinheimo. 2003. The Finnish forest fire index calculation system. pp. 645–648. *In*: J. Zschau and A. Kuppers (eds.). Early Warning System for Natural Disaster Reduction. Springer Verlag, Berlin, Germany.
Veteli, T. O., K. Kuokkanen, R. Julkunen-Tiitto, H. Roininen and J. Tahvanainen. 2002. Effects of elevated CO_2 and temperature on plant growth and herbivore defensive chemistry. Global Change Biology 8: 1240–1252.

Veteli, T. O., A. Lahtinen, T. Repo, P. Niemelä and M. Varama. 2005. Geographic variation in winter freezing susceptibility in the eggs of European pine sawfly (*Neodiprion sertifer*). Agricultural and Forest Entomology 7: 115–120.

Veteli, T., W. J. Mattson, P. Niemelä, R. Julkunen-Tiitto, S. Kellomäki, K. Kuokkanen et al. 2007. Do elevated temperature and CO_2 generally have counteracting effects on phenolic phytochemistry of boreal trees? Journal of Chemical Ecology 33(2): 287–296.

Virtanen, T., S. Neuvonen, A. Nikula, M. Varama and P. Niemelä. 1996. Climate change and the risks of *Neodiprion sertifer* outbreaks on Scots pine. Silva Fennica 30(2-3): 169–177.

Woods, A. J., D. Heppner, H. H. Kope, J. Burleigh and L. Maclauchlan. 2010. Forest health and climate change: A British Columbia perspective. The Forestry Chronicle 86(4): 412–422.

Zackrisson, O. 1977. Influence of forest fire on the north Swedish boreal forest. Oikos 29(1): 22–32.

PART VI

Management of Boreal Forests under Climate Change for the Adaptation and Mitigation of Climate Change

18

Management of Forests for Adaptation to Climate Change

ABSTRACT

Climate change effects can be direct, such as the response of growth to changes in temperature, precipitation and atmospheric CO_2, or indirect, such as losses due to abiotic and biotic damage related to changes in climate. The responses of forest ecosystems are specific to the properties of sites and the tree populations occupying sites. Management needs to be based on the careful selection of tree species and other operations in order to use the benefits and to avoid the problems, which future changes in climate may provide.

Keywords: adaptive management, adaptive measures in management, boreal forest climate change

18.1 Management to Meet Risk and to use Opportunities under Climate Change

Need for adaptive management

Climate warming is likely to modify the functioning and structure of forest ecosystems throughout the world. For example, Allen et al. (2010) and Zhao and Running (2010) claim that trees in some forest ecosystems are currently vulnerable to mortality in response to climate warming and to longer and more severe drought episodes. Based on global meta-analyses, Parmesan and Yohe (2003) estimated that climate warming has moved the ranges of different species northwards by 6.1 km per decade and moved spring 2.3 days earlier per decade. Choat et al. (2012) demonstrated that 70% of 226 forest species across the globe, even in wet forests, may face long-term reductions in productivity and survival. Based on permanent forest sample plots, Peng et al. (2011) showed that the mortality of trees in the Canadian boreal forests increased 4.7% in 1963–2008, as a result of increasing drought episodes (Wang et al. 2014). Similarly, water shortages may occasionally reduce tree growth and increase mortality in northern and eastern Europe (Mäkinen et al. 2000, 2001, Vygodskaya et al. 2002) under higher summer temperatures and longer droughts (Ma et al. 2012, Ge et al. 2014).

Adaptation in forestry

In forestry, damage refers to the difference between the expected and realized production of different ecosystem goods and services intended in management. In boreal conditions, the elevation of CO_2 and temperature, and changes in precipitation are likely to divert the forest growth to new successional paths. According to Johnston (2010), climate change causes changes in biotic/abiotic interactions, and has impacts on genetic diversity and selective pressure, species composition and ecosystem processes, and their annual cycle (phenology) (Fig. 18.1). The vulnerability of the ecosystem is the key issue in outlining the necessary management to increase adaptation to climate change. Vulnerability refers to the extent to which a natural or social system producing goods and services is susceptible to damage under climate change, including climate variability and extremes (Spittlehouse and Stewart 2003, Stankey et al. 2005).

In forestry or forest production system, vulnerability is dependent on exposure, sensitivity, resilience and adaptive capacity affecting potential and realized impacts on forest and necessary infrastructure (Fig. 18.1). Sensitivity refers to the degree to which forest-based production system is affected by specific changes in climate, including beneficial and harmful effects. On the other hand, resilience is the opposite of vulnerability; i.e., a resilient production system is not sensitive to climate variability and change, and it has the capacity to adapt to them. Adaptive capacity allows adjusting practices, processes or structures to moderate or offset the potential for damage, or take advantage of opportunities. Potentially, climate change has a wide set of impacts, but only a part

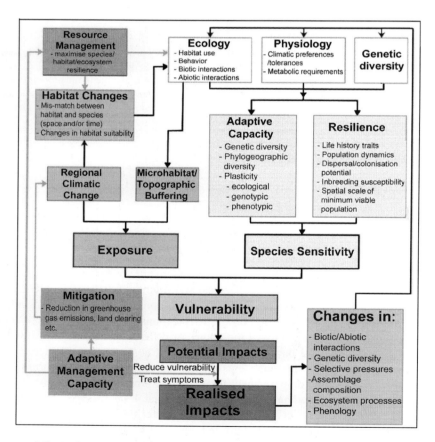

Fig. 18.1 Impacts of climate change on forest ecosystem and needs for adaptive management in the context of the vulnerability of ecosystem, as adopted from Johnston (2010).

of them is realized, e.g., abiotic and biotic factors with impacts on the potential producing different goods and services. Realized impacts modify the biotic/abiotic interactions in forest ecosystems, genetic diversity, phenology and ecosystem processes controlling the ecosystem performance (Johnston 2010).

In the vulnerability framework, adaptive management reduces negative impacts and improves adaptive capacity (Smit and Wandel 2006, Johnston 2010) for reducing future uncertainties in meeting management objectives under changing conditions (Fig. 18.1). Management affects the suitability of forest habitats to meet changes in biotic/abiotic interactions in populations/communities and their capacity for adaptive (e.g., genetic diversity, plasticity) and resilience processes (e.g., population dynamics, dispersal/colonization) controlling forest ecosystem dynamics. Subsequent changes in species vulnerability have potential impacts, which are realized depending on species sensitivity, climate change and management for reducing the vulnerability (Jandl et al. 2015).

According to Spittlehouse and Steward (2003), the adaptive process in forestry is based on monitoring the state of forests and recognizing when critical thresholds in their growth and development are reached (Fig. 18.2). Such adaptive needs are related to management objectives, depending on whether they are likely not to be realized or realized under warming climate. In the former case, the necessary management strategies are identified, including: (i) management plans; (ii) genotype management (tree species choice, provenance choice; tree breeding); (iii) protection of forests from insects and pests; (iv) silvicultural management (regeneration practices, pre-commercial thinning, thinning practices, rotation length, etc.); and (v) establishment and maintenance of the necessary technology (e.g., harvest technology) and infrastructure (e.g., forest roads). The choice of adaptive measures is the result of a process where a set of management operations are optimized against the management objectives. Necessary measures are applied to single populations of trees or sites, but adaptive management needs also cross-landscape integration in order to maintain the sustainability of timber and non-timber resources under climate change.

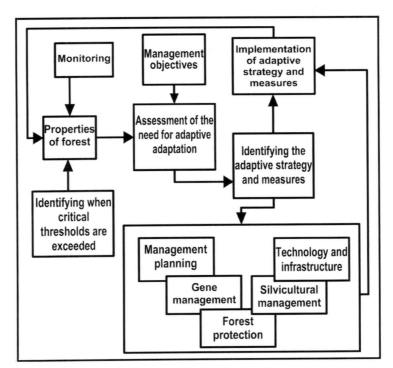

Fig. 18.2 Outline for identifying the needs of adaptive management and the measures required to meet changes in ecosystem structure and functioning under climate change (Spittlehouse and Steward 2003, Kellomäki and Leinonen 2003).

18.2 Adaptive Management in Boreal Forests

Potential damage and opportunities

The need for adaptive management varies from region to region and even from site to site, related to the degree of vulnerability of forests to climate change, depending on (Spittlehouse and Steward 2003): (i) climate change and climate variability; (ii) the impact or effects of climate change on forests (climate sensitivity); (iii) the extent to which forests may adapt autonomously; and (iv) the residual or net impact of actively adapting forests to climate change. The application of these criteria show that the northern edges of boreal forests are most vulnerable (Table 18.1) and lose much of their value for conservation, recreation and landscaping. At the same time, the timber producing capacity of these forests may increase substantially, thus providing more opportunities for forestry but fewer opportunities for non-timber production. On the southern edge of boreal forests, increased evaporation may reduce soil water in shallow soils in spring and early summer, causing problems for shallow rooted tree species. Furthermore, risks of abiotic and biotic damage are likely to increase throughout boreal forests, including the reduction of soil frost with problems in the harvest and logistics of timber and biomass.

Table 18.1 Assessing the need for adaptive management in boreal forests, with a focus on Finland (Kellomäki and Leinonen 2005).

Topic area	Factors affecting the need for adaptive management
Climate change and climate variability	• Higher temperatures and precipitation, with shorter duration of snow cover and soil frost, increasing risk of local thunderstorms with high wind, increasing risk of heavy snow fall with wet snow, wet soil with lower carrying capacity.
Impacts and climate sensitivities	• Enhanced regeneration and growth, increasing dominance of deciduous species thereby reducing fire risk, increasing risk of wind and snow damage, enhanced risk of insects and pathogens, enhanced input of carbon into soil, enhanced decomposition of soil organic matter.
Capacity to adapt autonomously	• Mainly native tree species are used, with much genetic variability to acclimate forests to elevating temperature and high variability in temperature. Enhancing seed production and the success of natural seeding forests. • Productivity of forests will enhance, with higher turnover of carbon in forest ecosystems. However, the low supply of nitrogen may limit the enhancement of growth, which is highest in the northern parts of boreal forests. • Currently, trees are well adapted to the local insects and pathogens, with low frequency of major outbreaks. Risk of invasion of alien species and local species further north, risk of major outbreaks of damaging insects and pathogens. Increasing wind and snow damage.
Vulnerability	• Increasing competition capacity of deciduous species may alter species composition, especially in southern boreal forests. • Timber line may move northwards and to higher altitudes, with a disappearance of current timber line forests. • Enhancing growth rates may reduce timber quality. • Likely increase of abiotic and biotic damage may result in major losses in the quantity and quality of timber, with more unscheduled cuttings and management. • Reducing soil frost along with higher precipitation may reduce the carrying capacity of soils, with problems in harvest and transportation of timber.
Need for planned adaptation	• Control of tree species composition to meet future needs and expectations. • Modifications of thinning practices (timing, intensity) and rotation length to meet the increasing growth and turnover of carbon, to maintain high quality of timber and resistance to abiotic and biotic damage. • Maintain infrastructure for forestry and non-timber forest production.

Current and future actions for adaptation

Under climate change, the future forest environment deviates from the current one but the changes are uncertain. Millar et al. (2007) emphasized "flexible approaches that promote reversible and incremental steps, and that favor ongoing learning and capacity to modify direction as situations change" under warming. Accordingly, proper adaptive management includes options: (i) to reduce the impact and protect forests (resistance); (ii) to improve the capacity of ecosystems to recover from disturbances (resilience); and (iii) to facilitate the transition of ecosystems from the current to new conditions (response). In this context, Spittlehouse and Steward (2003) break the current (up to 2050) and future (after 2050) actions into: (i) management planning; (ii) gene management; (iii) forest protection; (iv) silvicultural management; and (v) technology and infrastructure applicable in boreal forestry as detailed in Table 18.2.

Table 18.2 Examples of means to adapt forestry in boreal forests as modified from Kellomäki and Leinonen (2005).

Actions	Perspective before 2050	Perspective 2050–2100 and onwards
Management planning	• Include climate variables in growth and yield models for more specific predictions about the future development of forests. • Include risk management in management rules and plans. • Monitoring climate hazards.	• Plan forest landscape to resist high winds and fire hazards. • Plan forest landscape to minimize spread of insects and diseases. • Monitoring climate hazards, procedures and fighting climate disturbances.
Gene management	• Make choices about preferred tree species composition for future. • Identify suitable genotypes. • Breeding programs to increase the resistance of biotic damage.	• Plant alternative genotypes or new species. • Modify seed transfer zones and management of genetic resources. • Breeding for meeting climate change.
Forest protection	• Revise rules for importing fresh timber to reduce the risk of introducing alien species. • Breeding for reducing biotic hazards. • Conservation biodiversity for controlling forest health.	• Revise management rules to increase resistance of forest to abiotic and biotic damage. • Breeding for reducing biotic hazards. • Conservation biodiversity for controlling forest health.
Silvicultural management	• Revise management rules to consider effects of climate variability on regeneration, growth and mortality. • Prefer natural regeneration wherever appropriate. • Revise management rules for reducing climatic hazards and climate warming.	• Develop soil management to reduce influence of ground cover on success of regeneration. • Modify management rules to meet enhanced growth and turn-over of carbon and altered risks. • Change rotation lengths to meet increased turn-over of carbon and enhanced growth.
Technology and infrastructure	• Develop technology to use altered wood quality and tree species composition. • Develop infrastructure for timber harvest and transportation, and non-timber use of forests. • Technology procedures to fight climatic disturbances.	• Develop technology to use altered wood quality and tree species composition. • Develop infrastructure for timber harvest, transportation, and non-timber use of forests. • Technology and procedures to fight climatic disturbances.

18.3 Choice of Tree Species and their Provenance in Adaptation

Regeneration and tree species choice

Under climate warming, boreal tree species are probably growing faster with a more rapid life cycle and the increased turnover of tree populations (Zhao and Running 2010). From an evolutionary perspective, the preference for natural regeneration in management has great genetic potential for acclimating forests to climate change (Box 18.1). This choice may be realistic, since natural regeneration is a common practice in northern forestry, and even in forest plantations natural seedlings substantially affect the total success of reforestation. However, the preference for natural regeneration may in the long run lead to an inevitable shift of tree species composition on medium fertile and fertile sites to the dominance of deciduous tree species with a larger supply of hardwood timber if not controlled in pre-commercial and commercial thinnings (Kellomäki and Kolström 1992a,b). Regular and timely management and harvest make it possible to modify the growing conditions and properties of tree populations to allow easier adaption than that possible when excluding intentional management.

In northern Europe, Scots pine is currently successful on sites from low to high fertility, while Norway spruce and birch are successful on sites of medium and high fertility. Torssonen et al. (2015) showed that these differences are of large importance for successful choice of tree species for future conditions. They used a process-based ecosystem model (Sima) to study how sensitive Scots pine, Norway spruce and birch are to the warming climate throughout Finland, if preferred in plantations. These tree species were cultivated in three 30-year periods (2010–2099) in order to identify how different tree species perform in their early development under a gradually warming climate in order to have an early warning about long-term growth and development under climate warming.

Regardless of tree species and the climate change scenario (SRES B1, A1B and A2), growth increased on fertile and medium fertile sites throughout the country in the period 2010–2039, whereas on poor sites the growth of Norway spruce reduced, especially on the sites characterized by a higher current temperature sum (Torssonen et al. 2015). The increasing growth was also true for Scots pine and birch in the periods 2040–2069 and 2070–2099, while the growth of Norway spruce reduced slightly in the period 2040–2069 and further in the period 2070–2099. This was especially clear in the south (current temperature sum > 1,200 d.d.). Under the same conditions, the growth increase in Scots pine and birch remained small compared to that in medium fertile and fertile sites. In the north (current temperature sum < 1,000 d.d.), growth increased even on poor sites, regardless of tree species. The productivity in the north may be increased substantially, but it will be still less than that currently in the south. Torssonen et al. (2015) also showed that the climate change scenario

---

Box 18.1 Gradual warming may increase the adaptation to higher temperatures

In terms of epigenetics, Johnsen et al. (2005a,b) have shown that the adaptive traits of Norway spruce progenies are affected by the maternal temperatures during seed production. Epigenetics involves variation in the physiological traits not caused by changes in the DNA sequence but by external or environmental factors that turn genes on and off. Epigenetic changes may adapt offspring to the parental environment, thus facilitating cross-generation adaptation to transient warming (Saxe et al. 2001). Such acclimation is adaptive only if climate warming proceeds smoothly. The climatic limits of epigenetic acclimation are related to the plasticity of individual genotypes linked to a high degree of genetic variability. Genetically diverse populations that support a variety of phenotypes with a high degree of plasticity may therefore have a selective advantage during climate change (Saxe et al. 2001). Only two or three generations are needed for lodgepole pine (*Pinus contorta* Douglas) in northern latitudes to adapt to a warming climate while in southern latitudes adaptation needs six to 12 generations (Rehfeldt et al. 2001).

mattered to how the different tree species responded to climate change. A high temperature increase combined with a low increase in precipitation may reduce the growth of Norway spruce, whereas a moderate temperature increase with a high increase in precipitation may increase the growth. This is true especially in the south, whereas in the north a larger relative increase in temperature rather than precipitation increased the growth instead of reducing it.

Provenance choice

The proper choice of tree provenance is widely used in management to increase forest growth. Provenance refers to local populations of tree species, which are adapted to the prevailing conditions. Properties of trees in a particular provenance deviate from the mean value over the whole geographical area occupied by the species. Differentiation into provenances is driven mainly by the spatial variation in climatic conditions in the distribution area. The differences between provenances are inherent, and they are recognizable from generation to generation, but different provenances are able to breed with each other. The attribute differences between provenances are mainly physiological ones in response to the prevailing conditions of seed origin, with no clear differentiation between the provenances in eco-physiological responses to the changing climate (elevating CO_2, temperature) (Dang et al. 2008). This makes it possible to improve the fit of a species to the changes in growing conditions by using proper provenance choice in forest plantations.

Provenance trials have been used to investigate how climate change may affect the growth and mortality of trees. In provenance trials, trees are growing in a climate, which is different from that where they are originally adapted. Based on an old provenance experiment (established in 1920s), Beuker (1994) assessed the long-term effects of climate change on the growth of Scots pine and Norway spruce. The geographical origin of Scots pine was in the latitude range 46°–67° N and that of Norway spruce in the latitude range 59°–69° N. In these ranges, the altitude of origin varied from 50 m to 800 m above sea level, with a temperature sum range of 680–2,000 d.d. Beuker (1994) assumed that the change in stem wood growth (ΔPr, m³ ha⁻¹) during a given period is a function of the temperature sum at the original site of provenance (TS_{or} d.d.) and the difference between temperature sum at the original site and at the new site (ΔTS, d.d.):

$$\Delta Pr = C + b_1 \times TS_{or} + b_2 \times \Delta TS + b_3 \times TS_{or}^2 + b_4 \times \Delta TS^2 + b_5 \times TS_{or} \times \Delta TS \qquad (18.1)$$

where C is a constant [m³ ha⁻¹] and $b_1 - b_5$ are parameters. Based on Equation (18.1), Fig. 18.3 shows that a southward transfer of Norway spruce and Scots pine substantially increased the growth but it remained smaller than growth of local provenances. A warming climate may thus most increase growth on sites, where the temperature is currently most limiting, as in the northern boreal forests. On the other hand, climate warming is likely to increase growth to a lesser extent or even reduce it if the provenance is from a currently warmer climate than that on the original sites. This was especially the case in southernmost Finland, where the temperature only slightly limited the growth of Norway spruce and Scots pine (Beuker 1994).

In Beuker's (1994) study, the effects of climate warming on growth from varying provenances were assessed excluding the possible effect of photoperiod on growth at a new site. The effect of photoperiod, however, seems to be small compared to that of climate warming as found by Persson and Beuker (1997). Excluding the impact of altitude on climatic conditions, the growing season at latitude 60° N in the south is two months longer and the temperature sum 600 d.d. higher than at latitude 68° N in the north, where the amount of radiation is 40% smaller than in the south. Persson and Beuker (1997) showed that the mean annual growth of Scots pine increased linearly in relation to the increasing temperature sum, whereas a latitudinal transfer without the change in temperature sum had no effect on growth.

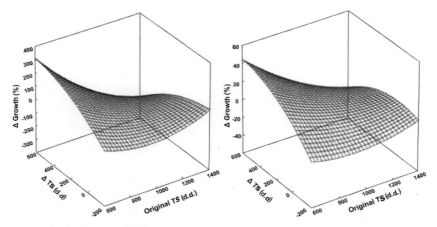

Fig. 18.3 Change in stem wood growth in Norway spruce (left) and Scots pine (right) provenances as a function of temperature sum at the original site of the provenances (Original TS, d.d.) and the difference between temperature sum at the original site and at the new site (ΔTS, d.d.) (Beuker 1994, Beuker et al. 1996). Courtesy of Taylor and Francis.

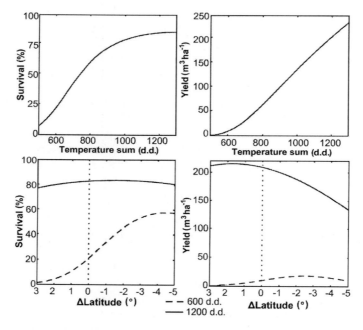

Fig. 18.4 Above: Survival and growth (yield) of local Scots pine provenances as a function of temperature sum. Below: Effect of latitudinal transfer of Scots pine provenances on survival and growth at the sites with temperature sum 600 d.d. and 1,200 d.d. Negative values of ΔLatitude indicate southward and positive northward transfer (Persson 1998). Courtesy of the Finnish Society of Forest Science and the Finnish Forest Research Institute.

Based on the same material as used in Persson and Beuker (1997), Persson (1998) further analyzed the interaction of warming and latitudinal transfer on the growth and survival of Scots pine. The material represented provenance trials in Sweden north of latitude 60° N. The survival and stem wood growth (yield) of local provenances increased as a function of temperature sum (Fig. 18.4). Survival in northern (temperature sum 600 d.d.) provenance increased substantially due to southward transfer, whereas the survival of southern Scots pines (temperature sum 1,200 d.d.) was fairly insensitive to the transfer. The yield of northern Scots pines increased due to southward transfer in relative terms, but remained absolutely below the yield of southern Scots pines. The southwards transfer clearly reduced the yield of southern Scots pines, however (Persson 1998).

Regarding climate change, Jääskeläinen (2014) used the calculation methods of Persson (1998) to simulate gradual climate warming and provenance transfer and to determine how it may affect the long-term survival and growth of Scots pine (Box 18.2). First, trees from a given provenance were transferred from more northern and more southern sites, where they grew under the current or warming climate. Under both climate, growth on the new sites was compared to those on the original site, with the current or warming climate of the original site. Second, trees of a given provenance were transferred to more northern and more southern sites as in the first case, but the comparison was made against the local provenance in the new site.

Box 18.2 Effects of provenance transfer on the dynamics of ecosystem model

In the Sima model (Kellomäki et al. 2008), stem volume growth is a function of radial and height growths. Height is based on the modified Näslund's (1937) model, which gives the height as a function of diameter and long-term temperature sum (TS, d.d.) at the site:

$$H = \left[\frac{Ts}{1000}\right]^c \times \left[1.3 + \frac{D^2}{(A + B \times D)^2}\right] \qquad (18.2)$$

where A, B and C are the species-specific parameters estimated from the forest inventory data. The model shows that trees with the same diameters are taller in southern (high temperature sum) than in northern (low temperature sum) sites (Fig. 18.5). Height growth followed by southward transfer thus remains smaller than that of local provenances: the temperature sum in the original site reduces the growth that is otherwise caused by higher temperature. Conversely, northward transfer increases height growth: higher temperature in the original site gives a larger growth than expected based on temperature on the new site.

The mortality of trees was assumed to be dependent on random reasons and the reduction of radial growth due to the excessive stand density and maturity of trees. Trees died for random reasons if an evenly distributed random number [0–1] had the value ≤ 0.368. Trees died due to reduced growth if an evenly distributed random number [0–1] had the value < L:

$$L = \left(BA \times \frac{\rho^{0.7}}{37}\right) / AgeMax \qquad (18.3)$$

where BA [m² ha⁻¹] is the basal area of trees in a stand, ρ is the density of the tree stand [trees per hectare] and AgeMax [yr] is the maximum age of tree species, e.g., 350 years for Scots pine. In both case, the value of L was scaled by the factor a:

$$a = \frac{S_0}{S_{prov}} \qquad (18.4)$$

where S_o is the survival of local provenance [0–1] and S_{prov} is the survival [0–1] in the new site. The value of a > 1 is for the transfer increasing survival and a < 1 for transfer reducing survival. The survival probability of local provenance (S_o) and the change in the survival probability (ΔS_o) is (Persson 1998):

$$\text{logit} S_o = \ln\left(\frac{S_o}{1 - S_o}\right) \qquad (18.5)$$

$$\text{logit}\Delta S_o = \ln\left(\frac{S_{prov}(1 - S_o)}{S_o(1 - S_{prov})}\right) \qquad (18.6)$$

The survival probability and the change in survival probability are related to the Temperature Sum (TS) and to the change in latitude (ΔLat) (Persson 1998):

$$S_0 = \exp(-75.69 - 0.009617 \times TS + 12.53 \times \ln TS) \qquad (18.7)$$

Box 18.2 contd....

Box 18.2 contd....

$$\Delta S_o = \exp(0.01510 - 1.406 \times \Delta Lat - 0.1458 \times \Delta Lat^2 + 0.001115 \times TS \times \Delta Lat + 0.000105 \times TS \times \Delta Lat^2)$$

Figure 18.6 demonstrates how the provenance transfer affects survival as a function of the temperature at the original site.

In the simulations, the current and the changing climate (SRES A2 scenario) were used for central Finland (62° N). In the period 1990–2100, the mean annual temperature increased 4°C and the mean annual precipitation 13%. The temperature increase implied an increase in temperature sum from 1,180 d.d. to 1,700 d.d. in the original site, where the local provenance of Scots pine was transferred northwards and southwards by one to three latitudes in both directions. The atmospheric CO_2 increased from 350 ppm to 840 ppm. The simulations were run for Scots pine planted at a density of 2,500 seedlings per hectare on a site of medium fertility (*Myrtilus* type, MT). Management was excluded over the 100-year rotation used in the simulations.

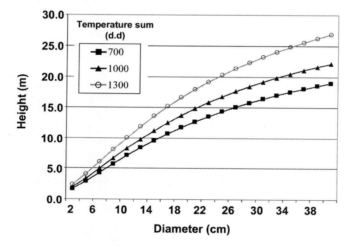

Fig. 18.5 Height of Scots pine as a function of the diameter and temperature sum on the site, where trees have grown throughout their life span (Kellomäki et al. 2005, 2008).

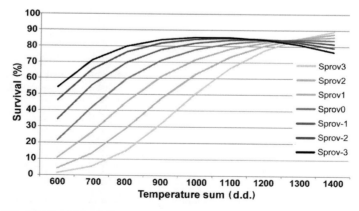

Fig. 18.6 Impact of the transfer of Scots pine provenance on survival as a function of the temperature sum, based on Persson (1998). Sprov0 indicates survival on the original site, whereas from Sprov1 to Spov3 indicates the northward and from Sprov-1 to Sprov-3 the southward latitude transfer.

Regarding the current climate, Jääskeläinen (2014) found that a southward transfer to a new site reduced the total growth by 6–14% compared to that at the original site, whereas northward transfer increased growth 1–4% (Table 18.3), most in a transfer over three latitudes. This pattern held with that found by Persson (1998). The same pattern was true under climate change, even though absolute value of stem wood production increased about 28–30% regardless of the latitudinal transfer. The effects of transfer on the total growth were further reflected in stocking: stocking at the end of rotation was 17–26% greater under climate warming than under the current climate. At the same time, climate warming reduced the stand density by 15–20% regardless of the latitudinal transfer compared to that under the current climate.

The effects of provenance transfer compared that of local provenances, where the transfer was done, is shown in Table 18.4. Under the current climate, the northward transfer across two latitudes reduced total growth 7% but increased stocking by 44% and increased mortality by 17%. A similar transfer southwards reduced total growth 21% and stocking 14% but reduced mortality by 10%. Under a warming climate, a northward transfer increased the total growth 23%. In a southward transfer the reduction of total growth (7%) was substantially smaller than under the current climate. The warming-induced changes in growth increased stocking substantially (13%) regardless of transfer. At the same time, mortality was reduced due to the southward transfer (29%) but increased due to the northward

Table 18.3 Total production of stem wood [m³ ha⁻¹ per 100 years], stem wood stocking [m³ ha⁻¹], and stand density (trees per hectare) of transferred provenances at the end of rotation as a function of latitudinal transfer under the current and warming climate compared to the same parameter values on the original site under the current and changing climate (Jääskeläinen 2014). Regardless of the transfer, the provenance was the same as in the Original site.

Transfer in latitudes	Current climate		Changing climate		Change in relation to current climate, %
Total production	m³ ha⁻¹	Change, %	m³ ha⁻¹	Change, %	
− 3 south	639	−13.9	820	−15.4	+28.2
− 2 south	682	−8.2	873	−10.0	+28.0
− 1 south	699	−5.8	912	−5.9	+30.4
Original site	743		969		+30.5
+ 1 north	752	+1.3	983	+1.4	+30.6
+ 2 north	755	+1.6	1,001	+3.2	+32.6
+ 3 north	774	+4.2	997	+2.8	+28.7
Stocking	m³ ha⁻¹	Change, %	m³ ha⁻¹	Change, %	
− 3 south	384	−12.6	451	−16.4	+17.5
− 2 south	418	−4.8	507	−6.1	+21.3
− 1 south	407	−7.3	520	−3.5	+27.9
Original site	439		539		+22.9
+ 1 north	442	+0.8	558	+3.4	+26.1
+ 2 north	446	+1.6	538	−0.2	+20.7
+ 3 north	439	+0.0	531	−1.6	+21.0
Stand density	Trees ha⁻¹	Change, %	Trees ha⁻¹	Change, %	
− 3 south	897	+2.9	705	−5.0	−21.4
− 2 south	910	+4.4	747	+0.7	−17.9
− 1 south	855	−1.9	739	−0.4	−13.6
Original site	872		742		−14.9
+ 1 north	869	−0.3	752	+1.3	−13.5
+ 2 north	841	−3.6	696	−6.2	−17.2
+ 3 north	824	−5.5	674	−9.2	−18.2

Table 18.4 Total stem wood growth [m³ ha⁻¹ per 100 years], stem wood stocking [m³ ha⁻¹] and stand density (trees per hectare) of transferred provenances at the end of rotation as a function of latitudinal transfer under the current and changing climate compared to the local provenances in new sites under the current and changing climate. The values in parentheses refer to the values in the original site (Jääskeläinen 2014). Regardless of the transfer, the provenance was the same as in the Original site, but the comparison was done against the local provenance.

Transfer in latitudes	Current climate		Changing climate		Change in relation to current climate, %
Total production	m³ ha⁻¹	Change, %	m³ ha⁻¹	Change, %	
− 2 south	867 (682)	−21.3	943 (873)	−7.4	+8.8
+ 2 north	813 (755)	−7.2	812 (1001)	+23.3	−0.2
Stocking	m³ ha⁻¹	Change, %	m³ ha⁻¹	Change, %	
− 2 south	486 (418)	−14.1	448 (507)	+13.2	−7.9
+ 2 north	310 (446)	+44.0	479 (538)	+12.5	+54.6
Stand density	Trees ha⁻¹	Change, %	Trees ha⁻¹	Change, %	
− 2 south	826 (910)	+10.2	580 (747)	+28.8	−29.8
+ 2 north	1,008 (841)	−16.6	777 (696)	−10.4	−22.9

transfer (10%). Assuming southward transfer, climate warming increased total growth 9% compared to growth under the current climate, whereas growth was not affected by the northward transfer. At the same time, warming increased stocking due to northward transfers but reduced it in southward transfers, while climate warming increased mortality (20–30%) regardless of transfer.

Tree species composition

Seidl et al. (2011) noted that the promotion of mixed stands of well adapted species to emerging warming is a successful way to reduce the vulnerability of forests. In northern Europe, mixtures of Norway spruce, Scots pine and birch are common on medium fertile and fertile sites, and, for example, Norway spruce-dominant mixtures comprise over 33% of the total forest resources in Finland. The effects of changing climate on the growth and dynamics of mixed boreal forests are still poorly known but the responses of trees growing in admixtures to resource availability are more diverse than in mono-species stands. In both cases, resource availability (water, nutrients, light, etc.) is tightly controlled by stand structure, as is canopy closure, but tree species composition is likely to make it different to mono-species stands.

Ge et al. (2011b) used a process-based ecosystem model (FinnFor) to determine whether climate change affects growth of Norway spruce, Scots pine and birch growing in mixtures compared to those growing in mono-species stands on the sites of medium fertility (*Myrtillus* type, MT). The main hypothesis was that tree species composition may change the availability of soil water and within-stand shading compared to that in mono-species stands. In the simulations, the success of tree species was indicated by the total net canopy photosynthesis and total stem wood growth. The initial tree species composition in mixtures (Norway spruce 50%, Scots pine and birch each 25% in regard to density) was the same regardless of the site. In both sites, the soil water content was 52% at the field capacity and 26% at the wilting point of the volume. The current climate was that for the period 1970–2000, with a constant CO_2 of 350 ppm. The changing climate was based on the SRES A2 emission scenario, with the increase of temperature up to 4–6°C, and atmospheric up to CO_2 840 ppm by 2099. Management was excluded from the analyses in order to identify how the natural stand dynamics affect the availability of soil water and the subsequent carbon uptake and growth of trees in different stand and climatic conditions.

On the southern site, the net canopy photosynthesis in mixed stands tended to be lower under the changing than under the current climate, with the exception of the early part of the simulation period (Table 18.5). However, climate warming reduced the total net canopy photosynthesis: 15%

Table 18.5 Cumulative net photosynthesis and total stem wood growth over 100 years in mixed and pure (mono-species) stands of Norway spruce, Scots pine and birch under the current and changing climate on southern and northern sites. Figures in parenthesis show the percentage change (%) under the changing climate compared to the current climate (Ge et al. 2011b).

Mixed stands	Net canopy photosynthesis, Mg C ha^{-1} per 100 years		Total stem wood growth, m^3 ha^{-1} per 100 years	
	Current	Changing	Current	Changing
Southern site (N 61°56', E 22°01'), mixed stand				
• Norway spruce	780	676 (−15)	667	610 (−9)
• Scots pine	78	69 (−12)	156	144 (−8)
• Birch	84	74 (−13)	133	127 (−5)
• Total	942	819 (−15)	956	881 (−8)
Southern site (N 61°56', E 22°01'), mono-species stand				
• Norway spruce	940	795 (−18)	834	738 (−13)
• Scots pine	807	926 (+13)	902	984 (+8)
• Birch	1,026	1,198 (+14)	1274	1,311 (+7)
Northern site (N 66°37', E 23°57'), mixed stand				
• Norway spruce	634	715 (+11)	394	441 (+10)
• Scots pine	63	80 (+21)	100	125 (+20)
• Birch	79	88 (+10)	76	83 (+8)
• Total	776	883 (+12)	569	649 (+12)
Northern site (N 66°37', E 23°57'), mono-species stand				
• Norway spruce	753	842 (+12)	507	562 (+11)
• Scots pine	685	896 (+24)	745	957 (+22)
• Birch	869	1,090 (+20)	865	1,044 (+17)

in Norway spruce, 12% in Scots pine and 13% in birch compared to that under the current climate. On the northern site, climate warming tended to increase the total net canopy photosynthesis: 11, 21 and 10% in Norway spruce, Scots pine and birch, respectively. The warming-induced changes in the total net canopy photosynthesis indicated similar changes in the total stem wood growth. On the southern site, growth in Norway spruce was 9% lower than under the current climate, whereas growth reduced by 8% for Scots pine and 5% for birch compared to that under the current climate. On the northern site, the total stem wood growth increased by 10, 20 and 8% for Norway spruce, Scots pine and birch when assuming a changing climate.

In general, the total net canopy photosynthesis and stem wood growth in mono-species stands were greater than in the mixed stands (Table 18.5). On the southern site, the carbon uptake (−18%) and growth (−13%) only reduced in Norway spruce when assuming climate change, whereas in Scots pine and birch the carbon uptake (+14%) and growth (+8%) increased. On the northern site, the values of net canopy photosynthesis and volume of stem wood were higher under the changing climate, regardless of tree species.

The differences between mixed and pure stands seemed to be related to the differences in water depletion (i.e., the total evapotranspiration from the site and tree stand), which was clearly greater in Norway spruce stands than in Scots pine and birch stands (Table 18.6). Climate change most increased the water depletion in Norway spruce stands, especially in the south. This implies that the proportion of total evapotranspiration to precipitation was higher in pure Norway spruce stands than in pure Scots pine and birch stands, regardless of site. The amount of water depleted was 11% greater under the changing climate than under the current climate. On the southern site, cumulative

Table 18.6 Share of total water evapotranspiration from precipitation and cumulative water use efficiency in the mono-species stands under the current and changing climate on the southern and northern sites over the 100-year simulation period. Figures in parenthesis show the percentage change (%) under the changing climate compared with the current climate (Ge et al. 2011b).

Mono-species stands	Proportion of evapotranspiration to precipitation, %		Water use efficiency, Mg C mm^{-1}	
	Current	Changing	Current	Changing
Southern site				
• Norway spruce	85	94 (+10)	0.043	0.049 (+14)
• Scots pine	66	67 (+1)	0.047	0.052 (+11)
• Birch	73	75 (+2)	0.054	0.060 (+11)
Northern site				
• Norway spruce	78	84 (+7)	0.058	0.061 (+5)
• Scots pine	66	66 (0.3)	0.062	0.062 (+8)
• Birch	70	70 (+0.5)	0.074	0.077 (+4)

water use efficiency increased by 11–14% under the changing climate compared to that under the current climate, most in the Norway spruce stand. In the northern site, the increase was 4–8%.

18.4 Thinning and Rotation Length in Adaptation

Thinning

Model-based studies show that Norway spruce is likely to be a loser under climate change compared to other tree species in northern (Bergh et al. 2005) and central Europe (Albert and Schmidt 2010). These claims are supported by experimental studies, such as that of Jyske et al. (2010). They found that preventing rainwater from replenishing soil water over 60–77 days from the beginning of growing season reduced the annual height and radial growth of Norway spruce in southern boreal conditions (61° N). The importance of soil moisture for Norway spruce is further demonstrated in model-based studies showing that even in middle boreal conditions water supply may be critical under a changing climate (Ge et al. 2010, 2011a, 2013b). On the other hand, Kohler et al. (2010) found that the effects of drought periods on Norway spruce could be decreased through thinning; i.e., increased growing space meant less growth reduction as a result of lower negative water balance (precipitation minus potential evaporation). This finding is in line with the model simulations of Briceño-Elizondo et al. (2006). They found thinning to increase the mean annual precipitation per remaining tree and soil moisture, along with increasing thinning intensity under boreal conditions. Consequently, growth of remaining trees was increased, which is consistent with spacing and thinning experiments.

Ge et al. (2013a) also studied how thinning affects soil water content and the consequent growth and development of Norway spruce. A process-based model (FinnFor) was employed in assessing how varying thinning intensity may affect the Gross Primary Production (GPP), total growth of stem wood and timber yield related to changes in temperature, precipitation, and atmospheric CO_2 and soil water content. The simulations represented Norway spruce in five different climatic zones, from southern to northern boreal conditions (61–67° N) over the south-north temperature gradient from the southern edge of the boreal forest to the northern edge against the Arctic tundra (Box 18.3).

Under climate warming, the mean annual evapotranspiration was 19–24% larger than that under current climate, if no thinning was assumed (Table 18.7). The proportion of evapotranspiration to precipitation increased from 5–14% under climate change. At the same time, the mean annual

Box 18.3 Simulation to study how thinning affects the growth and development of Norway spruce under climate change

Based on temperature sum (*TS*, d.d., threshold + 5°C), Finland was divided into five climatic zones: *TS* > 1050 d.d. for zone I (Z_I) and zone II (Z_{II}) in southern, 900 < *TS* < 1,050 d.d. for zone III (Z_{III}) in central Finland, and *TS* < 900 d.d. for zone IV (Z_{IV}) and zone V (ZV) in northern Finland (Fig. 18.7).

In each climatic zone, one Norway spruce stand was selected from the database of the Finnish National Forestry Inventory. The sample plots were of medium fertility (*Myrtillus* type, MT), with soil water moisture in the field capacity of 52% and the wilting point of 30% of the volume. The initial mass of organic matter in the soil was in the range of 60–70 Mg ha^{-1} for the sites, less at the northern than southern sites (Kellomäki et al. 2008). The initial stand density was 2,400 trees per hectare. Ten thinning regimes, with fixed proportions (5, 10, 15, 20, 25, 30, 35, 40, 45 and 50%) of the basal area removed from below, were used in thinning. Over the 100-year simulation period (2000–2099), three thinning events at an interval of 25 years were used. Clear cut was done at the end of the simulation period.

In the simulations, the current climate (CUR) was that for the period 1970–2000, and the Changing Climate (CC, 2000–2099) was based on the SRES A2 emission scenario (Ruoteenoja et al. 2005). In the period 2000–2099, the mean annual temperature increased 4°C in summer and 6°C in winter, and the atmospheric CO_2 increased from 350 ppm (mean for the period 1970–2000) to 840 ppm in the year 2099. The mean annual precipitation increased by 15%, mainly in winter.

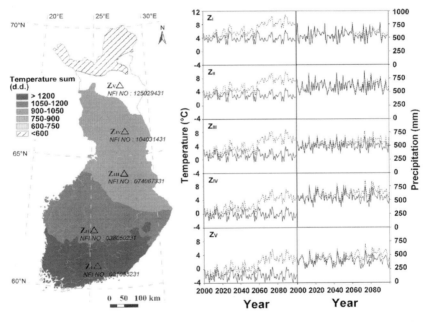

Fig. 18.7 Left: Location of the five Norway spruce sites in the climatic zones (temperature sum zones) used in the simulations on how Norway spruce is affected by climate change over boreal forests. Right: Annual variation in mean annual temperature and precipitation for the current (solid line) and changing (dashed line) climate at the sites (Ge et al. 2013a). Permission of Springer.

infiltration of water into the soil profile decreased by 24 and 13% at sites Z_I and Z_{II} in southern boreal and by 6% at site Z_{III} in middle boreal sites. In northern boreal conditions (sites Z_{IV} and Z_V), the increase in precipitation compensated for the increased evapotranspiration caused by climate warming.

On the southern boreal site (the site Z_I), thinning regimes of T05, T10, T15 and T20 increased gross primary production and total stem wood growth by 2–4% and 1–2% compared to those under no thinning (Table 18.8). High stocking through rotation thus gave the highest net primary growth

Table 18.7 Mean annual evapotranspiration, its share of precipitation, and water infiltration into soil over the 100-year simulation period under the current and changing climate, assuming no thinning. The numbers in parenthesis indicate the percentage (%) of infiltration under climate change of that under the current climate (Ge et al. 2013a). Explanation of sites, see Box 18.3.

Site	Climate scenarios	Evapotranspiration, mm per year	Evapotranspiration, % of precipitation	Water infiltration, mm per year
Z_I	Current	431.4	81	347.6
	Changing	533.6	92	327.2 (94)
Z_{II}	Current	454.6	76	375.5
	Changing	544.2	90	365.4 (97)
Z_{III}	Current	387.9	76	290.4
	Changing	463.4	83	284.0 (98)
Z_{IV}	Current	358.2	65	255.6
	Changing	441.9	71	272.0 (106)
Z_V	Current	301.5	68	190.4
	Changing	375.0	73	201.5 (106)

and increased timber yield compared to values under greater thinning intensity (Thornley and Cannell 2001). The increase in timber yield was up to 19% under thinning regimes T10, T15 and T20. However, heavy thinning (the regimes T30–T50) reduced gross primary production, total stem wood growth and the timber yield.

On the middle boreal site Z_{II}, the pattern was similar than in the south. The thinning regimes of T05, T10 and T15 increased the gross primary and total stem wood growth by 1–3%, compared to that under no thinning, and the regimes of T10, T15 and T20 gave the highest timber yield (increase by 16%). Under thinning regimes with a thinning intensity higher than 20%, the gross primary production, total stem wood growth and timber yield were lower than those under no thinning.

Further north on the site Z_{III}, gross primary production and total stem wood growth were slightly higher under thinning regimes T05 and T10 than that under no thinning. In this site, the timber yield was up to 13% larger than under thinning regimes T10, T15 and T20. Heavy thinning (T20–T50) again reduced gross primary production and the total stem wood growth and timber yield. Timber yield was also lower under very high thinning intensity (50%) than that under no thinning. This was true on the sites Z_{IV} and Z_V in the north, where any thinning regimes decreased gross primary production and total stem wood growth compared to those under no thinning. The reduction increased along with the increasing thinning intensity.

Rotation length and tree species choice

Based on model simulations, Kellomäki et al. (2008) used the rule-based management in reformulating management for adapting to climate change. There were two main tasks: to maintain the productivity of the forest ecosystems, especially the growth of Norway spruce if the current patterns of timber production are preferred in the future. First, the length of the rotation was reduced by making the terminal cut earlier than in conventional timber production but still aiming at producing saw timber and pulpwood. Second, Norway spruce was replaced by Scots pine or birch on sites of medium fertility, and Norway spruce was preferred in reforestation only on sites with high fertility if the species occupied the site prior to the terminal cut. Third, a more southern provenance of Norway spruce was used in reforestation on medium fertile and fertile sites, if Norway spruce previously occupied the site. In the analysis, the reference management with no modifications of the

Table 18.8 Changes (%) in evapotranspiration, water infiltration, total gross primary production (GPP), total stem wood growth, total timber yield, mean stem diameter and mortality due to varying thinning intensity on different bioclimatic zones (sites), as related to the values under no thinning, assuming climate change (Ge et al. 2013a).

Bioclimatic zone (site) and item	% change due to thinning in item values related to values under no thinning under climate change				
	T10	T20	T30	T40	T50
Site Z_I					
• Evapotranspiration	−1	−1	−2	−4	−8
• Water infiltration	+2	+4	+5	+7	+9
• Total GPP	+3	+3	−6	−16	−28
• Total stem wood growth	+2	+1	−1	−6	−14
• Total timber yield	+19	+19	+16	+10	+2
• Mean diameter	+7	+15	+24	+35	+47
Site Z_{II}					
• Evapotranspiration	−1	−2	−3	−6	−10
• Water infiltration	+2	+4	+6	+8	+10
• Total GPP	+3	−1	−8	−18	−30
• Total stem wood growth	+1	0	−5	−11	−16
• Total timber yield	+16	+16	+12	+6	−2
• Mean diameter	+7	+15	+25	+35	+47
Site Z_{III}					
• Evapotranspiration	−1	−2	−3	−5	−9
• Water infiltration	+2	+3	+5	+6	+7
• Total GPP	+1	−3	−10	−21	−32
• Total stem wood growth	+1	0	−3	−9	−16
• Total timber yield	+13	+13	+9	+4	−4
Mean diameter	+7	+15	+24	+34	+46
Site Z_{IV}					
• Evapotranspiration	−2	−4	−7	−12	−17
• Water infiltration	+2	+4	+6	+9	+10
• Total GPP	−1	−9	−18	−30	−42
• Total stem wood growth	0	−2	−6	−12	−19
• Total timber yield	+6	+4	+21	−6	−13
• Mean diameter	+6	+13	+21	+29	+39
Site Z_V					
• Evapotranspiration	−1	−3	−7	−10	−15
• Water infiltration	+3	+5	+7	+8	+10
• Total GPP	−7	−17	−29	−41	−51
• Total stem wood growth	−1	−5	−10	−16	−22
• Total timber yield	−2	−5	−10	−15	−21
• Mean diameter	+4	+9	+16	+20	+29

current management rules was used for detecting the impacts of different management strategies on adapting to the climate change.

The simulations showed that the reduction of rotation length clearly reduced the growth of Norway spruce (up to 16%) but increased total growth (up to 28%) (Table 18.9), as the growth of Scots pine and birch increased. The increase was largest in the south, where total growth increased up to 35%. This was much more than that obtained under the preference of Scots pine on sites of *Myrtillus* type (12%). The increased growth of Scots pine was not able to compensate for the reduction in the growth of Norway spruce and birch had done. In this case, the total growth increased most overall (38%). A proper choice of tree species and provenance seemed to maintain the productivity of forest land under the climate change. Furthermore, reduced rotation length with more rapid turnover of forest resources may help to maintain the productivity and makes it possible to modify management to meet climate-induced changes in site conditions (Kellomäki et al. 2008, Innes et al. 2009).

Table 18.9 Mean annual growth of different tree species in southern and northern Finland for the period 2070–2099 under climate change (SRES A2 scenario) and selected management regimes, with a percentage (%) of that under the management if no modification in management has been done (Kellomäki et al. 2008). Southern Finland is the area below 63° N, while northern Finland is above 63° N.

Management regime	Growth, m³ha⁻¹yr⁻¹ (% of that with no modifications in management)			
	Scots pine	Norway spruce	Birch	Total
Strategy 1: Management with no modifications of current management rules				
• South	2.81	0.26	3.62	6.69
• North	3.19	0.58	0.84	4.61
• Total	2.96	0.39	2.49	5.84
Strategy 2: Management with reduced rotation length				
• South	3.42 (+22)	0.24 (−8)	5.36 (+48)	9.02 (+35)
• North	3.70 (+16)	0.49 (−16)	1.07 (+27)	5.26 (+14)
• Total	3.54 (+20)	0.34 (−13)	3.62 (+45)	7.50 (+28)
Strategy 3: Preferring Scots pine on *Myrtillus* site if previously occupied by Norway spruce. Reduced rotation length				
• South	4.06 (+44)	0.17 (−35)	3.29 (−9)	7.53 (+13)
• North	3.99 (+25)	0.39 (−33)	0.70 (−17)	5.08 (+10)
• Total	4.03 (+36)	0.26 (−33)	2.24 (−10)	6.53 (+12)
Strategy 4: Preferring birch on *Myrtillus* site if previously occupied by Norway spruce. Reduced rotation length				
• South	3.12 (+11)	0.17 (−35)	6.79 (+88)	10.08 (+51)
• North	3.53 (+11)	0.49 (−16)	1.14 (+36)	5.16 (+12)
• Total	3.29 (+11)	0.30 (−23)	4.49 (+80)	8.08 (+38)
Strategy 5: Preferring Norway spruce of more southern ecotype. Reduced rotation length				
• South	3.04 (+8)	0.67 (+158)	5.56 (+54)	9.27 (+39)
• North	3.60 (+13)	0.44 (−24)	1.23 (+46)	5.27 (+14)
• Total	3.27 (+10)	0.57 (+46)	3.80 (+53)	7.64 (+31)

Reducing wind damage

Wind damage is related to the structure of both forest landscape and the single tree stands forming the landscape. Wind damage is most likely to occur in places where wind loading increases suddenly, not allowing trees to acclimate to the new situation. For example, stands adjacent to recent clear cut areas or stands recently thinned are prone to wind damage. When planning the spatial pattern of clear cuts in the landscape and the management of forest edges, the basic issue is how clear cuts affect the local speed and direction of airflow at the downwind edges of open areas. At the landscape scale, the risk of wind damage may be reduced, for example, by avoiding new edges in old stands and by cutting the most vulnerable stands, first old ones. After the closure of regenerated gaps, the risk of wind damage at the edges will reduce behind. Similarly, the probability of damage decreases over time, when thinning increases growth and the consequent strength of remaining trees to resist dynamic wind load.

Currently, the risk of wind damage is highest in southern Finland, where forests are in many places characterized by mature Norway spruces with shallow rooting most susceptible to wind damage. In northern Finland, damaging wind speeds are much higher (i.e., the risk is lower); because of Scots pines at the intermediate phases dominate northern boreal forests (Kellomäki et al. 2009). In both cases, the risk of wind damage may be reduced by adjusting the rotation length to make it short enough to reduce bending moment due to wind force. This is demonstrated in Fig. 18.8, which shows the current distribution of damaging wind speeds across Finland, and changes in the distribution, when the current rotation length is increased or decreased. The forests are most vulnerable in the south, and longer rotations make these forests even more vulnerable; i.e.,

Management of Forests for Adaptation to Climate Change   291

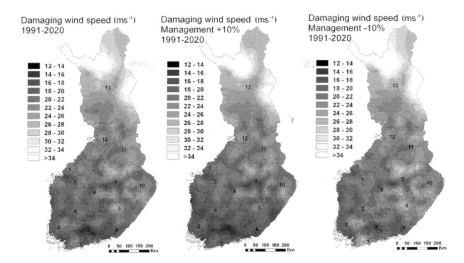

Fig. 18.8 Distribution of damaging wind speeds across Finland, currently (left) and if the rotation length is increased (+10%, with the increase in the minimum diameter indicating the maturity for clear cutting) or decreased (−10 %) (Kellomäki et al. 2009). Calculations are based on the forest inventory datasets for the whole country, and they represent the average critical wind speeds needed to uproot trees. A mechanistic wind damage model (Peltola and Kellomäki 1993) is used in calculations.

further maturing will increase the bending moment and thus reduce the damaging wind speed. A shorter rotation has the opposite effect, reducing vulnerability to wind damage.

Management and timber harvest affect the balance between timber production and risk of wind damage. For example, Zeng et al. (2007) used the length of vulnerable edges in the forest landscape to indicate this balance when using clear-cutting and planting in regeneration over three planning periods of 10 years (Table 18.10). The risk of wind damage was either minimized or maximized, with or without an even-flow target of harvested timber. Planning problems were focused only on minimizing the length of edges at risk (Problem 2), maintaining an even flow of timber (Problem 5), or combined both objectives (Problem 4). Problems 1 and 3 involved the maximum impact of forest harvest on the risk of wind damage, thus demonstrating how improper management may increase the risk of wind damage.

The simulations showed that the total length of vulnerable edges could be reduced by: (i) aggregating clear cuts (i.e., decreasing the total length of the edges); (ii) locating clear cuts at the edges of young stands (i.e., tree height < 10 m) or at the edges of stands with high critical wind speeds; and (iii) making the landscape smooth in terms of stand height. Clustering clear cut areas from the same period will further decrease the cost of logging operations. However, the requirement for an even flow of timber may limit the potential to decrease the risk of wind damage in terms of the total length of vulnerable edges (see Problem 4 compared to Problem 2). When minimizing the risk of wind damage without setting any specific harvest objectives, the amount of timber does not consistently increase when the critical wind speed criterion is decreased, and vice versa. At the forest unit level, the optimum clear cut regimes also depended on the age structure and spatial distributions of permanent gaps (e.g., fields, lakes) and old stands.

Table 18.10 Evaluation of wind-induced risk due to management focusing on timber production, assuming that damage occurs at the critical wind speed of < 20 m s⁻¹ (Zeng et al. 2007). The calculations refer to a forest area of 395 ha, representing central boreal forests (N 63°01', E 27°48'). The area was mostly dominated by Scots pine and Norway spruce.

Planning problems (395 ha of forest unit)	Problem 1: Max. risk edges	Problem 2: Min. risk edges	Problem 3: Max. risk edges, cut 12,000 m³	Problem 4: Min. risk edges, cut 12,000 m³	Problem 5: Cut 12,000 m³
Harvest, m³					
• 1st period	6,964	2,989	12,015	11,958	11,998
• 2nd period	9,307	8,752	12,006	11,932	11,999
• 3rd period	15,561	3,698	12,007	11,999	12,003
• Total	31,832	15,439	36,028	35,889	36,000
Final volume, m³	67,536	81,929	64,087	64,606	64,874
Number of gaps, with mean area in ha in the parenthesis					
• 1st period	17 (2.4)	10 (2.0)	26 (2.3)	20 (2.9)	20 (3.2)
• 2nd period	27 (1.9)	12 (3.6)	33 (1.7)	19 (3.1)	27 (2.0)
• 3rd period	45 (1.4)	9 (2.8)	24 (2.3)	25 (2.1)	23 (2.3)

18.5 Optimal Management under Climate Change

Stand-scale simulations

In general, management is an optimization problem, in which the preferences of the forest owner are balanced with constraints set by environmental conditions, legislation, economy, etc. Under stable climatic conditions, management rules are general recommendations for rotation length, timing and intensity of thinning, etc. Recommendations may be based on analysis, where the timing of thinning and the terminal cut are optimized to maximize timber production or the net present value. Under climate change, growth may deviate from that under the current climate, making the recommendations for the current climate sub-optimal if using the rule-based management excluding gradual changes in growing conditions. In this context, Zubizarreta-Gerendiain et al. (2015) used the simulation/optimization system in analyzing how climate change may change the optimal management at the stand scale under boreal conditions. In simulations, the growth provided by a growth and yield model was corrected to include the climate change impact on radial growth (Pukkala and Kellomäki 2012).

Figure 18.9 shows that climate warming made thinning earlier for Norway spruce and birch than under the current climate. At the same time, the final cut was advanced up to seven years earlier for Norway spruce and 12 years earlier for birch. In both cases, the advance was larger on fertile (*Oxalis-Myrtillus* type, OMT) than on medium fertile (*Myrtillus* type, MT) sites. In Scots pine, the response to climate warming tended to be opposite delaying thinning and the final cut. This was especially the case on poor (*Vaccinium* type, VT) sites, where the changing climate increased growth relatively more and longer than on fertile sites (Zubizarreta-Gerendiain et al. 2015).

Regardless of climate scenario and tree species, timber yield was smaller on medium fertile than on fertile site. Under the current climate, timber yield for Norway spruce was 6.6 and 8.2 m³ ha⁻¹ yr⁻¹ on the MT and OMT sites. At the same time, timber yield for Scots pine was 5.8 and 6.5 m³ ha⁻¹ yr⁻¹ on the VT and MT sites, and for birch 5.2 to 7.4 m³ ha⁻¹yr⁻¹ on the MT and OMT sites. The effect of site fertility on timber yield was of the same magnitude than climate warming effect on growth, thus increasing the timber yield for Scots pine by 27 and 21% on the VT and MT sites, and 14 and 16% for Norway spruce and 17 and 11% for birch on the MT site and OMT sites. Zubizarreta-Gerendiain et al. (2015) also found that the yield of saw logs increased more than the yield of pulp wood if

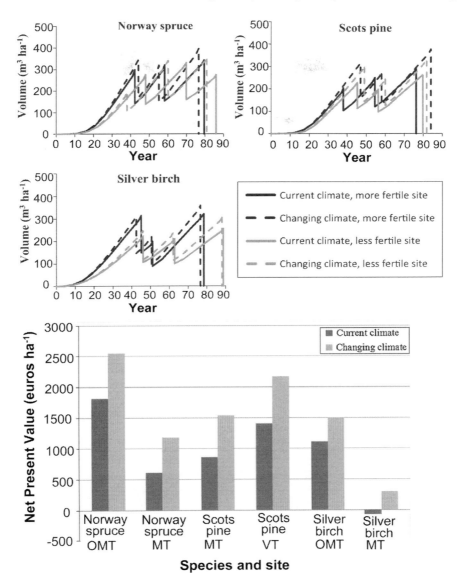

Fig. 18.9 Above: Stocking over time under optimized management (timing of thinning and final cut) per tree species as a function of site fertility and climate scenario, and Below: Net present value per trees species for optimized management as a function of site fertility and climate scenario for central Finland (62° N) (Zubizarreta-Gerendiain et al. 2015). Permission of Springer. The decision variables included the number of years since the previous thinning and the share of dominating species and other species removed in two thinning interventions. In the simulations, the mean annual temperature increased from the current 2.3 to 6.2°C by 2100, while the mean annual precipitation increased from 493 to 554 mm and atmospheric CO_2 from 363 to 635 ppm (SRES A1B emission scenario). The optimization for management maximized the net present value (NPV, 3% interest rate) of future income.

a warming climate was assumed. At the same time, the NPV increased regardless of species and site fertility. In Norway spruce, the NPV increased 40% on the MT site and 90% on the OMT site compared to the values under the current climate. In Scots pine, NPV increased even more: 78% on the MT site and 54% on the VT site.

Regional-scale simulations

At a regional scale, the maximum sustainable yield depends on the properties of sites (e.g., fertility, soil type and water holding capacity) and tree stands (e.g., tree species, maturity, stocking) throughout the region responding differently to climate change, with impacts on growth and risk of damage. In this context, Pukkala et al. (2013) investigated optimal management on the regional scale in a middle boreal forest landscape (62° N). The study area was 950 ha, representing managed forests dominated by even-aged stands of Scots pine, Norway spruce and birch distributed in 650 separate stands. A part of land (44%) is of low fertility (*Vaccinium* type, VT) occupied by Scots pine, whereas 32% of land is of medium fertility (*Myrtillus* type, MT) occupied all the species alone or in mixtures. The rest (24%) of land is of high fertility (*Oxalis-Myrtillus* type, OMT) occupied mainly by Norway spruce and birch alone or in mixtures. Scots pine represented 53%, Norway spruce 31% and birch 16% of total stocking of stem wood. The total stem wood stock was 233,500 m^3 distributed evenly in different age classes: 25, 39.4, 19.5 and 16.1% in the classes 0–19, 20–59, 60–99 and > 100 year.

The MONSU optimizing system was used, considering target variables such as timber production, carbon sequestration, biodiversity (dead wood) and the risk of damage due to high wind and snow load. Critical values of wind speed for uprooting trees were calculated for each stand based on Heinonen et al. (2010). The probability of wind damage (critical wind speed) was calculated every five years for each stand. In doing this, the occurrence of maximum wind speeds (of at least 10 minute duration during a 10-year period) was assumed to be normally distributed with a mean of 13.8 ± 3 m s^{-1}. Under the changing climate, the probability of wind-induced damage was further expected to increase by 0.17% per year until 2100 due to the gradually increasing period of unfrozen soil (Kellomäki et al. 2010), but not assuming any change in mean wind velocity.

Growth and development and management of each stand over 60 years were simulated by using five-year time step. Simulation design included the simulations both under current and changing climate, with and without considering the effects of wind damage on forest growth. Figure 18.10 shows that wind damage decreased the timber yield by 12% under the current climate and 13% in changing climate. The timber yield was, however, larger due to increased growth than that under the current climate thus compensating partly for the losses due to increased damage. The actual loss of timber may be even less since some of the wind thrown trees may be salvaged. On the other hand, the baseline management used in the simulations has no specific features with which to avoid the wind-induced damage likely to occur under the warming climate (Pukkala et al. 2013). The adaptation of regional forest structure to warming is likely to allow increasing the yields of timber and biomass in optimizing forest management at the regional scale.

18.6 Concluding Remarks

Adaptation is a long-term process including monitoring of growth, development and health of forests, and analysis of observations in order to plan to meet the future climate. The necessary actions are, however, very unclear, due to uncertainties about the development of the future climate and societal preferences, and they may involve a variety of strategies in order to respond to climate change in a suitable way (Lindner 2000, Ford et al. 2011). In this regard, Jandl et al. (2015) divide adaptive management and operations in management into those increasing the heterogeneity in forests and those affecting the exposure of forests to abiotic and biotic disturbances. From a strategic perspective, increasing heterogeneity distributes climate change risks throughout the landscape, including in forest sites and stands with different vulnerability. Such landscape structure is probably

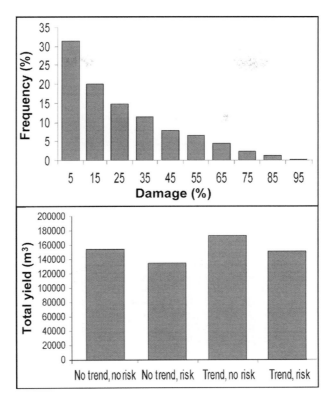

Fig. 18.10 Above: Assumed distribution of the severity of wind damage used in simulations. Below: Simulated timber yield from a forest area over 60 years using four different climate scenarios excluding/including wind damage under the current climate (no trend) and changing climate (trend) (Pukkala et al. 2013). In simulations, the current climate was that for the period 1960–2000, and the changing climate was based on the SRES A1B emission scenario with the atmospheric CO_2 increased from 363 to 635 ppm until 2100. The mean summer temperature (June–August) increased from 14 to 17°C, and the mean annual precipitation from 500 to 550 mm.

increasing the resistance and resilience of ecosystem dynamics in response to the increase in disturbances and losses that climatic warming is likely to result in for forests. Jandl et al. (2015) further note that adaptive management depends also on the needs that society expresses towards the provision of forest products and ecosystem services.

References

Albert, M. and M. Schmidt. 2010. Climate-sensitive modelling of site-productivity relationships for Norway spruce (*Picea abies* (L.) Karst.) and common beech (*Fagus sylvatica* L.). Forest Ecology and Management 259: 739–749.
Allen, C. D., A. K. Macalady, H. Chenchouni, D. Bachelet, N. McDowell, M. Vennetier et al. 2010. A global overview of drought and heat-induced tree mortality reveals emerging climate change risks for forests. Forest Ecology and Management 259: 660–684.
Bergh, J., S. Linder and J. Bergström. 2005. Potential production of Norway spruce in Sweden. Forest Ecology and Management 204: 1–10.
Beuker, E. 1994. Long-term effects of temperature on wood production of *Pinus sylvestris* L. and *Picea abies* (L.) Karst. in old provenance experiments. Scandinavian Journal of Forest Research 9: 34–45.
Beuker, E., P. Hari, J. Holopainen, T. Holopainen, H. Hypén, H. Hänninen et al. 1996. Metsät. pp. 70–106. *In*: E. Kuusisto, L. Kauppi and P. Heikinheimo (eds.). Helsinki University Press, Helsinki, Finland.

Briceño-Elizondo, E., J. Garcia-Gonzalo, H. Peltola, J. Matala and S. Kellomäki. 2006. Sensitivity of growth of Scots pine, Norway spruce and silver birch to climate change and forest management in boreal conditions. Forest Ecology and Management 232: 152–167.

Choat, B., S. Jansen, T. J. Brodribb, H. Cochard, S. Delzon, R. Bhaskar et al. 2012. Global convergence in the vulnerability of forest to drought. Nature 491: 752–755.

Dang, Q. -L., J. M. Maepea and W. H. Parker. 2008. Genetic variation of ecophysiological responses to CO_2 in *Picea glauca* seedlings. The Open Forest Science Journal 1: 68–79.

Ford, C. R., S. H. Laseter, W. T. Swank and J. M. Vose. 2011. Can forest management be used to sustain water-based ecosystem services in the face of climate change? Ecological Applications 21(6): 2049–2067.

Ge, Z. -M., X. Zhou, S. Kellomäki, K. -Y. Wang, H. Peltola, H. Väisänen et al. 2010. Effects of changing climate on water and nitrogen availability with implications on the productivity of Norway spruce stands in southern Finland. Ecological Modeling 221: 1731–1743.

Ge, Z. -M., S. Kellomäki, H. Peltola, X. Zhou, K. -Y. Wang and H. Väisänen. 2011a. Effect of varying thinning regimes on carbon uptake, total stem wood growth, and timber production in Norway spruce (*Picea abies*) stands in southern Finland under the changing climate. Annals of Forest Science 68: 371–383.

Ge, Z. -M., S. Kellomäki, H. Peltola, X. Zhou, K. -Y. Wang and H. Väisänen. 2011b. Impacts of changing climate on the productivity of Norway spruce dominant stands with a mixture of Scots pine and birch in relation to the water availability in southern and northern Finland. Tree Physiology 31: 323–338.

Ge, Z. -M., S. Kellomäki, H. Peltola, X. Zhou and H. Väisänen. 2013a. Adaptive management to climate change for Norway spruce forests along a regional gradient in Finland. Climatic Change 118(2): 275–289.

Ge, Z. -M., S. Kellomäki, X. Zhou, K. -Y. Wang, H. Peltola, H. Väisänen et al. 2013b. Effects of climate change on evapotranspiration and soil water availability in Norway spruce forests in southern Finland: ecosystem model based approach. Ecohydrology 6: 51–63.

Ge, Z. -M., S. Kellomäki, X. Zhou and H. Peltola. 2014. The role of climatic variability in controlling carbon and water budgets in a boreal Scots pine forests during ten growing seasons. Boreal Environmental Research 19: 181–194.

Heinonen, T., T. Pukkala, V. -P. Ikonen, H. Peltola, A. Venäläinen and S. Dupont. 2010. Integrating the risk of wind damage into forest planning. Forest Ecology and Management 258(7): 1567–1577.

Innes, J., L. A. Joyce, S. Kellomäki, B. Louman, A. Ogden, J. Parrotta et al. 2009. Management for adaptation. *In*: R. Seppälä, A. Buck and P. Katila (eds.). Adaptation of Forests and People to Climate Change. A Global Assessment Report. IUFRO World Series 22: 135–186.

Jääskeläinen, H. 2014. Männyn alkuperäsiirroista saatava hyöty muuttuvassa ilmastossa? M.Sc. Thesis, University of Eastern Finland, Joensuu, Finland.

Jandl, R., J. Bauhus, A. Bolte, A. Schindlbacher and S. Schüler. 2015. Effect of climate-adapted forest management on carbon pools and greenhouse emissions. Current Forestry Reports 1: 1–7.

Johnsen, Ø., O. G. Dæhlen, G. Østreng and T. Skrøppa. 2005a. Day length and temperature during seed production interactively affect adaptive performance of *Picea abies* progenies. New Phytologist 168: 589–596.

Johnsen, Ø., C. G. Fossdal, N. Nagy, J. Mølmann, O. G. Dæhlen and T. Skrøppa. 2005b. Climatic adaptation in *Picea abies* progenies is affected by the temperature during zygotic ambryogenesis and seed maturation. Plant, Cell and Environment 28: 1090–1102.

Johnston, M. 2010. Limited Report. Tree Species Vulnerability and Adaptation to Climate Change SRC Publication No. 12416-1E10. Saskatchewan Research Council, Saskatoon, Canada.

Jylhä, K., K. Ruoteenoja, J. Räisänen, A. Venäläinen, H. Tuomenvirta, L. Ruokolainen et al. 2009. Arvioita Suomen muuttuvasta ilmastosta sopetumistutkimuksia varten. ACCLIM- hankkeen raportti 2009. Raportteja 2009/4: 1–102.

Jyske, T., T. Hölttä, H. Mäkinen, P. Nöjd, I. Lumme and H. Spiecker. 2010. The effect of artificially induced drought on radial increment and wood properties of Norway spruce. Tree Physiology 30: 103–115.

Kellomäki, S. and M. Kolström. 1992a. Computations on the management of seedling stands of Scots pine under the influence of changing climate. Silva Fennica 26(2): 97–110.

Kellomäki, S. and M. Kolström. 1992b. Simulation of tree species composition and organic matter accumulation in Finnish boreal forests under changing climatic conditions. Vegetatio 102: 47–68.

Kellomäki, S. and S. Leinonen. 2005. Management of European forests under changing climatic conditions. University of Joensuu. Faculty of Forestry. Research Notes 163: 1–427.

Kellomäki, S., H. Strandman, T. Nuutinen, H. Peltola, K. T. Korhonen and H. Väisänen. 2005. Adaptation of forest ecosystems, forests and forestry to climate change. Finnish Environment Institute, FinAdapt Working Paper 4: 1–50.

Kellomäki, S., H. Peltola, T. Nuutinen, K. T. Korhonen and H. Strandman. 2008. Sensitivity of managed boreal forests in Finland to climate change, with implications for adaptive management. Philosophical Transactions of the Royal Society B363: 2341–2351.

Kellomäki, S., V. Koski, P. Niemelä, H. Peltola and P. Pulkkinen. 2009. Management of forest ecosystems. pp. 252–373. *In*: S. Kellomäki (ed.). Forest Resources and Sustainable Management. Gummerus Oy, Jyväskylä, Finland. Second edition.

Kellomäki, S., M. Maajärvi, H. Strandman, A. Kilpeläinen and H. Peltola. 2010. Model computations on the climate change effects on snow cover, soil moisture and soil frost in the boreal conditions over Finland. Silva Fennica 44(2): 213–233.

Kohler, M., J. Sohn, G. Nägele and J. Bauhus. 2010. Can drought tolerance of Norway spruce (*Picea abies* (L.) Karst.) be increased through thinning? European Journal of Forest Research 129: 1109–1118.

Lindner, M. 2000. Developing adaptive forest management strategies to cope with climate change. Tree Physiology 20: 299–307.
Ma, Z., C. Peng, Q. Zhu, H. Chen, G. Yu, W. Li et al. 2012. Regional drought-induced reduction in the biomass carbon sink in Canada's boreal forests. PNAS 109: 2423–2427.
Mäkinen, H., P. Nöjd and K. Mielikäinen. 2000. Climatic signal in annual growth variation of Norway spruce (*Picea abies*) along a transect from central Finland to the Arctic timberline. Canadian Journal of Forest Research 30: 769–777.
Mäkinen, H., P. Nöjd and K. Mielikäinen. 2001. Climatic signal in annual growth variation in damaged and healthy stands on Norway spruce [*Picea abies* (L.) Karst] in southern Finland. Trees 15: 177–185.
Millar, C. I., N. L. Stephenson and S. L. Stephens. 2007. Climate change and forests of the future: managing in the face of uncertainty. Ecological Applications 17: 2145–2151.
Näslund, M. 1936. Skogsförsöksanstaltens gallringsförsök I tallskog. Primärbearbetning. Meddelanden från Statens Skogsförsöksanstalt. Häfte 29: 1–179.
Parmesan, C. and G. Yohe. 2003. A globally coherent fingerprint of climate change impacts across natural systems. Nature 421(2): 37–42.
Peltola, H. and S. Kellomäki. 1993. A mechanistic model for calculating wind throw and stem breakage of Scots pine at stand edge. Silva Fennica 27(2): 99–111.
Peng, C., Z. Ma, X. Lei, Q. Zhu, W. Wang, S. Liu et al. 2011. A drought-induced pervasive increase in tree mortality across Canada's boreal forests. Nature Climate Change 1: 467–471.
Persson, B. and E. Beuker. 1997. Distinguishing between the effects of changes in temperature and light climate using provenance trials with *Pinus sylvestris* in Sweden. Canadian Journal of Forest Research 27: 572–579.
Persson, B. 1998. Will climate change affect the optimal choice of *Pinus sylvestris* provenances? Silva Fennica 32(2): 121–128.
Pukkala, T. and S. Kellomäki. 2012. Anticipatory vs. adaptive optimization of stand management when tree growth and timber prices are stochastic. Forestry 85(4): 463–472.
Pukkala, T., S. Kellomäki and H. Peltola. 2013. Wind risk in the case study for Finland. Unpublished internal report to the Motive project. University of Eastern Finland, Finland, Joensuu.
Rehfeldt, G. E., W. R. Wykoff and C. C. Ying. 2001. Physiologic plasticity, evolution, and impacts of a changing climate on *Pinus contorta*. Climatic Change 50: 355–376.
Saxe, H., M. R. G. Cannell, O. Johnsen, M. G. Ryan and G. Vourlitis. 2001. Tree and forest functioning in response to global warming. New Phytologist 149: 369–400.
Seidl, R., W. Rammer and M. J. Lexer. 2011. Adaptation options to reduce climate change vulnerability of sustainable forest management in the Austrian Alps. Canadian Journal of Forest Research 41: 694–706.
Smit, B. and J. Wandel. 2006. Adaptation, adaptive capacity and vulnerability. Global Environmental Change 16: 282–292.
Spittlehouse, D. L. and R. B. Stewart. 2003. Adaptation to climate change in forest management. BC Journal of Ecosystems and Management 4(1): 1–11.
Stankey, G. H., R. N. Clark and B. T. Bormann. 2005. Adaptive management of natural resources: theory, concepts, and management institutions. General Technical Reports PNW-GTR-654. Portland, OR: U.S. Department of Agriculture, Forest Service, Pacific Northwest Research Station, USA.
Thornley, J. H. M. and M. G. R. Cannell. 2001. Managing forests for wood yield and carbon storage: a theoretical study. Tree Physiology 20: 477–484.
Torssonen, P., H. Strandman, S. Kellomäki, A. Kilpeläinen, K. Jylhä, A. Asikainen et al. 2015. Do we need to adapt the choice of main boreal tree species in forest regeneration under the projected climate change? Forestry 88(5): 564–572.
Vygodskaya, N. N., E. -D. Schulze, N. M. Tchebakova, L. O. Karpachevskii, D. Kozlov, K. N. Sidorov et al. 2002. Climatic control of stand thinning in unmanaged spruce forests in the southern taiga in European Russia. Tellus 54B: 443–461.
Wang, Y., E. Hogg, D. T. Prince, J. Edwards and T. Williamson. 2014. Past and projected future changes in moisture conditions in the Canadian boreal forest. The Forestry Chronicle 90(5): 678–690.
Zeng, H., P. Pukkala and H. Peltola. 2007. The use of heuristic optimization in risk management of wind damage in forest planning. Forest Ecology and Management 241: 189–199.
Zhao, M. and S. W. Running. 2010. Drought induced reduction in global terrestrial net primary production from 2000 through 2009. Science 329: 940–943.
Zubizarreta-Gerendiain, A., T. Pukkala, S. Kellomäki, J. Garcia-Gonzalo, V. -P. Ikonen and H. Peltola. 2015. Effects of climate change on optimised stand management in boreal forests of central Finland. European Journal of Forest Research 134: 273–280.

Management of Forests for the Mitigation of Climate Change

ABSTRACT

Climate change mitigation involves reducing long-term risks, which may be the result of climate change. Mitigation includes reducing the sources or enhancing the sinks of greenhouse gases in order to limit the radiative forcing controlling global warming. In forestry, mitigation involves management, which enhances CO_2 absorption and elongates the CO_2 residence in forest ecosystems. Climate warming may be further reduced using forest biomass and timber to replace fossil fuels and to increase the use of wood-based materials to increase carbon density outside forests. Climate warming may also be slowed down if forest structures with high albedo are preferred in management.

Keywords: climate change mitigation, carbon sequestration, substitution of fossil fuels and materials, carbon neutrality of bioenergy, albedo, radiative forcing

19.1 Climate Change Mitigation in Forestry

Forest-based climate change mitigation

Climate change mitigation involves strategies to reduce the long-term risks and hazards to natural and human systems, which may be the result of climate change. Mitigation is focused on reducing and/or eliminating the causes behind climate change, whereas adaptation to climate change involves the adjustment of natural and/or human systems to the moderate harms caused by climate change, and/or exploits opportunities provided by the changing climate. Mitigation and adaptation have the same strategic framework, but mitigation involves anthropogenic interventions to reduce the sources or enhance the sinks of greenhouse gases (GHG) (IPCC 2007) for limiting the increase of radiative forcing and the potential global warming.

Forests and forestry contribute in several ways to mitigating climate change, including by retaining carbon in forest ecosystems, replacing fossil fuels and materials with biofuels and biomaterials and preferring a high albedo of forest cover in management. Through management, carbon stocks in the forest ecosystem are increased/decreased, with indirect effects on radiative forcing. At the same time, reduced/increased thinning intensity and longer/shorter rotation are likely to reduce/increase albedo, and more/less radiation is absorbed into the ecosystem, thus enhancing/reducing radiative forcing. Albedo is also altered by changes in tree species composition, with direct effects on radiative forcing. Management impacts on carbon stocks and albedo are generally opposite; i.e., increasing carbon stock tends to reduce the albedo, and reducing carbon stock to increase albedo. However, a high forest growth (and carbon sequestration) and processing wood for long-lived products is of importance in mitigating climate change (Eriksson et al. 2007, Bonan 2008, Klein et al. 2013).

Carbon uptake and emissions in climate change mitigation

Forest-based mitigation is affected by the CO_2 uptake and emissions in the forest ecosystem and in the forest product technosystem (Fig. 19.1). In the context of forest ecosystem, the Net Ecosystem Exchange (NEE) indicates the balance if carbon is emitted to the atmosphere or sequestrated in trees and soil. Forest product technosystem includes CO_2 emissions due to the management and harvest operations in producing timber and biomass, which is used in producing materials and fuels for replacing fossil carbon bound in materials like in concrete and coal. The net technosystem exchange contributes to net carbon emissions to the atmosphere and site, and it affects radiative forcing along with the contribution from forest ecosystem (Fig. 19.1).

There are several direct ways to mitigate climate change in forestry (Nabuurs et al. 2007): (i) to reduce deforestation and degradation of existing forests; (ii) to increase the forested land area in afforestation/reforestation; and (iii) to increase carbon density in existing forests. At the same time, timber and biomass can be used to avoid fossil-CO_2 emissions (Karjalainen 1996a,b): (iv) to replace fossil fuels with biomass-based fuels; and (v) to increase the carbon density outside existing forests in wood-based materials (Fig. 19.2). In forestry, the sink capacity in a forest ecosystem is maintained/increased. A forest product technosystem may reduce fossil-CO_2 emissions by substituting fossil fuels (e.g., coal, oil) and materials based on fossil carbon (e.g., concrete) when in cycling carbon in the atmosphere/biosphere/technosystem and slowing down the increase of atmospheric CO_2. Abandoned forest-based materials may further be used for raw material for new products, combusted for energy or abandoned in landfill.

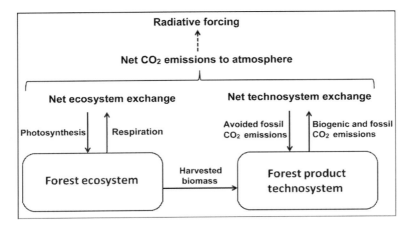

Fig. 19.1 Major flow of carbon in/from forest ecosystem and forest product technosystem, with radiative forcing related to the net emissions of CO_2 into the atmosphere from both the forest ecosystem and forest product technosystem (Sathre et al. 2013).

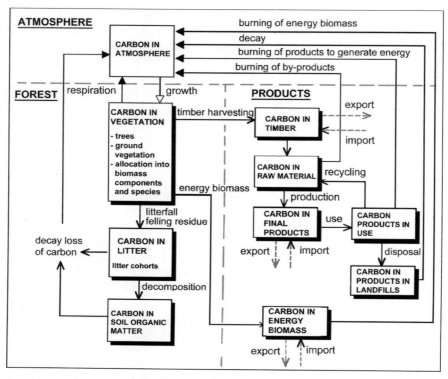

Fig. 19.2 Carbon in atmosphere/biosphere/techosystem. Boxes indicate the carbon stocks, and arrows the processes cycling carbon through the total system (Karjalainen 1996a,b).

19.2 Carbon in Forest Ecosystem

Carbon sink/source dynamics in forest ecosystem

A forest ecosystem is a carbon sink if more carbon is sequestrated than released in autotrophic and heterotrophic respiration. Conversely, a forest ecosystem is a carbon source if the carbon uptake in photosynthesis is less than emissions in respiration. The sink rate involves the carbon uptake rate (the carbon stock in the ecosystem increases), and the source rate involves the carbon emission rate (the carbon stock in the ecosystem reduces). Sink and source rates (carbon dynamics, carbon sink/source dynamics) vary over time depending on the factors driving the carbon uptake rate in Gross Primary Production (GPP), the carbon emission rate in Autotrophic (RA) and Heterotrophic (RH) Respiration, which affects the Net Ecosystem Exchange (NEE) (or Net Ecosystem Production NEP):

$$NEE = GPP - RA - RH = NPP - RH \tag{19.1}$$

where NPP refers to Net Primary Production in the ecosystem.

Net ecosystem exchange is commonly measured using eddy covariance technology (Baldocchi 2003, Box 19.1). In a forest canopy, if the flow velocity is high enough, the air flow is turbulent with eddies (Fig. 19.3). Eddies carry CO_2 (and H_2O) back and forth between atmosphere and canopy, whereas slower air flow is laminar with no/few eddies and only a small exchange of CO_2 (and H_2O) between canopy and atmosphere. The exchange of CO_2 occurs in the boundary layer of the leaves/canopy through the CO_2 uptake in photosynthesis and H_2O losses in transpiration. The CO_2

Box 19.1 Atmosphere/forest ecosystem exchange of carbon in eddy fluxes

The net ecosystem exchange of CO_2 (NEE) is the total carbon in eddy fluxes (F_e, g C m^{-2} ground d^{-1}) monitored above the canopy and the flux from the storage of CO_2 (F_s, g C m^{-2} ground d^{-1}) in the air layer below the canopy (Aubinet et al. 2000):

$$NEE = F_e + F_s \qquad (19.2)$$

The flux from the canopy (F_e) involves the mean covariance between the vertical wind velocity (w′) and the fluctuations in CO_2 density (c′) (Aubinet 2000, Ge et al. 2011b):

$$F_e = \left(\frac{\overline{pT}}{p_i T_a}\right)\left[\overline{w'c_i'} + \frac{}{}(\overline{c'}/\overline{\rho_a})\overline{w'\rho_{vi}'}\right] \qquad (19.3)$$

where p is pressure [Pa], T is temperature [K], ρ_{vi} and ρ_a are the mean density of water vapor. m_a/m_v is the ratio between the molecular mass of dry air and that of water vapor. The bars over factors refer to the time averages and primes to the instantaneous fluctuations around the mean quantities (Aubinet et al. 2000). Carbon storage below the canopy (F_s) is obtained by integrating the rate of temporal change in the profile of CO_2 concentration below the canopy:

$$F_s = \frac{v_a \Delta c_r}{m} \qquad (19.4)$$

where v_a is the volume [m³] of the air column below the height of eddy instrumentation, Δc_r is the change in CO_2 density per unit time at the canopy source height, and m is the molar volume of CO_2 (Aubinet et al. 2000).

Fig. 19.3 Schematic presentation of the CO_2 and H_2O exchange in a forest ecosystem if air flow through the canopy is laminar (left), and turbulent (middle). The exchange of CO_2 and H_2O takes place in the boundary layer of leaves (right).

content in eddies and three-dimensional wind velocity are used to calculate the exchange rate of CO_2. Uptake (photosynthesis) is a negative flux into the ecosystem from the atmosphere, while emission (respiration) is a positive flux from the ecosystem to the atmosphere. In day time, both fluxes occur, whereas at night time there is only respiration.

The diurnal variability in carbon eddy flux over several years is demonstrated in Fig. 19.4, based on measurements for a middle boreal Scots pine forest (62° N). The net ecosystem exchange measured in daylight ($NEE_{c,day}$, g C m^{-2} ground s^{-1}) is gross canopy photosynthesis (A_c, g C m^{-2} ground s^{-1}) minus the sum of respiration in canopy foliage ($R_{c,l}$, g C m^{-2} ground s^{-1}), the respiration of sapwood in stem and branches ($R_{c,w}$, g C m^{-2} ground s^{-1}), and carbon efflux from soil. Carbon efflux from soil involves all sources of autotrophic (e.g., coarse and fine roots) and heterotrophic respiration in the forest floor (R_s, g C m^{-2} ground s^{-1}) (e.g., decay of litter and humus, respiration of roots) (Wang et al. 2004):

$$NEE_{c,day} = A_c - \left(R_{c,l} + R_{c,w} + R_s\right) \qquad (19.5)$$

Similarly, the net ecosystem exchange in night ($NEE_{c,night}$, g C m^{-2} ground s^{-1}) is:

$$NEE_{c,night} = R_{c,d} + R_{c,w} + R_s \qquad (19.6)$$

where $R_{c,d}$ (g C m^{-2} ground s^{-1}) is dark respiration in foliage.

Figure 19.4 shows that the seasonal patterns of daily NEE values were similar throughout the monitoring period 2000–2002 (Wang et al. 2004, Zha et al. 2004). In winter, the values were slightly positive (carbon source) until the end of April. Thereafter, the values were negative, following the release of winter dormancy and increase of photosynthetic capacity from early May to August. In the growing season, the daytime values were −4.8, −5.4 and −6.6 g C m^{-2} d^{-1} at the maximum in 2000, 2001 and 2002, and the mean daytime values were −2.42 ± 0.83, −2.68 ± 0.97 and −2.88 ± 0.93 g C m^{-2} d^{-1}, respectively. In the growing season, nighttime maximum values were 1.8, 2.8 and 2.9 g C m^{-2} d^{-1}, and the mean values were 0.98 ± 0.23, 1.24 ± 0.32 and 1.63 ± 0.28 g C m^{-2} d^{-1}, respectively. From September to October carbon uptake declined until the trees fell into winter dormancy, and the site became a small carbon source again (0.28 ± 0.09 g C m^{-2} d^{-1}), when soil temperature dropped below 5°C.

The seasonality of carbon efflux from the forest floor was greatest in the growing season. The highest peak in mid-summer was up to 0.3 g C m^{-2} h^{-1}, while in winter the efflux was < 0.1 g C m^{-2} h^{-1} (Wang et al. 2004). The same magnitude of soil CO_2 efflux with similar annual and seasonal variability was shown by Niinistö et al. (2004, 2011) at the same site, using chamber measurements. In the snow-free period, soil temperature alone explained 70–80% of the variation in efflux, and root growth with the external mycelium of ectomycorrhizal fungi could explain most unexplained variability. Similarly, Boone et al. (1998) found that a large proportion of CO_2 efflux from soil is due to the root respiration and oxidation of rhizosphere carbon, which is more sensitive to temperature than respiration in bulk soil. Under boreal conditions, the effect of soil moisture on soil CO_2 efflux may often be negligible or nearly nonexistent (Zha et al. 2007).

Table 19.1 summarizes the modeled carbon uptake in canopy photosynthesis and losses in ecosystem respiration (see Fig. 19.4). Annual gross photosynthesis was −730, −817 and −991 g C m^{-2} yr^{-1} in 2000, 2001 and 2002, respectively. Similarly, ecosystem respiration was 565, 629 and 640 g C m^{-2} yr^{-1}, which accounted for 77, 77 and 65% of the annual gross photosynthesis. Carbon efflux from the forest floor contributed the most to ecosystem respiration. The share of the forest floor was 46–62%, while the share of foliage was 27–44% and wood (branches plus stem) 9–11%. The contribution of different components of ecosystem respiration varied more during the growing season than during the non-growing season, when the carbon efflux from the forest floor accounted for more than 77% of the total ecosystem respiration. Finally, the measured annual net carbon uptake (NEE) in ecosystem was −131, −210 and −258 g C m^{-2} yr^{-1} in 2000, 2001 and 2002. Modeling yielded values which were close to those measured (Wang et al. 2004). Measurements represented the peak NEE common for boreal tree stands at the ages varying from 10 to 60 years, declining thereafter along with the maturation of trees (Hyvönen et al. 2007).

The monitoring shown in Fig. 19.4 continued to the end of 2008. Ge et al. (2011b) analyzed the entire time series (1999–2008), which showed similar seasonal and annual variability as the early parts of the time series. During the growing season (May–September), the modeled NEE values varied from −189 to −263 g C m^{-2} yr^{-1} and the measured values from −195 to −268 g C m^{-2} yr^{-1}. Outside the growing season, the simulated values varied from 34 to 69 g C m^{-2} yr^{-1}, and the measured values fell in the range 27–67 g C m^{-2} yr^{-1}. On an annual basis, the modeled NEE was from −129 to −203 g C m^{-2} yr^{-1} and the measured values from −161 to −232 g C m^{-2} yr^{-1} over

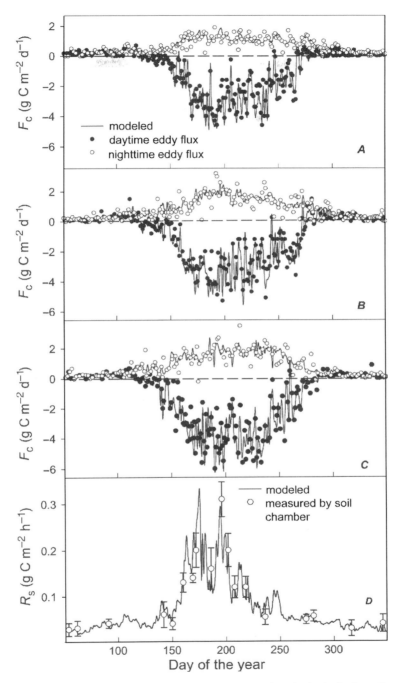

Fig. 19.4 Seasonal and annual variability in daily net ecosystem exchange and respiration in the forest floor in a Scots pine stand in middle boreal conditions (N 62°52', E 30°49', 145 m a.s.l.) in 2000–2002 (Wang et al. 2004). Permission of Oxford University Press. The points are measured, and the line indicates modeled values based on meteorological measurements. The monitoring site was on sandy soil with low fertility (*Vaccinium* type, VT) occupied by a pure Scots pine stand (a mean spacing of 1,175 trees ha^{-1}, a mean height of 12 m and a mean diameter of 11 cm). Climate at the site is characterized by cold winters and short growing season (140–175 days). In 1961–2000, the mean annual precipitation at the site was 700 mm, of which 38% was snow. The mean temperature in January was −10.4°C and in July 15.8°C.

Table 19.1 Modeled gross carbon uptake, ecosystem respiration and net ecosystem exchange [g C m^{-2} yr^{-1}] with the contribution from different sources (Wang et al. 2004).

Parameter	2000 Growing season	2000 Non-growing season	2001 Growing season	2001 Non-growing season	2002 Growing season	2002 Non-growing season
Gross carbon uptake	−718	−12	−808	−9	−980	−11
Ecosystem respiration	501	64	572	57	686	54
Share of ecosystem respiration (%) for						
• Needles	48	17	33	18	28	18
• Branches + stem	10	4	9	5	11	4
• Forest floor	42	79	58	77	61	78
Net ecosystem exchange (NEE)						
• Modeled	−217	52	−236	48	−294	43
• Measured	−194	63	−261	51	−316	58

the period 1999–2008. During the growing season, daily NEE values were mainly dependent on radiation, air temperature and vapor pressure deficit. Drought reduced NEE only in two summers out of 10 (Kellomäki and Wang 2000, Zha et al. 2004, Ge et al. 2011a, 2013, Box 19.2).

Climate change is likely to directly increase the soil CO_2 efflux, as was found in soil warming experiments (Eliasson et al. 2005). On the other hand, elevated CO_2 seems to increase root growth, mycorrhizal symbionts and root exudation as was found in the FACE experiments (Lukas et al. 2009). This implies a fast fine root turnover with an increase of 30–110% of soil carbon and 15–45% in soil CO_2 efflux. Fine roots were distributed under elevated CO_2 deeper in soil profiles than in the ambient conditions (Iversen 2010). Based on measurements in closed chambers over a snow-free season (May–October), Niinistö et al. (2004) found the elevated CO_2 alone to increase the efflux by 23–37% from poor sandy soil occupied by young Scots pines. At the same time, the increase for the elevated temperature alone was 27–43%. The increase was greatest, 35–59%, under the combined elevation of CO_2 and temperature compared to ambient conditions. Haimi et al. (2005) found, however, that elevated temperature had only a slight effect and CO_2 no effect on decomposing fauna in the chambers.

Carbon uptake, litter fall and carbon accumulation

Litter fall is a result of the death of whole trees, or part of the tissues and organs of living trees (foliage, branches, stems, roots) dying, transferring dead organic materials and nutrients into soil detritus. In coniferous forests, for example, the foliage litter is from the oldest foliage but it also includes younger needles (Helmisaari et al. 2009), while in deciduous forests a large part of the litter is from the current-year foliage. In both cases, the woody mass in dead branches declines gradually until the branches are decayed enough to be broken by wind and snow loads. Stem wood is transferred to the heterotrophic system after the death of a whole tree. Carbon in stems is accumulated in tens of decades, or even centuries, and is an important stock of carbon in the forest ecosystem. Nutrient-rich foliage and fine root litter are important short-term nutrient sources, but in the long term even decaying branches, coarse roots and stem wood provide nutrients for uptake (Mälkönen 1976, Hyvönen et al. 2000).

Box 19.2 Environmental control of NEE in a boreal forest ecosystem

A biochemical photosynthetic model (see Chapter 6) was used to identify how environmental conditions control variability in measured NEE values (Wang et al. 2004) (Fig. 19.5). The maximum NEE was −4.62 g C^{-2} d^{-1}. At the same time, the NEE response saturated at a flux density of 21.4 mol m^{-2} d^{-1}, while the initial slope of the response (Q_a < 10 mol m^{-2} d^{-1}) was 0.032. Carbon uptake increased more than 300% when mean temperature increased from 7 to 19°C. At the same time, the carbon flux to the atmosphere increased exponentially up to 4 g C m^{-2} d^{-1}. Throughout the growing season, the daytime NEE was only slightly responsive to the mean daily vapor pressure deficit (D_c) when D_c < 1.0 kPa but the daytime flux declined substantially when D_c > 1.0 kPa. The pattern for soil moisture was opposite, with a rapid increase in carbon flux along with the increase of moisture content (θ_s) from 0.08 m^3 m^{-3} to 0.25 m^3 m^{-3}. The night flux showed no clear pattern in response to soil moisture. During the non-growing season, the daytime flux only slightly responded to temperature < −10°C, above which the carbon flux increased slightly (Wang et al. 2004).

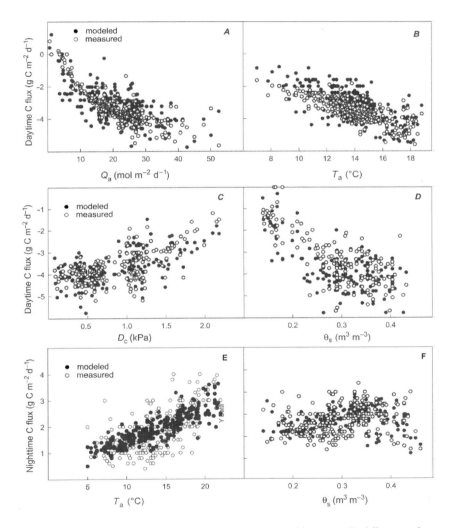

Fig. 19.5 Daytime NEE values as a function of A: daily radiation intercepted in canopy; B: daily mean air temperature; C: vapor pressure deficit of air; and D: water content in soil. Carbon efflux from soil is a function of E: mean daily air temperature; and F: soil water content in the monitoring period (2000–2002) in a middle boreal Scots pine stand described in Fig. 19.4 (Wang et al. 2004). Permission of Oxford University Press.

Litter fall is related to the tree growth, as shown by Matala et al. (2008). They found that in boreal and temperate forests litter fall is related to the growth following the allometric relationship between the growth and mass between different tree organs. Based on the available literature, Matala et al. (2008) related litter fall (the total above-ground litter, mainly foliage litter (Fig. 19.6)):

$$LF(t) = 343 \times \Delta V(t) \tag{19.7}$$

where $LF(t)$ is litter fall from a tree [kg tree^{-1} yr^{-1}] and $\Delta V(t)$ is the volume growth of a tree [m^3 tree^{-1} yr^{-1}]. The values of the regression coefficient varied from species to species, e.g., 362 for Scots pine and 284 for Norway spruce. Matala et al. (2008) applied the model across Finland using data from the Finnish National Forest Inventory. The results showed that litter fall varied between 340 kg ha^{-1} yr^{-1} in the northernmost to 2,300 kg ha^{-1} yr^{-1} in southernmost areas of boreal forests, following increasing forest growth from north to south. At the stand scale, the litter fall seemed to follow the temporal and spatial pattern of growth, whereas the litter in dying trees followed mortality resulting from a variety of reasons and reducing growth of maturing trees. Matala et al. (2008) further found a strong linear correlation between litter fall and the basal area and volume of a tree stand, and the latitude of the site, but the litter fall was not correlated with the age and density of the tree stand.

Fine roots also play an important role in contributing to soil carbon stock. Regarding northern boreal conditions (64° N, Sweden), Leppälammi-Kujansuu et al. (2014) found that in irrigated Norway spruce stands the amount of litter from foliage (1,340 kg ha^{-1} yr^{-1}) and from fine roots (1,020 kg ha^{-1} yr^{-1}) was of similar magnitude. In this experiment, nitrogen fertilization and soil warming alone increased the fine root litter substantially; i.e., 2,900 kg ha^{-1} yr^{-1} in both cases. Soil warming along with fertilization and irrigation, further increased fine root litter up to 4,500 kg ha^{-1} yr^{-1}. The ratio between below-ground/above-ground litter was 0.8–2.2, the lower limit representing the trees grown in the irrigated plots and the upper limit the trees grown in plots with soil warming and irrigation. The increased availability of nitrogen and/or increased soil temperature increased fine root growth but reduced the longevity of roots, thus substantially increasing the amount of fine root litter. The above-ground litter was increased only in the fertilized plots, suggesting that the increase in fine root litter was mainly due to increased root growth caused by soil warming and/or fertilization.

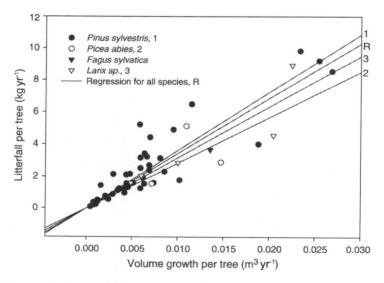

Fig. 19.6 Litter fall calculated per tree as related to volume growth per tree. The species-specific regression lines in Equation (19.7) are referred by the numbers upper right at the end of regression lines. The letter R at end of regression line indicates the regression line for all species combined in the analysis (Matala et al. 2008). Permission of Taylor & Francis.

In general, carbon stock in soil (SOM) is related to the accumulated balance between carbon input in litter fall and carbon output in litter decay. A litter fall event forms a litter cohort, which loses its original weight in decay over years. Matter in decaying litter in successive cohorts accumulates carbon until the addition of matter through litter fall and the loss of matter through decay reaches a balance (steady state), if assuming constant litter fall. According to Olson (1963), the remaining mass fraction of litter cohort and the mass fraction of accumulated litter compared to the amount of carbon at the steady state are:

$$\text{Fraction of litter still remaining from the original weight} = \frac{X}{X_o} = e^{-k \times t} \quad (19.8)$$

$$\text{Fraction of SOM from that at steady state} = \frac{X}{X_{ss}} = 1 - e^{-k \times t} \quad (19.9)$$

where is X_o is the mass of litter at the fall (t = 0), X is the remaining mass in the year t, X_{ss} is the mass of the accumulated litter at the steady state (Fig. 19.7).

The decay and accumulation trajectories under a given decay rate are inverse (Fig. 19.7; Table 19.2). The time for the decay of matter to the half of the original value is obtained from: $0.5 = e^{-k \times t}$, $t = -ln(0.5)/k = 0.693/k = 0.693/0.0625 = 11.1$ years. For example, at the decay rate k = 0.25 yr^{-1} the weight is reduced by 50% in 2.8 years, whereas at the rate k = 0.0625 yr^{-1} it takes 11.1

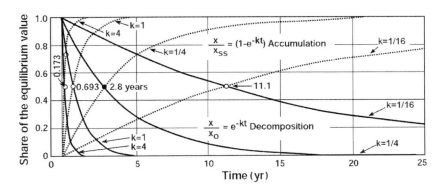

Fig. 19.7 Fraction of litter still remained from the original weight and the fraction of soil organic matter from that at the steady state under constant litter fall as a function of decay rate (k) and time (t) (Olson 1963). Permission of Wiley & Sons.

Table 19.2 Example of time for 50% and 95% mass loss of decaying matter using Equation (19.8) with given decay rates (k) assuming steady state of litter fall (Olson 1963).

Decay parameters		Decay time for 50% mass loss, 0.693/k	Decay time for 95% mass loss, 3/k
k = X/X$_{ss}$	1/k = X$_{ss}$/X		
4	0.25	0.173	0.75
2	0.5	0.346	1.50
1	1.0	0.693	3.0
0.693	1.442	1.000	4.33
0.5	2	1.386	6
0.25	4	2.772	12
0.125	8	5.544	24
0.0625	16	11.09	48
0.0312	32	22.21	96
0.0156	64	44.42	192
0.01	100	69.31	300
0.003	333	232.3	1,000
0.001	1,000	693.1	3,000

years for the same weight loss. The time period 3/k is needed to the weight 95% loss of the original weight (Olson 1963).

Equation (19.8) is most applicable to decaying litter such as deciduous foliage, but the constant decay rate is frequently used for other types of litter. Depending on tree species, the decay rate (k) for coniferous foliage is 0.20–0.40 yr^{-1} and for deciduous foliage 0.30–0.60 yr^{-1} (Valachovic et al. 2004). In the former case, 95% of the mass would be lost in 10–15 years and in the latter case 5–10 years, depending on species. The decay rate for fine roots is of the same magnitude as that for foliage, but the decay of fine root litter tends to be slower (Gholz et al. 2000). The decay rate for Scots pine twigs and branches may vary in the range of 0.128–0.188 yr^{-1} (95% weight loss in 15–25 years) and 0.066–0.127 yr^{-1} (95% weight loss in 25–45 years). The lower limit of k value represents northern boreal (66° N), and the upper limit the southern boreal (61° N) and hemiboreal (59° N) conditions (Vávrová et al. 2008).

In coniferous coarse wood debris (e.g., Scots pine, Norway spruce), the decay rate varies in the limits of 0.010–0.035 yr^{-1} (Laiho and Prescott 2004, Vávrová et al. 2009) implying the time range of 80–300 years for 95% weight loss. Similarly, Krankina and Harmon (1995) estimated that the decay rate for Norway spruces in north-western Russia (59° N) is 0.0033–0.034 yr^{-1}. This implies that 80–90 years are needed for 95% weight loss, which is close to that suggested by Mäkinen et al. (2006). They estimated that birch wood in middle boreal conditions disappears in 25–40 years, and that the full decay of coniferous wood occurs in 60–80 years. They further found that initial decay was slow, with a time lag of 5–10 years before fast decay with exponential weight loss, finally levelling off during the end phase before complete decay (see also Harmon et al. 2000, Tuomi et al. 2011). Russell et al. (2014) estimated that for temperate forest in eastern North America the residence time of dead wood is 57–120 years for conifers and 40–70 years for hardwoods. This implied a decay rate from 0.024 to 0.040 yr^{-1} for conifers, and from 0.043 to 0.064 yr^{-1} for hardwoods.

Carbon uptake and litter fall are closely related to each other during the growth and development of young stands, with a rapid accumulation of organic matter in soil as shown in Fig. 19.8. In this simulation, the carbon uptake culminated in 40–50 years after initiating the simulation, with most of the litter representing the foliage, branches, roots and stem wood of suppressed small trees. Larger trees began dying due to reducing space and maturing with declining growth, as indicated by the large variability in annual litter fall. The stocking in naturally developed mature stands represents only a small part of total growth during the life span of trees as in boreal forests. In these conditions, stem wood litter represents 20–50% of the total stem wood growth over longer periods (e.g., a 100-year period), depending on tree species, site fertility and location. The dying of maturing trees or self-thinning of tree stands is initiated earlier in birch stands than in Scots pine and Norway spruce stands, earlier on fertile than on poor sites, and earlier in southern than in northern boreal forests, thus following the growth rate and life span of different tree species in given conditions (Mazziotta et al. 2014).

In general, a warming climate increases growth and litter fall in boreal forests and enhances the decay of soil organic matter (Schlesinger and Andrews 2000) (Box 19.3; Fig. 19.9). This is demonstrated in soil warming experiments, which show the short-term increase of CO_2 efflux from the soil (Eliasson et al. 2005). In warming experiments, the CO_2 release declines, however, in the long term due to the reducing availability of substrate with fast decaying labile compounds (Knorr et al. 2005, Davidson and Janssens 2006). The increase of CO_2 emissions in soil warming experiments seems to be temporary, and in five to 10 years the emissions level off to the pre-warming level (Knorr et al. 2005, Ågren 2010). Eliasson et al. (2005) found that the combined effects of fast turnover and reduced substrate availability led to an annual increase in soil respiration by 60% in the first year after the temperature elevation, but decreased by 30% in a decade. The decay rate may further be indirectly affected by the climate change, which may change the properties of litter. Huttunen et al. (2009), for example, found that in naturally abscised birch (*Betula pendula* Roth) leaves the share of fast-decomposing compounds was small in the leaves grown in the elevated CO_2 and temperature, but large in the leaves grown in ambient conditions.

Management of Forest for the Mitigation of Climate Change   309

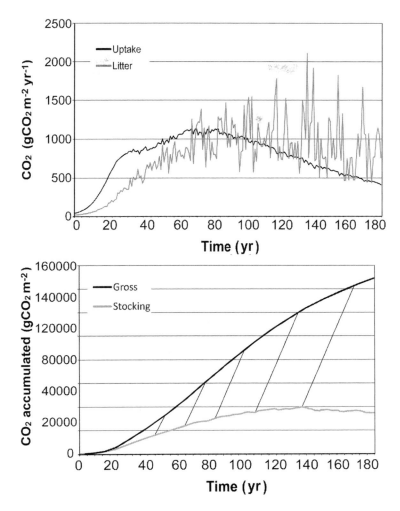

Fig. 19.8 Above: CO$_2$ uptake in growth and loss in litter fall in boreal forest ecosystem. Below: Total carbon accumulated in growth and remaining (stock) in stem wood over a longer period. The difference between the total accumulated growth and stock indicates accumulated litter. The simulations with the Sima model (Kellomäki et al. 2008) were for even-aged initial stands on medium fertile sites (*Myrtillus* type, MT) in middle boreal conditions (62° N). The initial stand density of Scots pine was 1,800 trees per hectare.

Box 19.3 Outlines of processes controlling the decay of soil organic matter

Decay of litter includes leaching, weathering and biological decomposition. Leaching and weathering are physical and chemical processes (Fig. 19.9), where several organic and inorganic substances are released. Biological decomposition involves fungal and bacterial activity, but several invertebrates also graze on soil organic matter. During the final phases of decay, the nutrients bound in organic matter are released, and the litter converts into humus colloids. Under boreal conditions, the decay rate is mainly controlled by temperature and seldom does low moisture limit the decay rate. High nitrogen content in litter indicates a high decay rate.

Box 19.3 contd....

Box 19.3 contd....

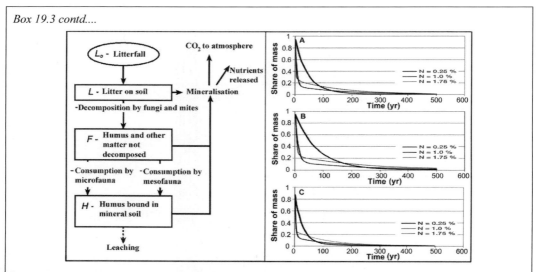

Fig. 19.9 Left: Schematic presentation of the conversion of litter into humus with the release of nutrients and emission of CO_2 under several decomposing processes, based on Chertov et al. (2001). Right: Fraction of original weight of litter cohort as a function of time, nitrogen content of litter and climate. Under A: current temperature; B: temperature 4°C < currently, and C: temperature 4°C > currently for boreal conditions (62° N). Simulated results are based on a process-based model (FinnFor) (Kellomäki and Väisänen 1997).

19.3 Carbon in Unmanaged Forest Ecosystem

Net ecosystem exchange in uneven-age and even-aged tree stands

In general, naturally growing forest landscape is a mixture of trees alone and stands in varying developmental phase, from small seedlings to large trees of full maturity (old-growth forest). In such a forest, the full decay of dead stem wood in large trees requires several decades, while the ingrowth of seedlings takes place in years. Luyssaert et al. (2008) claimed that "old-growth forests with tree losses do not necessarily become carbon sources, as has observed in even-aged plantations". They estimated that the Net Ecosystem Production (NEP) in natural stands of age ≥ 200 years was 2.4 ± 0.8 Mg C ha^{-1} yr^{-1}. Similarly, Desai et al. (2005) used an eddy covariance system to measure NEE values of –72 ± 36 and –147 ± 42 g C m^{-2} yr^{-1} in two successive years in hemlock [*Tsuga canadensis* (L.) Carrière] forests older than 300 years, while the values were –438 ± 49 and –490 ± 48 g C m^{-2} yr^{-1} for 70-year-old forests. The difference in NEE between these two sites is probably related to the difference between ecosystem respiration.

Janisch and Harmon (2002), Law et al. (2001), Law et al. (2003), Gough et al. (2008) and Law and Harmon (2011) also showed that old-growth forests intercept more carbon in photosynthesis than they emit in respiration. Despite this, carbon balance depends on the site fertility, tree species, maturity of dominant trees and the ingrowth of new tree generations. Luyssaert et al. (2008) concluded that carbon dynamics in old-growth forests are probably driven by the age structure of forest related to the mortality of mature trees and subsequent ingrowth of seedlings, as found by Sirén (1955) for northern boreal Norway spruce forests. Old-growth forests may be subject to natural disturbances such as fire or insect attacks, substantially increasing the release of carbon stored in boreal forest ecosystems (Kurz and Apps 1999).

Dynamics of trees in naturally growing and developing boreal forest involve regeneration, growth and mortality of trees, which changes the structure of tree populations/communities and carbon sequestration over time as is shown in Fig. 19.10. A gap-type growth and yield model (Sima)

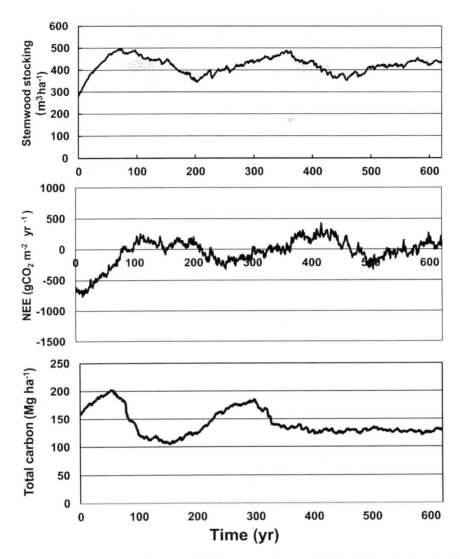

Fig. 19.10 Simulation of stem wood stocking (upper), Net Ecosystem Exchange (NEE, middle) and carbon stock (lower) in a naturally growing and developing Norway spruce stand on a medium fertile site (*Myrtillus* type, MT) in the middle boreal forest zone (62° N). The long-term mean temperature in the area was 2.2°C and precipitation 630 mm. The initial amount of litter and humus on the site was 67 Mg per hectare. In the simulations, the initial density of even-age stand was 2,500 seedlings per hectare. Simulations are based the Sima model (Kellomäki et al. 2008).

was used to simulate the stem wood stocking, Net Ecosystem Exchange (NEE) and the ecosystem carbon sequestration in the middle boreal conditions (62° N) over 640 years. The simulation was initiated by a managed even-aged tree stand, which was let to be converted to an uneven-aged, and used to initiate the simulation with no managed onwards. The values of NEE were maximized 20–25 years after launching the simulation. Thereafter, the mean NEE reduced gradually and stabilized to -40 ± 216 g CO_2 per ha per year from the year 200 onwards. At the same time, the mean stocking of trees stabilized to 420 ± 38 m^3 per hectare, with a small cycling driven by the dynamics of regeneration, growth and mortality of trees. The accumulation of carbon in the even-aged initial stand was thus replaced by the accumulation in the uneven-aged tree stand along with the maturing tree stand. During the transition between high and low accumulation rates, carbon accumulation

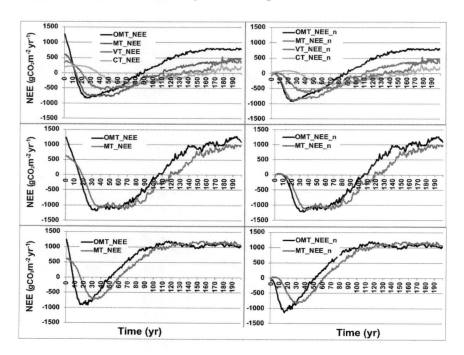

Fig. 19.11 Effect of site fertility on the values of Net Ecosystem Exchange (NEE) in unmanaged Scots pine (upper), Norway spruce (middle) and birch (lower), based the uptake and emissions of carbon including (left panels) and excluding (right panels) the old soil organic matter. The simulations of over 200 years are based on the Sima model (Kellomäki et al. 2008) for the sites in the middle boreal forests (62° N). For further details of simulation layout, see Fig. 19.10.

declined and carbon stocking stabilized at a lower level than the peak level. This pattern is similar to that found by Bormann and Likens (1979) for the northern hardwood forest. In the long term, the mean amount of carbon in the soil was fairly stable, 140 ± 25 Mg ha^{-1} from the year 150 onwards.

Carbon dynamics in unmanaged even-aged stands are mainly related to, e.g., site fertility tree species and maturity and stocking of trees, which control carbon uptake and emissions through ecosystem dynamics as shown in Fig. 19.11. The simulations are based on even-aged initial stands as typical in forest plantations or the management of forest landscape comprising of even-aged stands in varying developmental phases. They emit carbon (source) in the seedling and mature phases, and absorb (sink) carbon in the intermediate or thinning phases with the highest growth. The soil carbon originating from the previous cycle (old carbon) and that from the current cycle (new carbon) are separated in the simulations, the former representing the initial value of soil carbon. In the early phase of simulations, the old carbon dominates the carbon dynamics, depending on site fertility and tree species. However, the contribution of old carbon disappears in few decades allowing identification of how the growth and development of trees, and their management and harvest affect carbon dynamics over time.

In Scots pine, for example, the accumulation of new carbon is clear almost from the very beginning of the simulation on the fertile (OMT), medium (MT) and medium poor (VT) sites, while the emissions from the new soil organic matter on the poor site (CT) exceeded uptake in the first 30 years (Fig. 19.11). The fertile site was a carbon sink for 80 years and the medium fertile site for 110 years after initiating the simulation. On the medium poor site, the uptake exceeded emission a decade later, and the site was a carbon sink for 120 years after initiating the simulation. Similarly, the poor site was a sink for the next 130 years, beginning in year 30. During late rotation, the mean source strength was 700 g C m^{-2} yr^{-1} on the fertile (OMT) site, 400 g C m^{-2} yr^{-1} on the medium fertile (MT) and medium poor (VT) sites, and 200 g C m^{-2} yr^{-1} on the poor site (CT).

Table 19.3 Effect of tree species and site fertility on the length of periods with negative NEE values, the mean NEE and the amount of carbon in trees and soil excluding CO_2 emissions from old carbon (previous cycle) in unthinned stands.

Trees species and site	Years with negative NEE	Mean NEE in period with negative values, kg C ha^{-1}yr^{-1}	Mean amount of carbon in trees, Mg ha^{-1}	Mean amount of carbon in soil, Mg ha^{-1}
Scots pine				
• OMT	84	−1350	107	56
• MT	111 (+32)	−1300 (−4)	104 (−3)	31 (−45)
• VT	128 (+52)	−880 (−35)	89 (−17)	40 (−29)
• CT	131 (+56)	−530 (−61)	60 (−44)	23 (−59)
Norway spruce				
• OMT	95	−2100	139	66
• MT	107 (+13)	−2020 (−4)	145 (+4)	62 (−4)
Birch				
• OMT	51	−1610	86	67
• MT	52 (+2)	−1320 (−18)	81 (−6)	70 (+7)

The simulations were run over 200 years, based on the Sima model (Kellomäki et al. 2008) for sites in the middle boreal conditions (62°). The initial stand density was 1,800 seedlings ha^{-1}. Numbers in parenthesis indicate the percentage change (%) compared to the value for the most fertile site.

In Norway spruce, carbon uptake exceeded emission 15–20 years after the start of the simulation, earlier on the fertile (OMT) than on the medium fertile (MT) site (Fig. 19.11). The difference in the growth rate further affected the timing when the stand turned from a sink to a source; i.e., on the fertile (OMT) site at the age of 100 years and on the medium fertile site at the age of 120 years. In both cases, the source strength increased towards the end of simulation period, with the mean value 1200 g C m^{-2} yr^{-1} on the fertile (OMT) and 1,000 g C m^{-2} yr^{-1} on the medium fertile (MT) site. At the same time, the carbon sink in birch became a source at 50 years on the fertile (OMT) site and at 60 years on the medium fertile (MT) site, and it stabilized at 100–110 years to the same level as in Norway spruce.

Table 19.3 summarizes the main output of the simulations in Fig. 19.11, including only the effects of the new soil carbon on the current output. Carbon emissions exceeded the uptake during the first 10–12 years, regardless of site fertility and tree species, but the source phase was 35 years even for Scots pine on the poor (CT) sites. The period of carbon uptake (with negative NEE values) extended as the site fertility reduced, excluding birch; i.e., from 80 to 130 years in Scots pine, and from 95 to 110 years in Norway spruce. Following the reduction in site fertility, the mean NEE reduced from −1,350 to −530 kg C ha^{-1} yr^{-1} for Scots pine, from −2,100 to −2,000 kg C ha^{-1} yr^{-1} for Norway spruce and from −1,600 to −1,300 kg C ha^{-1} yr^{-1} for birch.

Carbon retention in unmanaged forest ecosystem

Carbon sequestrated in a forest ecosystem may retain in the ecosystem for a time period (the residence time (τ)) of years or even decades, depending on litter quality, management (an increase of tree growth, soil management), harvest (frequency, intensity) and the decay of litter and organic matter in soil. The residence time is obtained by dividing the store of carbon (storage, e.g., CO_2 kg ha^{-1}) in the ecosystem by the emission rate (q, e.g., kg CO_2 ha^{-1} yr^{-1}) for the ecosystem in the steady state (Govers et al. 2012):

$$\tau(t) = \frac{Capacity\ of\ a\ system\ to\ hold\ carbon(t)}{Rate\ of\ carbon\ flow\ through\ a\ system(t)} = \frac{Storage(t)}{q(t)} = \frac{Storage(t)}{k(t) \times Storage(t)} = \frac{1}{k(t)} \quad (19.10)$$

Table 19.4 Mean amount of soil carbon and residence of mean soil carbon in naturally growing and developing Scots pine, Norway spruce and birch stands as a function of site fertility in middle boreal zone (62° N) simulated with the Sima model (Kellomäki et al. 2008).

Species and site type	Mean amount of soil carbon, Mg ha⁻¹	Mean residence of soil carbon, yr
Scots pine		
• OMT	163	49
• MT	135	61
• VT	129	70
• CT	83	73
Norway spruce		
• OMT	206	48
• MT	207	55
Birch		
• OMT	153	35
• MT	151	36

Regardless of tree species, the initial even-aged stand density of 1,800 seedlings per hectare was used in simulations.

where k is the carbon turnover rate [yr⁻¹] representing the inverse value of carbon residence time. The residence time begins when carbon enters the system and ends when carbon leaves the system. In the boreal forests, residence time is about 90 years, but it varies substantially from 10s of years in sites of high fertility to more than 100 years in sites of low fertility (Raich and Schlesinger 1992, Mahli et al. 2002).

In boreal upland sites, the soil carbon stock depends on site fertility, tree species and stocking of trees. It is also fairly stable throughout succession cycle (McKinley et al. 2011, Kellomäki et al. 2013), which makes Equation (19.10) applicable through successive tree stands. Table 19.4 shows that in middle boreal zone (62° N) the mean amount of soil carbon reduced along with reducing site fertility, especially in sites dominated by Scots pine. In this case, the mean total soil carbon stock on the most fertile site (OMT) was 160 Mg ha⁻¹, whereas in Norway spruce and birch dominated sites the soil carbon stock was 200 Mg ha⁻¹ and 150 Mg ha⁻¹, respectively. At the same time, carbon residence in Scots pine stands increased from 49 years to 73 years, along with the reduction of site fertility. In Norway spruce, the residence time was about 50–55 years regardless of site fertility, and in birch it was 35 years.

19.4 Carbon in Managed Forest Ecosystem

Management/carbon interaction in forest ecosystem

In management, the choice of tree species and their genotypes for regeneration has long-term effects on the rate of carbon accumulation in trees and soils (Fig. 19.12). In both cases, high productivity in relation to site fertility maximizes the carbon storage rate. The share of growth and biomass in foliage, branches, stems and coarse and fine roots affects the mass in trees and the amount and quality of litter controlling the accumulation of matter in soil. In general, carbon residence in coniferous litter is longer than that in deciduous litter, but among coniferous or deciduous species residence time varies from species to species. In Norway spruce, for example, the accumulation of matter in trees and soil is faster than in Scots pine. In storing carbon on soil, Norway spruce and birch are preferable on fertile sites while Scots pine is successful even on poor sites, thus extending the role of Scots pine in storing carbon in forest ecosystems over the whole range of site fertility (Jandl et al. 2007).

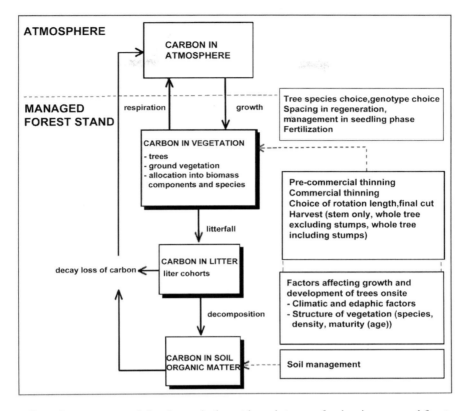

Fig. 19.12 Effects of management and disturbances in the uptake and storage of carbon in a managed forest ecosystem (Karjalainen 1996a).

Any measures to which trees are subjected, such as pre-commercial thinning and commercial thinning, reduce the carbon stored in them, but increase the carbon sequestration capacity of the remaining trees (Fig. 19.12). Consequently, the carbon absorption capacity of a thinned stand may recover in a few years to the level before thinning. Liski et al. (2001) showed that carbon accumulation in Norway spruce recovered in five years after thinning and in 10 years after planting in a clear cut area. The recovery of carbon accumulation may further be enhanced by nitrogen fertilizing, which increases both growth and litter fall (Johnson and Curtis 2001). For example, Mäkipää (1994) found that nitrogen addition increased carbon accumulation in soil in boreal coniferous stands by 14–115% at the thinning phase, mostly in Scots pine on poor sites (CT). The addition of nitrogen may also reduce the decay of organic matter indicated by the reduced CO_2 efflux from soil thus probably enhancing the accumulation of carbon in the soil (Martikainen et al. 1989, Nohrstedt et al. 1989, Olsson et al. 2005, Saarsalmi et al. 2014).

The removal of timber and residues in thinning and final cutting may reduce soil carbon in boreal forests but findings are highly dependent on, for example, species, site, harvest and logistics (Clarke et al. 2015). In general, harvest intensity may reduce soil organic matter but short term losses may be compensated over time. Based on experimental studies, Clarke et al. (2015) concluded that intensive harvest (i.e., whole-tree harvest in thinning or in a final cut) may or may not reduce soil carbon at some sites compared to non-intensive harvest (i.e., stem-only harvest). They suggest that the short-term losses of soil carbon are probably temporary, and even compensated for in a few decades by the natural ecosystem processes controlling regeneration and growth in the current and next production cycles (see also Egnell et al. 2015).

As in thinning, the removal of timber and harvest residues in clear cutting may reduce soil carbon depending on, for example, species, site, harvest and logistics. Pumpanen et al. (2004) found that the annual CO_2 efflux before clear cut in a boreal coniferous stand was 1,900 g CO_2 m^{-2} yr^{-1}. In three successive years after cutting, the efflux was 3,242, 2,845, and 2,926 g CO_2 m^{-2} yr^{-1}, when letting logging residues decay in the soil. In the first year after the cut, the CO_2 emission from residues was 1,423 g CO_2 m^{-2}; i.e., 23% of the carbon in the logging residues was released. The remaining carbon would probably be retained in decaying residues over the regeneration period to the phase when growing seedlings turn the site into a carbon sink. Site preparation prior to planting, however, may substantially affect carbon residence. For example, Pumpanen et al. (2004) found that the CO_2 efflux in mounds with a mixture of organic matter and mineral soil was greater than on sites where only logging residues were present.

Net carbon exchange in uneven-aged and even-aged tree stands

In using uneven-aged (or selection cutting) management, a fraction of large, medium and small trees is harvested in 10- to 20-year intervals. The basic assumption behind the system is that the ingrowth of new trees is based on natural regeneration in gaps created by removing dominant and subdominant trees (thinning above), mimicking autogenetic development of natural forest. Thus, forest landscape is compiled by trees in different development phase from seedlings to full maturity.

Impacts of uneven-aged management on carbon dynamics in boreal forest has been investigated only a little compared to impacts of even-aged management. In this respect, the study by Nilsen and Strand (2013) is exceptional when comparing the carbon storages and fluxes in uneven-aged and even-aged Norway spruce stands on fertile sites in maritime southern boreal condition in Norway (59° N, 11° E, 80 m a.s.l.). They showed that in a long-term experiment (81 years), the mean carbon storage in trees including roots was 210 Mg C ha^{-1} in the even-aged and 76 Mg C ha^{-1} in the uneven-aged stand. At the same time, the amount of soil carbon was 178 and 199 Mg C ha^{-1}, respectively. The mean total amount of carbon (in trees plus soil) in even-aged stand was 388 Mg C ha^{-1}, which was 30% larger than in uneven-aged stand (275 Mg C ha^{-1}). In the 81-year period, the total net carbon sequestered in trees under even-aged management accounted 16 Mg C ha^{-1} more than under uneven-aged management. Nilsen and Strand (2013) found further that over the monitoring period the timber production in the uneven-aged stand was 95% of that in the even-aged stand. During the same period, the net carbon sequestration under even-aged management was 37 Mg C ha^{-1} larger than under uneven-aged management. Nilsen and Strand (2013) concluded that in long run uneven-aged management is likely allowing less carbon being sequestrated in Norway spruce stand than even-aged management.

In even-aged management, a fraction of small, medium and large trees (thinning below) is removed letting the rest of trees, mainly dominant and subdominant, for future harvest, e.g., in 10- to 20-years or longer intervals until final cut at the end of rotation for repeating the new cycle of regeneration and thinnings to final cut for regeneration, etc. This mimics the allogenetic development of natural forests, where abiotic disturbances create large gaps allowing even-aged regeneration. Thus, forest landscape is compiled by tree stands in different development phase from young to full maturity.

Figure 19.13 shows the time series of the Net Ecosystem Exchange (NEE) in thinned Scots pine, Norway spruce and birch in even-aged stands as a function of site type. In Scots pine, three to four thinnings were made, the earliest on a fertile site (OMT) and the latest on a poor site (CT). Every thinning reduced the sink strength immediately after thinning, but it recovered following the recovery of growth. This was especially true for the first thinning, while later the sink strength recovered only partially, or the thinning even turned the ecosystem from a sink to a source. On the poorest (CT) site, the NEE values remained negative late in the rotation, and they were more negative (sink) than on more fertile sites, which lost their sink capacity earlier in the rotation. Before

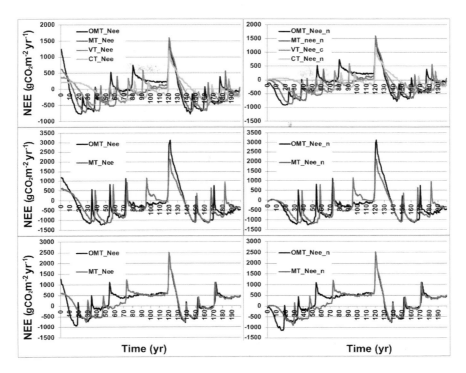

Fig. 19.13 Effect of site fertility on the values of Net Ecosystem Exchange (NEE) in managed Scots pine (upper), Norway spruce (middle) and birch (lower) in even-aged stands, based on the uptake and emissions of carbon, including (left panels) and excluding (right panels) emissions from the old carbon. The simulations over 200 years with the rotation of 120 years, i.e., high emission peak) are based on the Sima model (Kellomäki et al. 2008) for sites in the middle boreal forests (62° N). The initial stand density for all the trees species was 1,800 seedlings ha^{-1}.

the final felling, the positive NEE values were about a half of those in unthinned stands at the same time. This pattern held for Norway spruce, which was thinned three times on the fertile (OMT) site and four times on the medium fertile (MT) site. Even in the first thinning, the sink turned to a source for a while. Thinned Norway spruce lost their sink capacity more rapidly than in unthinned stands as was the case for birch.

Table 19.5 summarizes the main output of carbon in thinned stands shown in Fig. 19.13. The period with negative NEE values was 60–100 years for Scots pine, 100–110 years for Norway spruce and 50–55 years for birch. At the same time, the mean NEE was from –600 to –1,000 kg C ha^{-1} yr^{-1} for Scots pine, from –1,000 to –1,400 kg C ha^{-1} yr^{-1} for Norway spruce and from –1,000 to –1,300 kg C ha^{-1} yr^{-1} for birch, the lower limit representing low fertility and the upper limit high fertility. The mean amount of carbon in trees and soil increased along with the increasing site fertility, as did the mean NEE. The comparison to the values for unthinned stands (Table 19.3) show that thinning reduced by 10–30% of the period with negative NEE values, less on fertile than on the poor site. At the same time, the mean NEE reduces by 5–30%, less on the poor than the fertile site. This implied further that the mean amount of carbon in trees in thinned stands was approximately a half of that in unthinned stand, while carbon in soil in thinned stand was 50–80% of that in unthinned stand. As expected, the mean amount of carbon in soil was less sensitive to thinning than the amount of carbon in trees.

Table 19.5 Effect of tree species, site fertility and thinning on the length of periods with negative NEE values, the mean NEE and the amount of carbon in trees and excluding CO_2 emissions from old carbon (previous cycle) in soil.

Trees species and site	Years of period with negative NEE	Mean NEE in period with negative values, kg C ha^{-1})	Mean amount of carbon in trees, Mg ha^{-1}	Mean amount of carbon in soil, Mg ha^{-1}
Scots pine				
• OMT	77	−963	50	28
• MT	99 (+29)	−950 (−1)	53 (+6)	21 (−25)
• VT	80 (+4)	−812 (−16)	34 (−32)	26 (−7)
• CT	89 (+16)	−508 (−47)	25 (−50)	19 (−32)
Norway spruce				
• OMT	107	−1,346	69	31
• MT	108 (0)	−1,165 (−13)	64 (−7)	29 (−6)
Birch				
• OMT	50	−1,215	48	36
• MT	50 (0)	−992 (−18)	43 (−10)	40 (+11)

The simulations were run over 200 years with a rotation of 120 years, based on the Sima model (Kellomäki et al. 2008) for sites in the middle boreal conditions (62° N). The initial stand density was 1800 seedlings ha^{-1}. Logging residues were left on the site after thinning. Numbers in parenthesis indicate the percentage change (%) compared to the value for the most fertile site.

Carbon sequestration vs. timber production in managed forest landscape

Under even-aged management, a forest landscape is a mosaic of single stands in varying phases of growth and development, from clear cut areas to mature stands. In wide spatial and long temporal scales, carbon stocks in the landscape are stable, while in single stands the variability of carbon stocks mainly follows the variability of carbon bound in growing trees (McKinley et al. 2011). Soil carbon tends to reduce but it is more stable than that bound in trees. In the landscape scale, the variability of carbon stocks become smaller when there are more stands. The mean amount of carbon in the landscape is also affected by the intensity of management and harvest and forest structure. Thus, thinning in short intervals with low remaining stocking and short rotation reduces the mean carbon stocking, while thinning in long thinning intervals with high remaining stocking and long rotation increases the mean carbon stocks in the landscape scale.

Based on rule-based management, Garcia-Gonzalo et al. (2007a,b) found that under no thinning (UT (0,0)), with only the final cut, the mean carbon storage over a boreal forest area was 154 Mg C ha^{-1}, of which 48% was in trees (73 Mg C ha^{-1}) and 52% in the soil (81 Mg C ha^{-1}) (Box 19.4, Table 19.6). Under basic thinning (BT (0,0)), the total carbon stock was 45% lower than that under no thinning (UT (0,0)) but the reduced thinning intensity increased the stocking of timber and biomass, with an increase in ecosystem carbon. For example, the increase in the upper basal area limit triggering thinning was increased by 15% (BT (15,0)) and 30% (BT (30,0)), increased carbon stock 3 and 6% compared to that under basic thinning (BT (0,0)). If the remaining basal area (BT (30,30)) also increased, carbon stock in the ecosystem increased by up to 11%. The carbon stock in the forest ecosystem was, however, greatest if no thinning was used before the final cut. Regardless of management, carbon sequestration in Norway spruce stands was clearly greater (15–20%) than in Scots pine and birch stands.

A similar analysis was made under the changing climate using the ECHAM4 and HadCM2 emission scenarios (Garcia-Conzalo et al. 2007a,b). In the simulation area, the current annual mean temperature and precipitation were 3.1°C and 478 mm, increasing by 4.2–5.5°C and 85–113 mm with the elevating atmospheric CO_2 from 350 ppm to 650 ppm towards the end of this century. In general, climate change increased the amount of carbon in the ecosystem substantially more without thinning than in other cases, and in some cases the carbon stock even reduced slightly. For example,

Table 19.6 Effect of thinning intensity on carbon sequestration in the boreal forest landscapes (62° N), where Norway spruce occupies 64% of the area (1,451 ha), Scots pine 28% and birch 7%. Most of the stands were on the sites of medium fertility (*Myrtillus* type, MT) (Garcia-Gonzalo et al. 2007a,b). For thinning regimes, see Fig. 13.17.

Tree species and management	Carbon in trees		Carbon in soil		Carbon in ecosystem	
	Mg ha^{-1}	Change, %	Mg ha^{-1}	Change, %	Mg ha^{-1}	Change, %
Scots pine						
• BT(0,0)	31	-	37	-	68	-
• BT(15,0)	32	6	38	2	70	3
• BT(30,0)	34	12	39	3	73	7
• BT(15,15)	34	12	39	3	73	7
• BT(30,30)	37	22	41	7	78	14
• UT(0,0)	45	47	50	31	95	38
Norway spruce						
• BT(0,0)	46	-	82	-	128	-
• BT(15,0)	48	4	83	1	131	2
• BT(30,0)	52	12	84	2	135	6
• BT(15,15)	51	10	84	3	136	6
• BT(30,30)	56	20	86	5	142	11
• UT(0,0)	90	94	98	20	188	47
Birch						
• BT(0,0)	22	-	34	-	56	-
• BT(15,0)	23	6	35	2	58	3
• BT(30,0)	25	12	35	3	60	7
• BT(15,15)	25	13	36	5	61	8
• BT(30,30)	27	25	37	9	64	15
• UT(0,0))	35	62	45	32	80	44
Whole area						
• BT(0,0)	40	-	66	-	106	-
• BT(15,0)	42	5	67	1	109	3
• BT(30,0)	45	12	68	3	112	6
• BT(15,15)	44	11	68	3	112	6
• BT(30,30)	48	21	70	6	118	11
• UT(0,0)	73	83	81	23	154	45

BT (0,0) indicates current thinning rules (basic thinning), and, e.g., BT(15,0) indicates a thinning regime with a 15% increase in the upper limit of stocking and 0% change in the remaining stocking compared to that under basic thinning (BT(0,0)), while UT(0,0) indicates no thinning before the terminal cut (Fig. 13.17).

Box 19.4 Effects of thinning and structure of the forest landscape on carbon

In simulating the effects of age class distribution on timber production and carbon sequestration in the scale of a forest landscape, Garcia-Gonzalo et al. (2007a) divided the tree stands per tree species into age classes: 0–20, 21–40, 41–70 year and > 70 year. Four age class distributions were created: (A) distribution dominated by intermediate age classes (normal distribution); (B) distribution dominated by no age class (equal distribution); (C) distribution dominated by young age classes (left-skewed distribution); and (D) distribution dominated by old age classes (right-skewed distribution). Stem wood growth, timber yield (saw logs and pulp wood) and carbon sequestration were calculated for the 100-year simulation period over the area, applying different age class distributions, management regimes and climates. The cost of carbon sequestration was calculated based on indirect pricing: costs for reducing timber when increasing carbon sequestration (Garcia-Gonzalo et al. 2007a).

the carbon stock increased by 5–6% under no thinning (UT (0,0)) under climate warming, but the response depended on tree species. The increase was the largest in Scots pine for both climate change scenarios. Similarly carbon stock in birch increased by 1–2% under climate warming, regardless of the thinning regime. In Norway spruce, the response was variable remaining virtually constant under basic thinning (BT (0,0)), but reduced under the thinning regime BT (30,30) regardless of the climate change scenario.

Garcia-Gonzalo et al. (2007a,b) noted that the structure of a forest landscape (e.g., species and age class distribution) is among the key factors affecting carbon stocks and yields of timber and biomass. In seedling stands, carbon is lost for several years until the uptake in tree growth exceeds carbon loss from the soil. Over time, carbon uptake reduces in maturing trees, whereas carbon losses from soil increase due to the increasing amount of soil organic matter. At the same time, the amount and value of timber in maturing trees increases. In single stands, the amount of timber and carbon increases concurrently indicating no trade-off between timber production and carbon sequestration.

In the scale of forest landscapes, the distribution of young and mature tree stands has a strong effect on ways to meet needs to sequestrate carbon in managed forests but still produce timber (Kilpeläinen et al. 2012) (Box 19.4). As expected, the timber yield was affected by both the initial age class distribution and management, but the effects of age class distribution on carbon stocks were small (Fig. 19.14). Assuming the current climate and basic thinning (BT (0,0)), the largest values of timber and carbon stock (687 m^3 ha^{-1}, 106 Mg carbon ha^{-1}) were obtained if the initial landscape was dominated by mature stands. Conversely, the initial age distribution dominated by young stands gave the smallest values (573 m^3 ha^{-1} of timber and 103 Mg carbon ha^{-1}). Under the changing climate, the patterns remained the same, but climatic warming increased the timber yield up to 18% and carbon stocks up to 6%, depending on the management and landscape structure. Regardless of the initial age class distribution, any increase in tree stocking tended to increase the timber yield, the Net Present Value (NPV) and the carbon stocks. The timber yield and the NPV were the smallest under no thinning except in the final cutting (UT (0,0)). Conversely, carbon stocks were the highest under no thinning (UT (0,0)), being 45% greater than under basic thinning (BT (0,0)).

Figure 19.14 shows further that the maximum timber yield and maximum carbon stock in the forest ecosystem are not compatible. The timber yield and the amount of carbon increased if the stocking of trees was increased, compared to that under basic thinning (BT (0,0)). This implies that under the current climate the potential cost (potMC) per additional Mg of carbon was from 32 to 41 € Mg^{-1} depending on the age class distribution, if maximizing carbon stock (UT (0,0)) or maximizing the timber yield (BT (30,30)) in management. This implies that an additional 32 Mg ha^{-1} (normal distribution) to 36 Mg C ha^{-1} (left distribution) will be stored in the forest ecosystem at a cost of 32 to 41 € Mg^{-1} depending on the landscape structure. Under the changing climate, the additional amount of carbon that can be stored if using no thinning (UT (0,0)) ranged from 39 to 42 Mg ha^{-1}. In this case, the cost varied from 31 to 41 € Mg^{-1} (interest rate 3%). The increasing discount rate decreased the marginal cost for carbon sequestration (Garcia-Gonzalo et al. 2007a,b).

Under the curMC approach, the shift from basic thinning (BT (0,0)) to no thinning (UT (0,0)) under the current climate enhanced carbon stock 44–47 Mg ha^{-1} depending on the initial age class distribution, and under the changing climate the increase was 50–53 Mg ha^{-1}. This implied that the cost increased 27 € Mg^{-1} under the current climate and 28 € Mg^{-1} at most under climate change. The real option approach (roMC) showed that carbon stock was increased when shifting from the management regime BT (0,0) to BT (30,30) with no loss in NPV regardless of the climate scenario. In this case, the thinning regime maximizing NPV clearly increased carbon stock compared to basic thinning (BT (0,0)). This choice would increase carbon stock 11–12 Mg C ha^{-1} in the forest ecosystem depending on the initial landscape structure (Garcia-Gonzalo et al. 2007a,b).

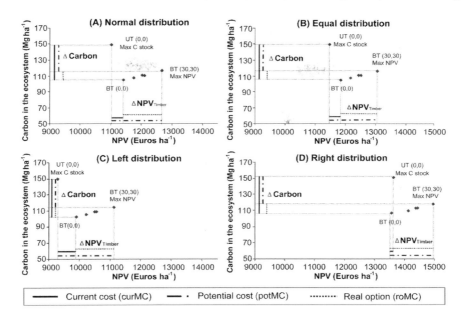

Fig. 19.14 Relationship between the carbon in the ecosystem and the Net Present Value (NPV) of timber (discount rate 3%) when using varying thinning regimes (Garcia-Gonzalo 2007a). Permission of Elsevier. Legend: Potential marginal cost of carbon sequestration (potMC) involves the differences in carbon stock and in the NPV of timber, representing management regimes maximizing carbon stock and NPV. The current marginal cost of carbon sequestration (curMC) indicates the differences in carbon stock and NPV of timber production, when shifted from the current management to one maximizing carbon stock. The real option marginal cost of carbon sequestration (roMC) involves the differences in carbon stock and in NPV when shifted from the current management to management increasing both carbon stock and the NPV of timber production compared to current management. For thinning regimes, see Table 19.6.

Carbon retention in managed forest ecosystem

Under management, a forest is unbalanced, with varying amounts of carbon in the ecosystem depending on the rate of carbon uptake and emission control through ecosystem development. Figure 19.15 shows the accumulation of new carbon in the ecosystem as a function of the initial stand density. At the end of a 100-year rotation, the amount of carbon under basic thinning (130 Mg ha^{-1}) was a half of that under no thinning (250 Mg ha^{-1}) if the initial stand density of 1,800 trees per hectare was used in the simulations (Kellomäki et al. 2013). At the same time, carbon emissions temporarily exceeded emissions under thinning, but in long run they stabilized to less than those under no thinning. If the stocking under thinning is higher than under basic thinning (the basal area before and after thinning remains higher than that under basic thinning), the stabilized carbon emission converged to those under no thinning.

Figure 19.15 shows further that depending on the frequency and intensity of thinning and rotation time, management has an effect on the residence time of carbon in the forest ecosystem. In young stands, not yet thinned or left unthinned for longer, carbon is retained in the ecosystem for 20–30 years. At the culmination of stem wood growth, residence time stabilized towards the end of rotation in the range 30–50 years under no or low thinning intensity. Under thinning, the residence time varied from 60 years just before thinning, to 30 years after the first thinning, with low initial density, but up to 20 years under high initial density. By the end of rotation, the residence time in unthinned stands was 55–60 years, the lower limit representing the high, and the upper limit the low initial density. Over the whole rotation, the mean residence time was about 50 years without thinning, regardless of the initial stand density (Table 19.7). Thinning reduced the mean residence time up to 15% compared to that under no thinning, and was highest under basic thinning.

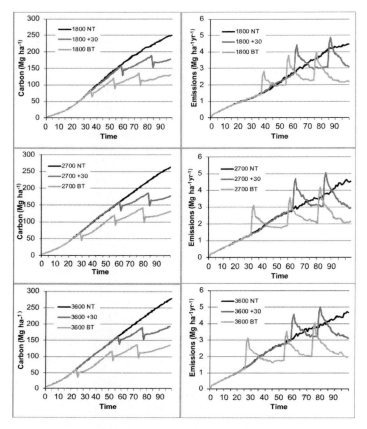

Fig. 19.15 Stocking (trees and soil) of new carbon (left) and carbon emission (right) as a function of initial stand density (1,800, 2,700 and 3,600 seedlings per hectare). In management, no thinning (NT), thinning in practical management (basic thinning, BT), and thinning with stocking 30% higher than that under basic thinning were used (Kellomäki et al. 2013). The simulations were for a site of medium fertility (*Myrtillus* type, MT) in the middle boreal conditions (62° N) applying a 100-year rotation. Simulations are based on the Sima model.

Table 19.7 Mean amount of carbon and mean carbon residence time in the ecosystem as a function of the thinning regime (Kellomäki et al. 2013).

Management regime	Mean amount of carbon over rotation, Mg ha^{-1}	Mean residence time of carbon over rotation, yr
Initial density 1,800 trees per ha		
• No thinning	127 -	51 -
• Basic thinning	83 (65)	45 (88)
• Basic thinning + 30%	107 (84)	49 (96)
Initial density 2,700 trees per ha		
• No thinning	135 -	52 -
• Basic thinning	86 (64)	45 (87)
• Basic thinning + 30%	113 (84)	47 (90)
Initial density 3,600 trees per ha		
• No thinning	139 -	51 -
• Basic thinning	85 (61)	45 (88)
• Basic thinning + 30%	118 (85)	47 (92)

The simulations are for a site of medium fertility (*Myrtillus* type, MT) in the middle boreal conditions (62° N) applying a 100-year rotation length. Details of the simulations are given in Fig. 19.15. Number in the parentheses indicate the percentage (%) of the value of that for no thinning.

19.5 Carbon in Forest-based Production System and Products

Flow of carbon through forest-based production

In forest ecosystems, carbon is bound in the growth of trees but emitted in autotrophic respiration and heterotrophic respiration through the detritus cycle, where litter, harvest residues and humus are decayed (Fig. 19.16). Carbon is also emitted in management/logistics operations producing and harvesting timber and biomass for materials and energy. The carbon path can also include emissions from processing timber and biomass for materials and energy, including the raw material provided by the recycling of carbon back to processing abandoned products. Carbon is further emitted into the atmosphere in the combustion of biomass-based fuels and materials abandoned in landfills.

Liski et al. (2001) simulated carbon stocks in a Scots pine dominated boreal forest ecosystem, and carbon in materials based on timber using the 90-year rotation in management. The Gross Primary Production (GPP) was 4.0 Mg C ha^{-1} yr^{-1} yielding 1.20 Mg C ha^{-1} yr^{-1} of Net Primary Production (NPP) (Fig. 19.17; Box 19.5). One third of this was harvested in timber, and two thirds were lost in litter and residue. Considering the losses of timber in harvest/logistic operations and in manufacturing, one quarter of the NPP was bound in wood-based materials. At the same time, 8% of the NPP was in recycled materials. On average, trees, soil and materials in use bound 38, 83 and 7.4 Mg ha^{-1} of carbon over a 300-year simulation including three 90 year cycles with two to three thinnings. The carbon in products increased immediately after cutting but reduced over time. Liski et al. (2001) calculated that the average residence time of carbon was 15 years in trees, 49 years in soil and nine years in products.

Carbon in forest-based production has a direct link to management, especially the rotation length, which substantially affects the timber yield and the share of how timber is allocated to pulp wood and saw logs (Kaipainen et al. 2004). The elongation of rotation increases the share of saw logs. This implies that stem diameters more frequently exceed the threshold for saw logs, thus increasing the share of saw logs and the value of timber yield. The reduction of rotation has the opposite effect when increasing the share of pulp wood in the total harvest (Fig. 19.18). The total carbon bound in trees, soil and wood-based materials is, however, fairly insensitive to the rotation length (Mäkipää et al. 1999).

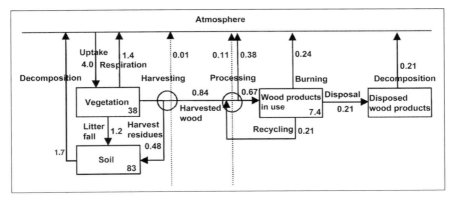

Fig. 19.16 Carbon cycle between forest ecosystem/forest product technosystem and atmosphere in a 90-year rotation (Liski et al. 2001). Permission of Canadian Science Publishing. Numbers in the boxes show the stocks [Mg ha^{-1}], solid arrows show fluxes of carbon [Mg ha^{-1} yr^{-1}] from forest ecosystems and processing forest-based materials and energy. Broken arrows indicate carbon fluxes of fossil carbon in management and process.

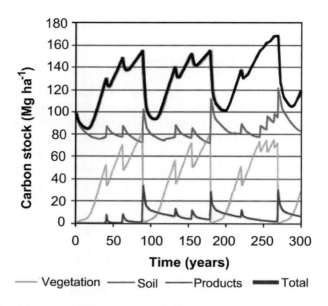

Fig. 19.17 Carbon bound in trees, soil and wood-based materials in use over a 300-year simulation including three 90-year management cycles with two thinnings in the first and second rotations and one in the third rotation (Liski et al. 2001). Permission of Canadian Science Publishing. The simulations were run using the Sima model for the middle boreal forests (62° N) for medium fertile sites (*Myrtillus* type, MT) using the recommended management rules in the early 1990s (Metsäkeskus Tapio 1994).

Box 19.5 Processing timber for wood-based materials

Liski et al. (2001) used the wood-product model developed by Karjalainen et al. (1994) in tracking carbon bound in timber through the forest product technosystem. Timber may be used in several production lines, e.g., sawmilling and manufacturing paper. Over time, the share of original material bound in products reduces (Karjalainen et al. 1994):

$$\text{Share of original material in use}(t) = \frac{PU(t)}{PU(0)} = d - \frac{a}{1 + b \times e^{-cxt}} \quad (19.11)$$

where PU (0) is the amount of material at the year 0, PU(t) the amount of material still in use at the year t, and a, b, c and d are parameters specific for a given wood-based material. According to Row and Phelps (1990), wood-based materials may be divided into short (e.g., chemical pulp), medium-short (e.g., mechanical pulp), medium-long (e.g., plywood) or long-term (e.g., sawn timber) lifespans (Fig. 19.18).

Box 19.5 contd....

Box 19.5 contd....

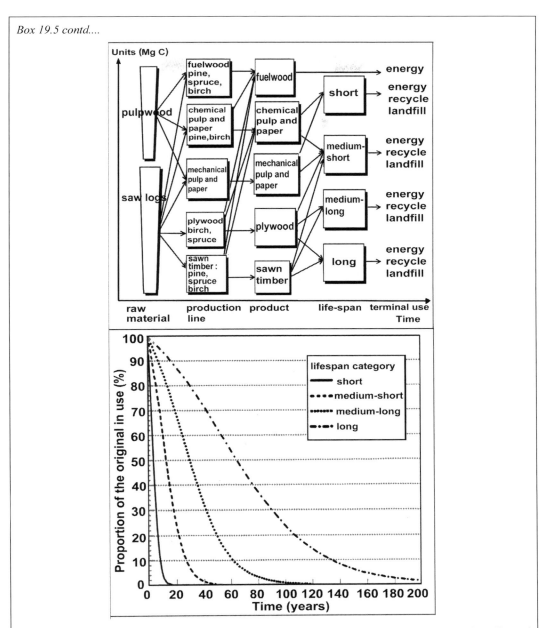

Fig. 19.18 Above: Example of production lines using timber and biomass for materials and energy. Below: Share of original amount still in use in materials with varying life span as a function of time (Karjalainen et al. 2004). Courtesy of the Finnish Society of Forest Science and the Finnish Forest Research Institute.

Life cycle of carbon in producing timber and energy biomass

In the ecosystem/technosystem/atmosphere cycle, carbon is removed in tree growth and stored in pools within the ecosystem (tree biomass, soil organic matter) and outside the ecosystem in stocks of wood-based materials in use or abandoned. Carbon taken up in growth and emitted in different phases of manufacturing and use has a lifespan which may be analyzed using the environmental Life Cycle Assessment (LCA). LCA involves a technique to assess environmental impacts of forest-based

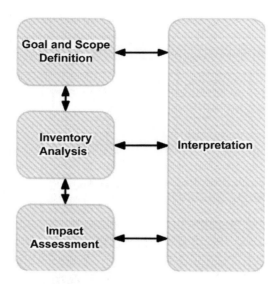

Fig. 19.19 Schematic presentations of different phases in life cycle assessment (LCA) (ISO 2006).

materials in any phase of their life span (i.e., from "cradle to grave") (ISO 2006). Environmental impacts may refer, for example, to radiative forcing due to the net CO_2 emissions from the forest ecosystem and forest technosystem, thus providing necessary information for climate change mitigation. Carbon flow through the ecosystem/technosystem/atmosphere cycle is fundamentally different from fossil carbon flows, which include extracting carbon from long-term geologic carbon stocks and accumulating carbon, e.g., only in the atmosphere when using fossil carbon for energy.

In LCA, the goal and scope define the context of the assessment (Fig. 19.19). They include: (i) the functional unit defining the quantities of goods and services produced in the system and the reference system to which the input and output of the production system can be related; (ii) the system boundaries, assumptions and limitations behind the assessment; (iii) the methods partitioning the environmental load of a process when several products or functions share the same process; and (iv) impacts included in the assessment. In LCA, the inventory involves, for example, the input and output of water, energy, raw materials, and releases to air, land, and water, including all activities within the system boundaries and from the supply chain relevant in relation to the functional unit defined in the goal and scope. The significance of potential environmental impact is evaluated in the impact assessment, based on the impact categories selected and measured for LCA. The results from the inventory, analysis and impact assessment are summarized for a set of conclusions and recommendations. The sustainable choice of production system is that with the least negative impact on land, water and air resources (ISO 2006).

In a forestry context, LCA includes the production system combining the forest ecosystem and the forest product technosystem. Radiative forcing may be used to assess the atmospheric impacts of the combined systems and optional ways to mitigate climate change by modifying the structure and function of the forest ecosystem and the forest product technosystem. LCA makes it possible to identify how carbon uptake in growth meets the carbon emissions in a forest-based production system, including the emissions in management, harvesting and logistic operations. The assessment may further include the use of timber and biomass for materials and energy, thus including the uptake and emissions throughout both forest ecosystem and forest product technosystem, for comparing different ways to manage forests and use timber and forest biomass in the manufacturing and energy industry.

Kilpeläinen et al. (2011) developed the LCA tool for estimating the net CO_2 exchange between the forest ecosystem, technosystem and atmosphere, when producing timber and energy biomass used for wood-based materials and energy. The system boundary included forestry producing timber and biomass for materials and energy, as a substitute for fossil-intensive materials and energy. In the analysis, fossil- and bio-systems can be compared with a focus on: (i) the carbon balance in the ecosystem and technosystem; (ii) retention of carbon in the ecosystem and technosystem; (iii) substitution and climate impacts; and (iv) performance of carbon in the ecosystem and technosystem under varying management. To calculate climate impact, the cumulative net CO_2 exchange (C_{net}) combines both bio- and fossil-systems, including the flow of carbon between the pools in an ecosystem ($C_{eco-bal}$) and technosystem ($C_{techno-bal}$):

$$C_{net} = C_{ecol-bal} + C_{techno-bal} \tag{19.12}$$

$$C_{ecol-bal} = C_{seq} + C_{decomp} \tag{19.13}$$

$$C_{techno-bal} = C_{man} + C_{bio} + C_{fos} + C_{seq-con} \tag{19.14}$$

Carbon dynamics in the ecosystem ($C_{eco-bal}$) includes carbon uptake in growth (C_{seq}) and carbon in emissions from decomposing litter and humus (C_{decomp}). In the technosystem, carbon is emitted in management, manufacturing and logistic operations (C_{man}), and in processing timber and biomass for materials and fuels (C_{bio}). Carbon is further emitted from fossil materials and energy (C_{fos}), and bound, e.g., in concrete ($C_{seq-con}$) used in buildings. C_{net} gives the net climate impact, indicating the amount of carbon emissions avoided when fossil-carbon intensive materials and energy were replaced by forest-based materials and energy (Box 19.6).

Box 19.6 Example of the phases of LCA in producing timber and energy biomass

In a forest ecosystem, the assessment may begin from regeneration, such as a clear cut area prepared for planting with seedlings grown in a nursery (Fig. 19.20). Before and during planting, carbon is emitted, for example, in site preparation, commuter traffic and the transport of machinery and seedlings to the site. Later in the production cycle (rotation), carbon is emitted in management, harvest and logistic operations (e.g., thinning, final cut). Logistic operations include both the short- and long-distance transportation of timber and biomass to road side, and further to the sites used in processing wood-based products and energy.

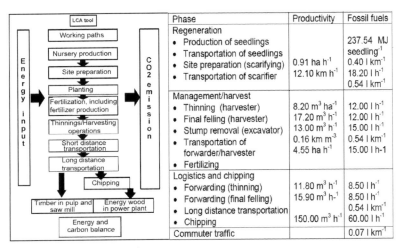

Fig. 19.20 Phases of LCA in producing timber and energy biomass (Kilpeläinen et al. 2011). Left: Working path in producing. Right: Productivity of phases of work and use of fossil fuels in operations used in implementing working path.

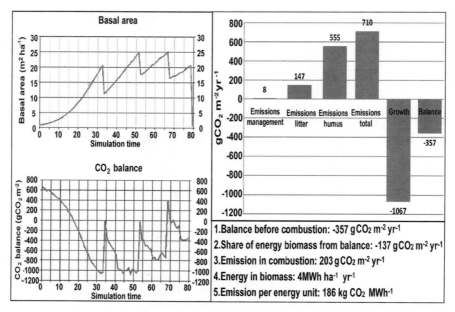

Fig. 19.21 Simulation of growth and dynamics (upper left), carbon balance (lower left, NEE), and carbon uptake and emissions in/from (right) a Scots pine dominated forest ecosystem managed for timber and energy biomass (logging residues). Simulation was run with the Sima model (Kellomäki et al. 2008) for a middle boreal forest (62° N) on a site of *Myrtillus* type (MT). The initial stand density was 2,500 seedlings per hectare, and the initial amount of organic matter in the soil was 67 Mg ha^{-1}. During an 80-year rotation length, timber and biomass were harvested in three thinnings and in the clear cut at the end rotation.

Figure 19.21 demonstrates how carbon is taken up and emitted in managing Scots pine for producing timber and energy biomass. During the first 15 years, the site was a source of CO_2 due to low uptake and high emission of CO_2. The high source strength was mainly related to the large amount of organic matter in soil originating from the previous management cycle. Thereafter, the site was a CO_2 sink. The sink strength varied following the reduction of biomass stocking in thinning and the increase of logging residue and litter on the soil, which increased the CO_2 emission from soil. The sink strength recovered in 10–15 years due to the enhanced growth of the remaining trees and due to the reduction of CO_2 emissions from decaying logging residues originating from the previous thinning. The sink strength culminated at the age of maximum growth rate (i.e., at the age of 40–50 years), and declined with the maturation of trees. The site finally became a CO_2 source due to reduced growth and increased CO_2 emission from the soil. Carbon balance covering the forest ecosystem and the management and logistic operation was −357 g CO_2 m^{-2} yr^{-1}. The share of energy biomass from the carbon balance was −137 g CO_2 m^{-2} yr^{-1}. The use of biomass for energy emitted 186 kg CO_2 MWh^{-1}, which was substantially smaller than emitted (340 kg CO_2 MWh^{-1}) if using coal (reference energy raw material) for energy.

Management for the combined production of timber and energy biomass

Pyörälä et al. (2012) used a process-based forest ecosystem model (Sima) to simulate the growth and development of managed boreal Norway spruce to produce timber and energy biomass. The carbon balance of production scenarios included the uptake of CO_2 in growth, CO_2 emissions in decaying litter and humus in the soil, and the emission of CO_2 from management, harvest and logistics, and the combustion of biomass for energy. In the first management regime, thinning was excluded and only the final cut (R0) was made (Fig. 19.22). In the second regime, thinning was also

Management of Forest for the Mitigation of Climate Change   329

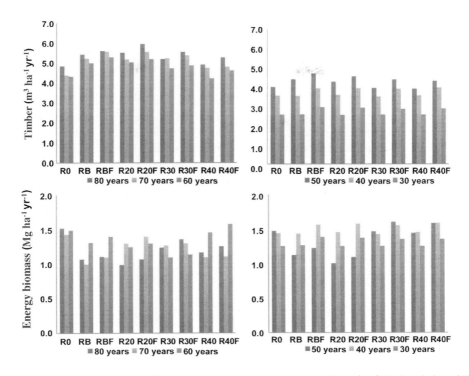

Fig. 19.22 Mean annual yield of timber (m³ ha⁻¹ yr⁻¹) (above) and energy biomass (Mg ha⁻¹ yr⁻¹) (below) in boreal Norway spruce under varying management regimes as a function of rotation length (30, 40, 50, 60, 70 and 80 years) and thinning regime. Simulations were run with the Sima model in the middle boreal conditions (62° N) on medium fertile site (*Myrtillus* type, MT) for Norway spruce using an initial density of 2,400 seedlings per hectare (Pyörälä et al. 2012). Permission of Elsevier. Legends: F = nitrogen fertilization (150 kg N ha⁻¹, once to twice over the rotation), R0 = only final felling, RB = thinning based on current management, R20 = increase of the basal area before and after thinning by 20%, R30 = increase of the basal area before and after thinning by 30%, and R40 = increase of the basal area before and after thinning by 40%, compared to that under RB.

included (RB). Three other regimes were used, similar to the second one, but thinning intensity was selected to maintain higher stocking over the rotation. Nitrogen fertilization (150 kg N ha⁻¹) was further used with long rotations (60–80 years) twice, and once with short rotations (30–50 years). Logging residues (stem top, branches and needles), stumps and coarse roots were harvested for energy biomass only in final cut.

In general, the use of long rotations for Norway spruce increased the mean annual timber yield compared to that of energy biomass. For example, on the medium fertile site (*Myrtillus* type, MT), the timber yield was 4–6 m³ ha⁻¹yr⁻¹ for long rotation, the largest values representing the current thinning with nitrogen fertilizing (Fig. 19.22) (Pyörälä et al. 2012). Using short rotation, the timber yield remained lower; i.e., 2–5 m³ ha⁻¹ yr⁻¹ without and 3–5 m³ ha⁻¹ yr⁻¹ with nitrogen fertilization. At the same time, the mean annual yield of energy biomass varied in the range of 1.0–1.5 Mg ha⁻¹ yr⁻¹ for long and 1.2–2.0 Mg ha⁻¹ yr⁻¹ for short rotation, regardless of nitrogen fertilization. On fertile sites (*Oxalis-Myrtillus* type, OMT), yields of timber and energy biomass were higher (10–20%) than on medium fertile site. However, the production patterns in relation to rotation length, thinning intensity and fertilization were similar than on medium fertile site. On the other hand, high stocking increased the yields of timber and energy biomass yields without and with nitrogen fertilization, regardless of the rotation length and site fertility.

Substitution of fossil fuels and the carbon neutrality of forest-based energy

In general, the CO_2 uptake most exceeds the loss of CO_2 if high stocking and long rotation was used in management. Pyörälä et al. (2012), for example, found that the carbon balance for boreal Norway spruce under long rotation was from −355 to −662 g CO_2 m^{-2}. Nitrogen fertilization slightly increased the uptake in relation to emissions. Under short rotation, the carbon balance was from −165 to −556 g CO_2 m^{-2}, with larger values under fertilization. The CO_2 emission per energy unit for the long rotations was 49–242 and 92–183 kg CO_2 MWh^{-1} without and with fertilization, on the fertile site (OMT). When using short rotations, the emission varied from 169 to 422 kg CO_2 MWh^{-1} and from 141 to 380 kg CO_2 MWh^{-1} without and with fertilization (Fig. 19.23). Regardless of the rotation length, the CO_2 emission per energy unit increased if energy biomass originated from the medium fertile sites (MT). In this case, the CO_2 emissions ranged from 80 to 219 kg CO_2 MWh^{-1} for the long rotations without fertilization and from 135 to 184 kg CO_2 MWh^{-1} with fertilization. The carbon emission per energy unit increased substantially if using the short rotations, the values being 230–564 and 188–485 kg CO_2 MWh^{-1} without and with fertilization, respectively.

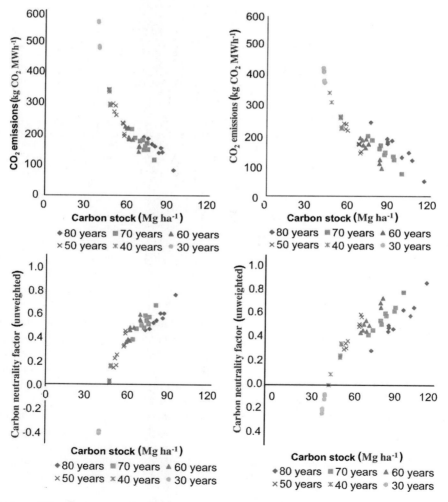

Fig. 19.23 Above: CO_2 emissions per unit of energy (kg CO_2 MWh^{-1}), and Below: Carbon neutrality, both as a function of mean carbon stock in the forest ecosystem (Mg ha^{-1}) and rotation length on medium (left panels, *Myrtillus* type, MT) and fertile (right panels, *Oxalis-Myrtillus* type, OMT) (Pyörälä et al. 2012). Permission of Elsevier.

Similarly, Routa et al. (2011a,b) and Routa et al. (2012) found that the lowest CO_2 emissions per unit of energy for boreal conifers were obtained if the pre-commercial stand was high and the density/stocking of stand was high through the rotation, and nitrogen fertilization was repeated three times during the rotation. This suggests that intensive management for timber and energy biomass clearly decreases CO_2 emissions in energy production over the energy supply chain compared to that based on coal. The forest bio-energy supply chains were effective, and the energy consumption in management, harvest and logistics was only 2–3% of the energy obtained from forest biomass.

Pyörälä et al. (2012) used further the carbon neutrality factor (CN (t)) developed by Schlamadinger et al. (1995) to indicate a reduction of CO_2 emissions when replacing coal with biomass in energy production. The carbon neutrality factor is the ratio between the net reduction/increase of CO_2 emissions from the use of forest biomass compared to CO_2 emissions from the use of optional energy raw material as coal (the reference energy system), which forest biomass substitutes:

$$CN(t) = \frac{[E_r(t) - E_n(t)]}{E_r(t)} = 1 - \frac{E_n(t)}{E_r(t)} \tag{19.15}$$

where $E_r(t)$ is the CO_2 emissions from the energy system based on coal and $E_n(t)$ is the CO_2 emissions from the energy system based on forest biomass between the years 0 and t, including the emissions from management and harvest/logistic operations. The following cases can be separated: (i) CN < 0, if the emissions from the bioenergy system are higher than the emissions from the fossil system; (ii) CN = 0, if the emissions from the bioenergy system are equal to the emissions from the fossil system; (iii) 0 < CN < 1, if the emissions from the bioenergy system are smaller than from the fossil system; (iv) CN = 1, if the bioenergy system is fully carbon neutral. The value CN > 1 indicates the full reduction of CO_2 emissions and the concurrent sequestration of carbon to the ecosystem (Zanchi et al. 2012).

Pyörälä et al. (2012) found that the carbon neutrality of the bioenergy production chain is greater than that based on coal, except if the rotation length was short and the mean stocking was low through the rotation (Fig. 19.23). The increase in the rotation length increased the carbon neutrality of the supply chain most if a high mean stocking was maintained through the rotation. Pyörälä et al. (2014) noted that "the positive effects of longer rotation length on carbon neutrality are due to the fact that in the early phase of rotation the CO_2 emissions from the soil will substantially exceed the CO_2 uptake of young trees, unlike in the later phase of rotation". The nitrogen fertilization further increased the carbon neutrality regardless of site fertility or the management regime used in the simulations. This held even when CO_2 emissions in manufacturing nitrogen fertilizers were included in the calculations. Maximizing the net present value (NPV) and carbon neutrality of the production chains simultaneously was, however, not possible in general; i.e., higher carbon sequestration and carbon stocks in forest ecosystem provide higher carbon neutrality, but not higher NPV, and vice versa.

19.6 Management of Forests for Reducing Radiative Forcing

Radiative forcing and mitigating climate change

The earth and atmosphere systems are in radiative balance, where a proportion of the incoming radiation is intercepted and a proportion is reflected into space. The reflection is indicated by the albedo; i.e., the ratio between the radiation reflected and the incidental radiation on surfaces. Albedo is scaled from zero for no reflection from a perfectly black surface, to one for perfect reflection from a white surface. Land use including management and the harvest of forest resources, changes the albedo and thus the interception of short-wave radiation in forest ecosystems. In boreal forests, forest management preferring high coniferous stocking is likely to reduce albedo, thus partly offsetting a part of the climate benefits, which carbon sequestration in forest ecosystems is likely to provide (Betts 2000).

The balance between the income and outcome of energy is indicated by radiative forcing (RF, W m^{-2}), which measures the flow rate of energy entering/departing the earth system (Sathre et al. 2013). The positive balance indicates warming (positive forcing) and the negative balance cooling (negative forcing). When added over a given time, the accumulated energy yields the Cumulative Radiative Forcing (CRF), which is a measure of the total excess of energy trapped in the earth system during the given period. In this context, the mitigation of climate change involves limiting the increase of greenhouses gases (GHGs) in the atmosphere affecting the intensity of radiative forcing and thus potential global warming (IPCC 2001). The ability of the main GHGs to trap heat (radiative efficiency) varies in such a way that the ability of nitrous oxide (N_2O) is greater than that of methane (CH_4) and carbon dioxide (CO_2). The concentration of GHGs in the atmosphere declines gradually in natural processes; i.e., once emitted, any GHG continues trapping heat as long it remains in the atmosphere specified by its life (Fig. 19.24) (IPCC 2007, Sathre et al. 2013) (Box 19.7).

Box 19.7 Decay of emission pulse of selected GHGs and radiative forcing

Let $(CO_2)_0$ be the mass carbon dioxide, $(N_2O)_0$ nitrous oxide and $(NH_4)_0$ methane emitted in the year 0. The mass of these GHGs in the atmosphere at the year t since emitted is (Fig. 19.24):

$$(CO_2)_t = (CO_2)_0 \times \left(0.217 + 0.259 \times e^{\frac{-t}{172.9}} + 0.338 \times e^{\frac{-t}{18.51}} + 0.186 \times e^{\frac{-t}{1.186}} \right) \tag{19.16}$$

$$(N_2O)_t = (N_2O)_0 \times e^{\frac{-t}{114}} \tag{19.17}$$

$$(CH_4)_t = (CH_4)_0 \times e^{\frac{-t}{12}} \tag{19.18}$$

The time series of CO_2, N_2O and CH_4 are converted into the time series of atmospheric concentrations based on their molecular mass and the molecular mass of air (28.95 g mol^{-1}) and the total mass of the atmosphere (5.148 x 10^{21} g). The change in the instantaneous radiative forcing relative to the changes in the CO_2, N_2O and CH_4 concentrations is (Sathre et al. 2013):

$$F_{CO2} = \frac{3.7}{\ln(2)} \times \ln\left(1 + \frac{\Delta CO_2}{CO_{2ref}}\right) \tag{19.19}$$

$$F_{N_2O} = 0.12 \times \left[\sqrt{\Delta N_2O + N_2O_{ref}} - \sqrt{N_2O_{ref}} \right] - f(M,N) \tag{19.20}$$

$$F_{CH_4} = 0.036 \times \left(\sqrt{\Delta CH_4 + CH_{4ref}} - \sqrt{CH_{4ref}} \right) - f(M,N) \tag{19.21}$$

where F_{CO2}, F_{N2O} and F_{CH4} are the instantaneous values of radiative forcing [W m^{-2}] related to the concentration of GHGs and their concentration change of ΔCO_2, ΔN_2O and ΔCH_4. On the volume bases, the dimension for CO_2 is [ppmv], and for N_2O and CH_4 [ppbv]. The reference values can be chosen, e.g., CO_{2ref} = 383 ppmv, N_2O = 319 ppbv and CH_{4ref} = 1774 ppbv (Sathre et al. 2013). The function (M, N) compensates for the overlap of spectral absorption between N_2O and CH_4 (IPCC 2007). The annual cumulative radiative forcing (CRF, Ws m^{-2} or J m^{-2}) of a pulse of GHG gas is obtained by integrating the instantaneous forcing over a year.

Box 19.7 contd....

Box 19.7 contd....

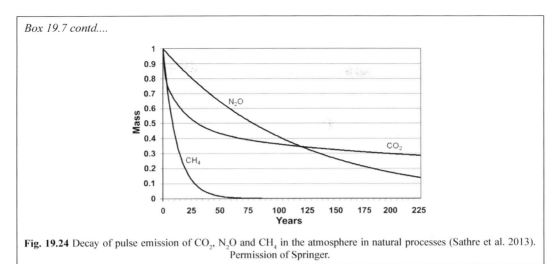

Fig. 19.24 Decay of pulse emission of CO_2, N_2O and CH_4 in the atmosphere in natural processes (Sathre et al. 2013). Permission of Springer.

Impacts of management of carbon storage and albedo on radiative forcing

Forestry with management, harvest and logistics affects climate by altering fluxes of heat, momentum and moisture exchanges between forest surface and atmosphere. Until now, biogeophysical forcing has only seldom been included in forestry and even carbon sequestration is still gaining importance in management for climate change mitigation (Bright et al. 2015). Based on carbon orientation in management, reduced thinning intensity, longer rotation and nitrogen fertilization increase carbon stocks and enhance negative forcing with cooling. At the same time, the albedo of forest canopy is, however, reduced and more radiation is absorbed in the ecosystem, thus enhancing positive forcing with warming (Caiazzo et al. 2014). Management and harvest impacts through changes in carbon stocks and albedo are almost the opposite, as increasing carbon stocks tends to decline albedo and vice versa (Schwaiger and Bird 2010, Kirschbaum et al. 2011). This is especially important in boreal regions, where changes in albedo are likely more important than carbon storages in controlling radiative forcing (Betts 2000, Kirschbaum et al. 2011).

Forcing due to carbon stocks is related to changes in land-use or changes in management. In the boreal forests, soil carbon stock is fairly stable through the management and harvest cycle, while tree carbon stocks increase along with the maturation of trees. This is especially the case if even-aged management or plantation forestry are used in producing timber and biomass. Radiative forcing (ΔRF_{CSC}, J m^{-2} d^{-1}) per unit change of atmospheric CO_2 concentration ($\Delta[C]$, ppm) is attributable to, for example, the change in thinning intensity related to the reference thinning (Kirschbaum et al. 2011):

$$\Delta RF_{CSC} = 86400 \times 5.35 \times \ln(1 + \Delta[C]/[C]) \tag{19.22}$$

where 84,400 is a day in seconds, the factor 5.35 [W m^{-2}] converts the unit of carbon dioxide to radiative forcing, and [C, ppm] is the background atmospheric CO_2 concentration (Box 19.8).

The albedo of forest cover is related to the properties and structure of tree canopy (e.g., density foliage cover, tree species) and ground vegetation, and seasonal changes in trees and ground vegetation. The presence/absence of snow and the seasonal angular distribution of irradiation results in a pronounced variability in albedo. In general, snow cover, on both soil and the canopy, increases albedo; i.e., its value is 0.8 for new snow and 0.20 for treeless ground cover regardless of site fertility with no snow (Ni 2000, Lukeš et al. 2013). The effect of snow on the balance of incoming and reflected short-wave radiation is especially strong in late winter, when irradiation is already high in boreal conditions.

> **Box 19.8 Converting carbon stock change to change in atmospheric CO$_2$ concentration and radiative forcing**
>
> Change of the carbon stock (ΔC, Mg C) is converted to the change of atmospheric CO$_2$ concentration ($\Delta[C]$, ppm): one ppm of CO$_2$ concentration in the whole atmosphere equals to 2.123 Gton of C (Kirschbaum et al. 2011):
>
> $$\Delta[C] = \Delta C / 2.123 \times 10^9 \qquad (19.23)$$
>
> The total radiative forcing over a year for the whole surface of earth is (ΔR_E, J m^{-2} yr^{-1}):
>
> $$\Delta R_E = \Delta RF_{CSC} \times 510 \times 10^{12} \times 365 \qquad (19.24)$$
>
> where 510 × 10^{12} m^2 (510 million km^2) is the surface of earth and 365 is the number of days per a year. If assuming a background CO$_2$ concentration of 390 ppm, the total radiative forcing of the earth over a year per the removal of one ton of carbon gives:
>
> $$\Delta R_E = 86400 \times 5.35 \times \ln\left(1 + \frac{\Delta C/(2.123 \times 10^9)}{390}\right) \times 510 \times 10^{12} \times 365 \qquad (19.25)$$
>
> with a reduction of radiative forcing by –104 GJ (Mg C)$^{-1}$ yr^{-1}.

In managed forests, the daily radiative forcing (ΔR_d, J m^{-2} d^{-1}) due to the change in albedo is related to a given management and harvest regime, such as tree species choice, thinning intensity, clear cutting, etc. (Kirschbaum et al. 2011):

$$\Delta R_d = Q_s \downarrow \times \Delta a \times (1 - \alpha_{atm}) \qquad (19.26)$$

where $Q_{s\downarrow}$ is the total daily downward radiation, Δa is the difference in albedo between two different management regimes and α_{atm} is the share of short-wave radiation intercepted by the atmosphere, the average global value of 20%. The annual change in radiative forcing on a hectare basis (ΔR_a, J ha^{-1} yr^{-1}) related to albedo change is (Kirschbaum et al. 2011):

$$\Delta R_a = \sum_{day=1}^{365} 10000 \times \Delta R_d \qquad (19.27)$$

In general, the value of albedo over a forested area is related to the tree species composition, stocking and age class distribution of tree stands affecting the canopy cover. Rautiainen et al. (2011) showed that the shortwave albedo in boreal coniferous forest reduced if the tree stand became older and the stocking increased. Similarly, Lukeš et al. (2013) found that species-specific albedo in the middle boreal forests (61° N) is correlated to the forest structure (Fig. 19.25), especially to the Leaf Area Index (LAI).

Contribution of negative forcing of carbon stocks exceeds the positive forcing of albedo in fairly mature phases of stand development depending on the location and properties of site and tree species. Kirschbaum et al. (2011) analyzed how afforestation with Monterey pine (*Pinus radiate* D. Don) affected to radiative forcing of a pasture at low latitude in the southern hemisphere (S 38°) in New Zealand (Fig. 19.26). The carbon accumulation in a stand increased rapidly after canopy closure since five years of the establishment of plantation, with high increase in negative (cooling) radiative forcing due to carbon accumulation. In 30 years, the biomass in a pine stand increased up to 200 Mg C ha^{-1}, while albedo reduced from 0.2 to 0.13, with positive (warming) radiative forcing. Thinning in the early development of the plantation declined the carbon accumulation, but albedo reduced only slightly showing only a temporarily reduction in canopy closure. By the end

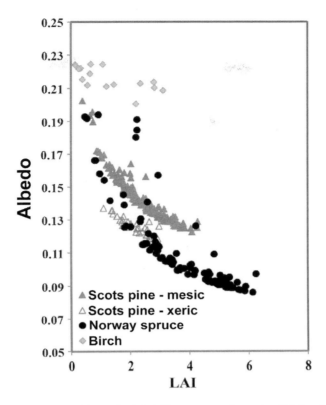

Fig. 19.25 Albedo of forest as a function of Scots pine (pine), Norway spruce (spruce) and birch and leaf area index (LAI) in middle boreal conditions (61° N) (Lukeš et al. 2013). Measurements are for solar zenith angle of 40°. Permission of Elsevier.

of monitoring period, the mitigation effect of carbon storing (carbon effect in stand) reduced by one third when assessing carbon benefits in the context of global carbon cycle (carbon effect with carbon cycle in the context of land-use change). Carbon storage was still important in compensating for declining albedo when assessing the benefits of afforestation in mitigating climate warming. Kirschbaum et al. (2011) estimated that the declining albedo reduced, averaged over the monitoring period, by 17–24% the climate benefits from increasing carbon storage.

Impacts of replacing fossil fuels and fossil materials on radiative forcing

Climate warming may further be reduced by using forest biomass to replace fossil fuels alone or combining the use of energy biomass with the use of wood-based materials to increase carbon density outside forests. Figure 19.27 shows the profile of Cumulative Radiative Forcing (CRF) over 240 years following the pulse of GHG due to logging residue left on site or burnt in the year 0 when replacing coal and gas in energy production (Sathre and Gustavsson 2011). The decay of logging residues on site increased radiative forcing during the first 10 to 20 years, whereas the use to energy immediately reduced radiative forcing. Burning reduced the radiative forcing more than if left on the site to decay and store carbon in the soil. The reduction in radiative forcing is less for gas than for coal used as a reference; i.e., the GHG emissions per energy unit for gas are substantially smaller than those for coal. The mitigation efficiency of forest biomass is thus closely related to the fossil fuel, which is substituted in energy production (McKechnie et al. 2011, Sathre and Gustavsson 2010, 2011, 2012, Zetterberg and Chen 2014).

336 *Managing Boreal Forests in the Context of Climate Change*

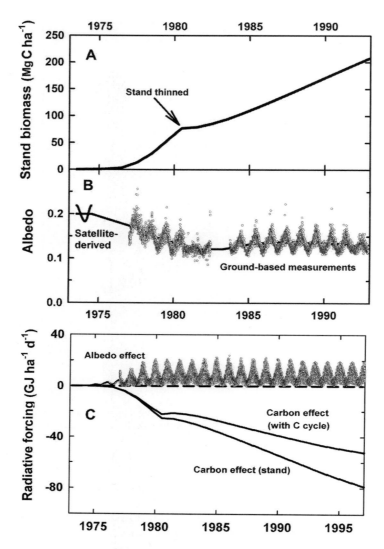

Fig. 19.26 Time series on carbon accumulation (upper), development of albedo (middle) and radiative forcing (lower) in a Monterey pine (*Pinus radiata* D.Don) plantation used to afforest a pasture at low latitude in the southern hemisphere (S 38°) in New Zealand, based on Kirschbaum et al. (2011). Courtesy of Miko Kirschbaum.

Sathre et al. (2013) used further thinning or thinning with nitrogen fertilization in studying how management affected radiative forcing in substituting for coal and non-wood materials in energy production and construction. The net climate benefit (net cumulative radiative forcing) related to management regimes is obtained from the difference between benefits provided by managed and unmanaged forests. In the latter case, the reduction in cumulative radiative forcing is only dependent on the accumulation of carbon in trees and soil, whereas in the former case the benefits further represent the replacement of fossil fuels in energy production and non-woody materials in construction. Figure 19.28 shows that cumulative radiative forcing is reduced substantially using intensive management (thinning plus nitrogen fertilization) than non-intensive management (thinning only). Thus, intensive management for manufacturing wood-based materials and energy biomass provided the greatest potential for reducing net carbon emissions in forestry (Sathre et al. 2010, Sathre and Gustavsson 2011, Sathre et al. 2013).

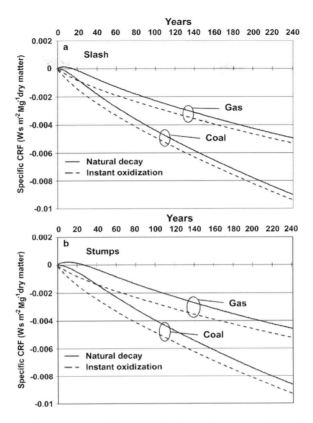

Fig. 19.27 Cumulative radiative forcing per unit of harvest residue (above) and stumps (below) when left to decay in forest (solid line, natural decay) or used to replace gas and coal (dotted line) in energy production (Sathre and Gustavsson 2011, Sathre et al. 2013). Permission of Springer.

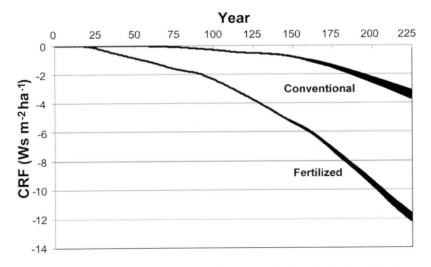

Fig. 19.28 Cumulative radiative forcing when thinning only (conventional) and thinning with nitrogen fertilization (intensive management) is used over a forest landscape in substituting coal in energy production and non-woody materials in construction (Sathre et al. 2013). Permission of Springer. In both management cases, the shaded area indicates the difference between the cases where high and low carbon levels are assumed for equilibrium carbon stock in unmanaged (natural) forest.

The stand-based analysis demonstrates that early in the rotation the atmospheric impact was fairly similar for both management regimes (Sathre et al. 2013) (Fig. 19.28). In late rotation, the atmospheric impact was increased much more rapidly if fertilization was used to increase growth. In this case, the cumulative radiative forcing was nearly double compared to conventional management (Sathre and Gustavsson 2011). The greatest effect was obtained if the increased growth was used to substitute for non-wood materials (e.g., concrete) in construction. Similarly, radiative forcing was substantially reduced if harvest residue from thinning, final cut and abandoned wood-based materials were used to substitute for coal in energy production. This is also true for a forest landscape with net climate benefits due to the difference between those under management and without management. Sathre et al. (2013) found that under thinning and fertilizing, the net CO_2 emissions avoided were double compared to management with thinning only.

19.7 Concluding Remarks

The management and harvest of forests provides renewable raw material for wood-based products and energy for reducing carbon in the ecosystem/technosystem/atmosphere cycle (Eriksson et al. 2007, Klein et al. 2013). Sustainable management and harvest, with a long-term balance between growth and harvest, provides a solid basis by which to obtain climate benefits when the climate/forest integrations are combined into management and harvest. Management is a means to control the forest structure, including the biophysical properties of forest cover such as albedo. For example, Otto et al. (2014) found that canopy albedo is driven by tree species in the early succession, but the effect of tree species decreases along with the maturation of trees. They found that the tree species effects declined further related to thinning towards the end of rotation, where thinning explains up to 70% of the variance in canopy albedo. Thus, it may be possible to obtain albedo benefits by shortening the rotation and using regular thinning (Bonan 2008, Bright et al. 2011).

Carbon stocks are maximized in naturally growing and developing forests if devastating disturbances (e.g., wild fire, strong wind) are excluded (Karjalainen 1996c,d, Luyssaert et al. 2008). When converting natural forests to managed ones, the carbon bound in trees and soil reduces. The new balance between uptake and emissions, and subsequent carbon stocks, is related to the intensity of management and harvest in maintaining high carbon stocks in trees and soil in forests and outside forests in wood products. Liski et al. (2001) concluded that longer rotation for Scots pine and Norway spruce are probably the most preferable in the Nordic countries for enhancing carbon residence outside the atmosphere. In this respect, novel management systems may be needed to combine the production of timber and energy biomass in an appropriate way to maintain or even increase carbon stocks both within and outside ecosystems, considering biogeophysical benefits provided for climate change mitigation (Eriksson et al. 2007, Sathre et al. 2010, Sathre and Gustavsson 2011).

References

Ågren, G. I. 2010. Microbial mitigation. Nature Geoscience 3: 303–304.
Aubinet, M., A. Grelle, A. Ibrom, Ü. Rannik, J. Moncrieff, T. Foken et al. 2000. Estimates of the annual net carbon and water exchange of European forests: the EUROFLUX 612 methodology. Advances in Ecological Research 30: 113–175.
Garcia-Gonzalo, J., H. Peltola, A. Zubizarreta-Gerendiain and S. Kellomäki. 2007a. Impacts of forest landscape structure and management on timber production and carbon stocks in the boreal forest ecosystem under changing climate. Forest Ecology and Management 241(1-3): 243–257.
Garcia-Gonzalo, J., H. Peltola, E. Briceño-Elizondo and S. Kellomäki. 2007b. Changed thinning regimes may increase carbon stock under climate change: a case study from a Finnish boreal forest. Climatic Change 81: 431–454.
Ge, Z. -M., S. Kellomäki, X. Zhou, K. -Y. Wang and H. Peltola. 2011a. Climate, canopy conductance and leaf area development controls on evapotranspiration in a boreal coniferous forests over a 10-year period: a united model assessment. Ecological Modelling 222: 1626–1638.

Ge, Z. -M., S. Kellomäki, X. Zhou, K. -Y. Wang and H. Peltola. 2011b. Evaluation of carbon exchange in a boreal coniferous stand over a 10-year period: an integrated analysis based on ecosystem model simulations and eddy covariance measurements. Agricultural and Forest Meteorology 151(2): 191–203.

Ge, Z. -M., S. Kellomäki, X. Zhou and H. Peltola. 2013. The role of climatic variability in controlling carbon and water budgets in a boreal Scots pine forest during ten growing seasons. Boreal Environment Research 129: 181–194.

Gholz, H. L., D. A. Wedin, S. M. Smitherman, M. E. Harmon and W. J. Partons. 2000. Long-term dynamics of pine and hardwood litter in contrasting environments: towards a global model of decomposition. Global Change Biology 6: 751–765.

Gough, C. M., C. S. Vogel, H. P. Schmid and P. S. Curtis. 2008. Controls on annual forest carbon storage: lessons from the past and predictions for the future. BioScience 58(7): 609–622.

Govers, G., R. Merck, K. van Oost and B. van Wesemael. 2012. Soil organic carbon management for global benefits: a discussion paper presented at the workshop "Soil Organic Carbon Benefits": A Scope Study, 10th–12th September 2012. Nairobi. Workshop organized by the Scientific and Technical Advisory Panel of the Global Environmental Facility. http://www.staggef.org. Visited in Mach 3, 2015.

Haimi, J., J. Laamanen, R. Penttinen, M. Räty, S. Koponen, S. Kellomäki et al. 2005. Impacts of elevated CO_2 and temperature on soil fauna in boreal forests. Applied Soil Ecology 30: 104–112.

Harmon, M. E., O. M. Krankina and J. Sexton. 2000. Decomposition vectors: a new approach to estimating wood detritus decomposition dynamics. Canadian Journal of Forest Research 30: 76–84.

Helmisaari, H. -S., I. Ostonen, K. Lohmus, J. Derome, A. -J. Lindroos, P. Merilä et al. 2009. Ectomycorrhizal root tips in relation to site and stand characteristics in Norway spruce and Scots pine stands in boreal forests. Tree Physiology 29(3): 445–456.

Huttunen, L., P. J. Aphalo, T. Lehto, P. Niemelä, K. Kuokkanen and S. Kellomäki. 2009. Effects of elevated temperature, elevated CO_2 and fertilization on quality and subsequent decomposition of silver birch leaf litter. Soil Biology and Biochemistry 41: 2414–2421.

Hyvönen, R., B. A. Olsson, H. Lundkvist and H. Staaf. 2000. Decomposition and nutrient release from *Picea abies* (L.) Karst. and *Pinus sylvestris* L. logging residue. Forest Ecology and Management 126: 97–112.

Hyvönen, R., G. I. Ågren, S. Linder, T. Persson, M. F. Cotrufo, A. Ekblad et al. 2007. The likely impact of elevated $[CO_2]$, nitrogen deposition, increased temperature and management on carbon sequestration in temperate and boreal forest ecosystems: a literature review. New Phytologist 173: 463–480.

IPCC. 2001. Climate Change 2001: Impacts, Adaptation and Vulnerability. Contribution of Working Group II to the Third Assessment Report of Intergovernmental Panel of Climate Change. Cambridge University Press, Cambridge, UK.

IPCC. 2007. Climate Change 2007: Mitigation of Climate Change. Contribution of Working Group III to the Fourth Assessment Report of the Intergovernmental Panel on Climate Change. Cambridge University Press, Cambridge, UK.

ISO (International Organization of Standardization). 2006. ISO 14040:2006. Environmental management – life cycle assessment – principles and framework. http://www.iso.org/iso/. Visited January 14, 2015.

Iversen, C. M. 2010. Digging deeper: fine-root responses to rising atmospheric CO_2 concentration in forested ecosystems. New Phytologist 186: 346–357.

Jandl, R., M. Lindner, L. Vesterdal, B. Bauwens, R. Baritz, F. Hagedorn et al. 2007. How strongly can forest management influence soil carbon sequestration? Geoderma 137: 253–268.

Janisch, J. E. and M. E. Harmon. 2002. Successional changes in live and dead wood carbon stores: implications for net ecosystem productivity. Tree Physiology 22: 77–89.

Johnson, D. W. and P. S. Curtis. 2001. Effect of forest management on soil C and N storage: meta analysis. Forest Ecology and Management 140: 227–238.

Kaipainen, T., J. Liski, A. Pussinen A. and T. Karjalainen. 2004. Managing carbon sinks by changing rotation length in European forests. Environmental Science & Policy 7: 205–219.

Karjalainen, T., S. Kellomäki and A. Pussinen. 1994. Role of wood-based products in absorbing atmospheric carbon. Silva Fennica 28(2): 67–80.

Karjalainen, T. 1996a. Dynamics of carbon flow through forest ecosystem and the potential of carbon sequestration in forests and wood products in Finland. University of Joensuu, Faculty of forestry. Research Notes 40: 1–31.

Karjalainen, T. 1996b. Dynamics and potentials of carbon sequestration in managed stands and wood products in Finland under changing climatic conditions. Forest Ecology and Management 80: 113–132.

Karjalainen, T. 1996c. Model computations on sequestration of carbon in managed forests and wood products under changing climatic conditions in Finland. Journal of Environmental Management 47: 311–328.

Karjalainen, T. 1996d. Carbon sequestration potential of unmanaged forest stands in Finland under changing climatic conditions. Biomass and Bioenergy 10(5/6): 313–329.

Kellomäki, S. and H. Väisänen. 1997. Modelling the dynamics of the boreal forest ecosystems for climate change studies in the boreal conditions. Ecological Modelling 97(1,2): 121–140.

Kellomäki, S. and K. -Y. Wang. 2000. Short-term environmental controls on carbon dioxide flux in a boreal coniferous forest: model computations compared with measurements by eddy correlation. Ecological Modelling 128: 63–88.

Kellomäki, S., H. Peltola, T. Nuutinen, K. I. Korhonen and H. Strandman. 2008. Sensitivity of managed boreal forests in Finland to climate change, with implications for adaptive management. Philosophical Transactions of the Royal Society 363: 2341–2351.

Kellomäki, S., A. Kilpeläinen and A. Alam. 2013. Effects of bioenergy production on carbon sequestration in forest ecosystems. pp. 125–158. *In*: S. Kellomäki, A. Kilpeläinen and A. Alam (eds.). Forest BioEnergy Production. Springer Science + Business Media, New York, USA.

Kilpeläinen, A., A. Alam, H. Strandman and S. Kellomäki. 2011. Life cycle assessment tool for estimating net CO_2 exchange of forest production. Global Change Biology, Bioenergy 3: 461–471.

Kilpeläinen, A., S. Kellomäki and H. Strandman. 2012. Net atmospheric impacts of bioenergy production and utilization in Finnish boreal forests Global Change Biology, Bioenergy 4: 811–817.

Kirschbaum, M. U. F., D. Whitehead, S. m. Dean, P. M. Beets, J. D. Shephard and A. -G. Ausseil. 2011. Implications of albedo changes following afforestation on the benefits of forests as carbon sinks. Biogeosciences 8: 3687–3696.

Klein, D., S. Höllert, M. Blaschke and C. Schulz. 2013. The contribution of managed and unmanaged forest to climate mitigation–a model approach at stand level for main tree species in Bavaria. Forests 4: 43–69.

Knorr, W., I. C. Prentice, J. I. House and E. A. Hollan. 2005. Long-term sensitivity of carbon turnover to warming. Nature 433: 298–301.

Krankina, O. N. and M. E. Harmon. 1995. Dynamics of the dead wood carbon pool in Northwestern Russian boreal forests. Water, Air and Soil Pollution 82: 227–238.

Kurz, W. A. and M. J. Apps. 1999. A 70-year retrospective analysis of carbon fluxes in the Canadian forest sector. Ecological Applications 9(2): 526–547.

Laiho, R. and C. E. Prescott. 2004. Decay and nutrient dynamics of coarse woody debris in northern coniferous forests: synthesis. Canadian Journal of Forest Research 34: 763–777.

Law, B. E., P. E. Thornton, J. Irvine, P. M. Anthoni and S. van Tuyl. 2001. Carbon storage and fluxes in ponderosa pine forests at different developmental stages. Global Change Biology 7: 755–777.

Law, B. E., O. J. Sun, J. Campbell, S. van Tuyl and P. E. Thornton. 2003. Changes in carbon storage and fluxes in a chronosequence of ponderosa pine. Global Change Biology 9: 510–524.

Law, B. E. and M. E. Harmon. 2011. Forest sector carbon management, measurement and verification, and discussion of policy related to climate change. Carbon Management 2(1): 73–84.

Leppälammi-Kujansuu, J., M. Salemaa, D. Berggren Kleja, S. Linder and H. -S. Helmisaari. 2014. Fine root turnover and litter production of Norway spruce in a long-term temperature and nutrient manipulation experiment. Plant and Soil 374: 73–88.

Liski, J., A. Pussinen, K. Pingoud, R. Mäkipää and T. Karjalainen. 2001. Which rotation length is favourable to carbon sequestration? Canadian Journal of Forest Research 31: 2004–2013.

Lukas, M., A. Lagomarsino, M. C. Moscatelli, P. De Angelis, M. F. Cotrufo M and D. L. Godbold. 2009. Forest soil carbon cycle under elevated CO_2—a case of increased throughput? Forestry 82: 75–86.

Lukeš, P., P. Stenberg and M. Rautiainen. 2013. Relationship between forest density and albedo in the boreal zone. Ecological Modelling 261-262: 74–79.

Luyssaert, S., E. -D. Schulze, A. Börner, A. Knohl, D. Hessenmöller D., B. E. Law et al. 2008. Old-growth forests as global carbon sinks. Nature 455(11): 213–215.

Mahli, Y. P. Meir and S. Brown. 2002. Forests, carbon and global change. Philosophical Transactions A 360: 1567–1591.

Mäkinen, H., J. Hynynen, J. Siitonen and R. Sievänen. 2006. Predicting the decomposition of Scots pine, Norway spruce and birch stems in Finland. Ecological Applications 16(5): 1865–1879.

Mäkipää, R. 1994. Effect of nitrogen fertilization on the humus layer and ground vegetation under closed canopy of boreal coniferous stands. Silva Fennica 28(2): 81–94.

Mäkipää, R., T. Karjalainen, A. Pussinen and S. Kellomäki. 1999. Effects of climate change and nitrogen deposition on carbon sequestration of a forest ecosystem in the boreal zone. Canadian Journal of Forest Research 29: 1490–1501.

Mälkönen, E. 1976. Effect of whole tree harvest on soil fertility. Silva Fennica 10(3): 157–164.

Martikainen, P. J., T. Aarnio, V. -M. Taavitsainen, L. Päivinen and K. Salonen. 1989. Mineralization of carbon and nitrogen in soil samples taken from three fertilized pine stands: long-term effects. Plant and Soil 114: 99–106.

Matala, J., S. Kellomäki and T. Nuutinen. 2008. Litter fall in relation to volume growth of trees–analysis based on literature. Scandinavian Journal of Forest Research 23: 194–202.

Mazziotta, A., M. Mönkkönen, H. Strandman, J. Routa and S. Kellomäki. 2014. Modeling the effects of climate change and management on the dead wood dynamics in boreal forest plantations. European Journal of Forest Research 133: 405–421.

McKechnie, J., S. Colombo, J. Chen, W. MacBee and H. L. MacClean. 2011. Forest bioenergy or forest carbon? Assessing trade-offs in greenhouse gas mitigation with wood-based fuels. Environmental Science and Technology 45(2): 789–795.

McKinley, D. C., M. G. Ryan, R. A. Birdsey, C. P. Giardina, M. E. Harmon, L. S. Heath et al. 2011. A synthesis of current knowledge on forests and carbon storage in the United States. Ecological Applications 21(6): 1902–1924.

Metsäkeskus Tapio (Tapio). 1994. Luonnonläheinen metsänhoito: metsänhoitosuositukset. Metsäkeskus Tapio julkaisu 6/1994.

Nabuurs, G. J. O., K. Masera, P. Andrasko, R. Benitez-Ponce, M. Boer, E. Dutschke et al. 2007: Forestry. pp. 542–584. *In*: B. Metz, O. R. Davidson, P. R. Bosch, R. Dave and L. A. Meyer (eds.). Climate Change 2007: Mitigation. Contribution of Working Group III to the Fourth Assessment Report of the Intergovernmental Panel on Climate Change, Cambridge University Press, Cambridge, UK.

Ni, W. 2000. Effect of canopy structure and the presence of snow on the albedo of boreal conifer forests. Journal of Geophysical Research 105(D9): 11879–11888.

Niinistö, S. M., J. Silvola and S. Kellomäki. 2004. Soil CO_2 efflux in a boreal pine forest under atmospheric CO_2 enrichment and air warming. Global Change Biology 10(8): 1363–1376.

Niinistö, S. M., S. Kellomäki and J. Silvola. 2011. Seasonality in a boreal forest ecosystem affects the use of soil temperature and moisture as predictors of soil CO_2 efflux. Biogeosciences 8: 3169–3186.

Nilsen, P. and L. T. Strand. 2013. Carbon stores and fluxes in even- and uneven-aged Norway spruce stands. Silva Fennica 47: 1–15.

Nohrstedt, H. O., K. Arnebrant, E. Bååt and B. Söderström. 1989. Changes in soil carbon content, respiration, ATP content, and microbial biomass in nitrogen-fertilized pine forest soils in Sweden. Canadian Journal of Forest Research 19: 323–328.

Olson, J. S. 1963. Energy storage and the balance of producers and decomposers in ecological systems. Ecology 44(2): 322–331.

Olsson, P., S. Linder, R. Giesler and P. Högberg. 2005. Fertilization of boreal forest reduces both autotrophic and heterotrophic soil respiration. Global Change Biology 11: 1745–1753.

Otto, J., D. Berveiller, F. -M. Bréon, N. Delpierre, G. Geppert, G. Granier et al. 2014. Forest summer albedo is sensitive to species and thinning: how should we account for this in Earth system models? Biogeosciences 11: 2411–2427.

Pumpanen, J., C. J. Westman and H. Ilvesniemi. 2004. Soil CO_2 efflux from a podzol forest soil before and after forest clearcut and site preparation. Boreal Environment Research 9: 199–212.

Pyörälä, P., S. Kellomäki and H. Peltola. 2012. Effects of management on biomass production in Norway spruce stands and carbon balance of bioenergy use. Forest Ecology and Management 275: 87–97.

Pyörälä, P., H. Peltola, H. Strandman, A. Kilpeläinen, A. Asikainen, K. Jylhä et al. 2014. Effects of management on economic profitability of forest biomass production and carbon neutrality of bioenergy use in Norway spruce stands under the changing climate. Bioenergy Research 7: 279–294.

Raich, J. W. and W. H. Schlesinger. 1992. The global carbon dioxide flux in soil respiration and its relationship to vegetation and climate. Tellus 44B: 81–99.

Rautiainen, M., P. Stenberg, M. Mõttus and T. Manninen. 2011. Radiative forcing simulations link boreal forest structure and shortwave albedo. Boreal Environment Research 16: 91–100.

Routa, J., S. Kellomäki, A. Kilpeläinen, H. Peltola and H. Strandman. 2011a. Effects of forest management on the carbon dioxide emissions of wood energy in integrated production of timber and energy biomass. Global Change Biology, Bioenergy 3: 483–497.

Routa, J., S. Kellomäki, H. Peltola and A. Asikainen. 2011b. Impacts of thinning and fertilization on timber and energy wood production in Norway spruce and Scots pine: scenario analyses based on ecosystem model simulations. Forestry 4(2): 159–175.

Routa, J., S. Kellomäki and H. Peltola. 2012. Impacts of intensive management and landscape structure on timber and energy wood production and net CO_2 emissions from energy wood use of Norway spruce. Bioenergy Research 5: 106–123.

Row, C. and R. B. Phelps. 1990. Tracing the flow of carbon through U.S. forest product sector. Presentation at the 19th World Congress, IUFRO. Montreal Canada, August 5–11, 1990.

Russell, M. B., C. W. Woodall, S. Fraver, A. W. D'Amato, G. M. Domke and K. E. Skog. 2014. Residence times and decay rates of downed woody debris biomass/carbon in eastern US forests. Ecosystems 17: 765–777.

Saarsalmi, A., P. Tamminen and M. Kukkola. 2014. Effects of long-term fertilisation on soil properties in Scots pine and Norway spruce stands. Silva Fennica 48(1): 1–18.

Sathre, R., L. Gustavsson and J. Bergh. 2010. Primary energy and greenhouse gas implications of increasing biomass production through fertilization. Biomass and Bioenergy 34: 572–581.

Sathre, R. and L. Gustavsson. 2011. Time-dependent climate benefits of using forest residue to substitute fossil fuels. Biomass and Bioenergy 35: 2506–2516.

Sathre, R. and L. Gustavsson. 2012. Time-dependent radiative forcing effects of forest fertilization and biomass substitution. Biogeochemistry 109: 203–218.

Sathre, R., L. Gustavsson and S. Haus. 2013. Time dynamics and radiative forcing of forest bioenergy systems. pp. 185–206. *In*: S. Kellomäki, A. Kilpeläinen and A. Alam (eds.). Forest BioEnergy Production. Springer Science+Business Media, New York, USA.

Schlamadinger, B., J. Spitzer, G. H. Kohlmaier and M. Lüdeke. 1995. Carbon balance of bioenergy from logging residues. Biomass and Bioenergy 8: 221–234.

Schlesinger, W. H. and J. A. Andrews. 2000. Soil respiration and global carbon cycle. Biogeochemistry 48: 7–20.

Schwaiger, H. P. and D. N. Bird. 2010. Integration of albedo effects caused by land use change into the climate balance: should we still account in greenhouse units? Forest Ecology and Management 260: 278–286.

Sirén, G. 1955. The development of spruce forest on raw humus sites in Northern Finland and its ecology. Acta Forestalia Fennica 62(4): 1–408.

Tuomi, M., R. Laiho, A. Repo and J. Liski. 2011. Wood decomposition model for boreal forests. Ecological Modelling 222: 709–718.

Valachovic, Y. S., B. A. Caldwell, K. Cromack, Jr. and R. P. Griffiths. 2004. Leaf litter chemistry controls on decomposition of Pacific Northwest trees and shrubs. Canadian Journal of Forest Research 34: 2131–2147.

Vávrová, P., T. Penttilä and R. Laiho. 2009. Decomposition of Scots pine fine woody debris in boreal conditions: implementations for estimating carbon pools and fluxes. Forest Ecology and Management 257: 401–412.

Wang, K. -Y., S. Kellomäki, T. Zha and H. Peltola. 2004. Component carbon fluxes and their contribution to ecosystem carbon exchange in a pine forest: an assessment based on eddy covariance measurements and an integrated model. Tree Physiology 24: 19–34.

Zanchi, G., N. Pena and N. Bird. 2012. Is woody bioenergy carbon neutral? A comparative assessment of emissions from consumption of woody bioenergy and fossil fuel. Global Change Biology Bioenergy 4: 761–772.

Zetterberg, L. and D. Chen. 2014. The time aspect of bioenergy–climate impacts of solid biofuels due to carbon dynamics. Global Change Biology, Bioenergy 7: 785–796.

Zha, T., S. Kellomäki, K. -Y. Wang and I. Rouvinen. 2004. Carbon sequestration and ecosystem respiration for 4 years in a Scots pine forest. Global Change Biology 10: 1492–1503.

Zha, T., S. Niinistö, Z. Xing, K. -Y. Wang, S. Kellomäki and A. G. Barr. 2007. Total and component carbon fluxes of a Scots pine ecosystem from chamber measurements and eddy covariance. Annals of Botany 99: 345–353.

PART VII
Managed Boreal Forests under Climate Change
Summary and Perspectives

Climate Change and Managed Boreal Forests

ABSTRACT

The productivity of managed boreal forests increases with climatic warming but the risk of abiotic and biotic damage also increases. Under climate warming, the opportunities exceed the risks in the short term (before 2050), but in the long term (beyond 2050) the risks increase more rapidly than the opportunities. Opportunities include the increased cutting potential for the wood processing and energy industries. The regular and sustainable management and harvest of timber and biomass provides opportunities to redirect the growth and development of forests to meet, step by step, climate warming and consequent changes in other ecological conditions.

Keywords: climate change, opportunities, risks, management

20.1 Climate Change and Forest-based Ecosystem Services

The supporting services of boreal forests ecosystem involve the basic structure and functioning of ecosystems: the interaction between genotypes and environment, cycling water, nutrients and carbon and maintaining biological diversity. In these limits, supporting services produce provisioning (e.g., timber, biomass, and ground water), regulating (e.g., erosion control) and cultural services (e.g., recreation values). Climate change has direct and indirect impacts on supporting services and also on provisioning and regulation services. Seppälä et al. (2009) noted that forest-based services are of global importance for human well-being, and they are vulnerable to climate change. In this context, the vulnerability is related to the pattern, magnitude and rate of climate change and climate variability: (i) how forests and forest-based functions in societies are exposed (exposure); (ii) how sensitive the forests and forest-based systems (sensitivity) are to climate change and variability; and (iii) how large adaptive capacities (resilience) forests and forest-based societies have (Seppälä et al. 2009). In boreal conditions, climate change is likely to provide benefits, but also various risks are likely to increase.

Table 20.1 Likely impact of climate change on managed boreal forests and forestry in Finland from short (2010–2039), medium (2040–2069) and long (2070–2099) term perspectives as modified from Peltola et al. (2012).

Impact of climate change on forests and possible ways to adapt to climate change	2010–2039	2040–2069	2070–2099
Opportunities			
• Potential forest growth	↑	↑↑	↑↑↑
• Potential cutting drain	↑	↑↑	↑↑
• Potential climate change mitigation	↑	↑↑↑	↑↑
Risks			
• Reduced wood quality	↑	↑	↑
• Snow damage	↑↓	↓↓	↓↓↓
• Wind damage	↑	↑↑	↑
• Fire damage	↑	↑↑	↑↑↑
• Insect attacks	↑	↑↑	↑↑↑
• Fungi attacks	↑	↑↑	↑↑↑
• Invasion of alien organisms	↑	↑↑	↑↑↑
• Reduced biodiversity	↑	↑↑	↑↑↑
• Reduced carrying capacity of soil	↑	↑↑	↑↑

The assessment is based mainly on model simulations using the FinAdapt SRES A2 climate scenarios, assuming the atmospheric CO_2 to rise from 350 ppm to 840 ppm, with a mean annual temperature rise of 5°C by 2100. Legend for opportunities and losses: small increase ↑, large increase ↑↑, very large increase ↑↑↑ in opportunities, small reduction ↓, large reduction ↓↓, very large reduction ↓↓↓ in risk.

20.2 Opportunities

Increasing forest growth and the potential to produce timber and biomass

Throughout the boreal forests in Finland, the gradual increase of 4–6°C in the mean annual temperature implies a likely increase of stem wood growth up to 10% by 2020, 20% by 2050 and 40% by 2100, compared to that expected under the current climate if applying current management rules and harvest intensity (Table 20.1). The increase in growth is relatively greater in northern (above 63° N) than in southern boreal forests (below 63° N). This applies for Scots pine, Norway spruce and birch, but the growth of Norway spruce may even reduce in the southern boreal forests. However, climate change may increase the potential cutting drain by 5% by 2020, 50% by 2050 and 80% by 2100. The increase is relatively larger in the northern than in southern boreal forests. At the same time, forest carbon storage is expected to increase by 5% by 2020, 20% by 2050 and 30% by 2100. However, the accumulation of carbon in forest is dependent, e.g., on tree species composition, growth and management. Stendahl et al. (2010), for example, found that the accumulation of carbon in soil in managed Norway spruce forests (5.7 kg C m^{-2}) was 22% greater than in managed Scots pine forests (4.7 kg C m^{-2}), as simulated over three 100 year rotations. This difference was related to differences in the productivity and consequent litter fall between these species.

Increasing potential to mitigate climate change

In the ecosystem/atmosphere cycle, carbon is removed from the atmosphere and retained in several ways in the ecosystem and outside the ecosystem (Fig. 20.1); i.e., (i) to enhance regular reforestation after final cut; (ii) to reduce deforestation and degradation of forests; (iii) to increase forested land area through afforestation; (iv) to increase the carbon density in existing forests; (v) to increase the use of biomass to replace fossil fuels in energy production; and (vi) to increase the use of biomass-based products to increase carbon density outside existing forests. For example, Sathre

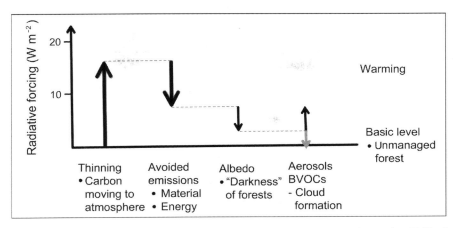

Fig. 20.1 Schematic presentation showing how managed forests may affect climatic warming (Valsta 2012). Courtesy of Lauri Valsta.

et al. (2010) found that proper management with intensive nitrogen fertilization may facilitate a reduction of the total greenhouse gas (GHG) emissions over the whole of Sweden by up to 28%, if the increased forest growth is used to substitute for fossil-intensive materials and energy. Even very intensive nitrogen fertilization may be used without excess nitrogen leaching outside the sites (Bergh et al. 2008).

Sustainable management maintains/increases the carbon stored in the forest ecosystems, thus reducing carbon emissions in short- and medium-long time horizons. However, carbon sink in forest ecosystems (tree biomass, soil organic matter) is limited, while the substitution benefits of forest-based materials and fuels are cumulative. The management and use of forest resources provides an attractive means in mitigating climate change over longer time horizons (Fig. 20.1). A key question is whether the sustainable management and harvest of timber and biomass for wood-based fuels and materials reduce CO_2 emissions more than leaving forests outside management and harvest for storing carbon. In this context, the largest single mechanism for reducing carbon emissions is to use wood and biomass in the construction and energy industries for replacing fossil-CO_2-rich materials and energy. Based on a meta-analysis, Sathre and O'Connor (2010) concluded that each Mg of carbon in wood products substitutes for 2.1 Mg of carbon in non-wood products.

Management and harvest affect climate by altering the fluxes of heat, momentum and moisture in the exchange between forest and atmosphere. Albedo of forest cover plays an important role in controlling heat fluxes and consequent warming (Bright et al. 2015). The albedo of forest canopies is related to the structure of forest ecosystems, which controls the interception of short-wave radiation (Fig. 20.1). Management impacts on carbon stocks and albedo are partly opposed to each other. For example, albedo is enhanced by the high share of regeneration areas with seedling stands before canopy closure, especially if deciduous species are preferred. Albedo in coniferous forests is low, with small seasonal variability, compared to albedo in open forest land or land dominated by deciduous species. The albedo of Scots pine is somewhat higher than that of Norway spruce (Kuusinen et al. 2012, Kuusinen 2014).

Novel/revised management systems may be needed to combine the production of timber and energy biomass in an appropriate way in order to maintain or even to increase carbon stocks and control albedo, compared to that under management optimized only for the production of timber and biomass (Schwaiger and Bird 2010, Kirschbaum et al. 2011). The carbon-only management of boreal forests is likely to be suboptimal in mitigating climate change, as demonstrated by Bright et al. (2014). Relative to the difference in albedo, they found a small cooling in deciduous sites (−0.13°C) and clear cut areas (−0.25°C) compared to coniferous sites. They concluded that the preference of deciduous species in plantations or let them invade in pure coniferous plantations is likely to reduce warming due to an increase in summer and winter albedo.

Management and harvest, in general, cause changes in albedo and enhance the capacity of the forest sector to mitigate climate change. Cherubini et al. (2011, 2012) note that the climate impact of management and harvest is affected by tree species, local climate, time horizon and albedo, integrating biogeochemical (GHG emissions) and biogeophysical (water and energy fluxes) climate impacts. This affects options to mitigate climate change. Sjølie et al. (2013), and Lutz and Howarth (2014) suggested that forest policies that only consider greenhouse gas fluxes in management and harvest without considering changes in albedo, will not lead to the optimal use of the forests and forest sector for climate change mitigation. Jackson et al. (2008) and Naudts et al. (2016) similarly claimed that including biogeochemical and biogeophysical interactions in management may enhance the potential of forests and forestry for climate change mitigation. For example, forests affect particles in the atmosphere, thus reflecting short-wave radiation to space (Fig. 20.1). At the same time, forests emit aerosols and volatile organic compounds (BVOCs) which increase the formation of clouds and reflectance of short-wave radiation, thus slowing climate warming (Spracklen et al. 2008). All these factors are related to the forest structure, including tree species composition, foliage mass and the dynamics of forest ecosystems.

20.3 Risks

Decline in wood quality

Physical and chemical properties of wood are closely linked to growth, but the climatic impact on properties is small compared to the changes in growth. The temperature elevation alone may increase the fiber length in some boreal conifers, while elevated CO_2 may reduce fiber length. Under elevated CO_2, the cellulose content may be reduced but the lignin content increased. Furthermore, the increased thickness of growth rings may limit the use of such wood in the joinery industry, which needs wood with enough thin and homogenous annual rings and thus high density and strength.

Increasing abiotic damage

Risks caused by climate change to forests and forestry are mainly related to changes in abiotic and biotic damage which interacts (Table 20.1): trees fallen down and damaged by strong wind, for example, provide breeding sites for many damaging insects. The role of abiotic damage in triggering biotic damage was demonstrated in the 1960s in the large-scale damage due to bark beetles in Sweden. This destroyed 6 million m³ of Norway spruce forest after wind catastrophes that provided large amounts of breeding material in which bark beetle populations could expand (Eidman 1992).

In boreal conditions, the mean annual velocity of wind is unlikely to increase under a warming climate. However, wind damage may increase due to the shorter duration of soil frost in autumn and spring coinciding with episodes of the highest wind speed. This is especially the case for mature Norway spruce, whose superficial rooting does not give the same support as the deeper rooting of Scots pine and birch. On the other hand, the risk of wind damage and the amount of damaged trees is likely to increase due to increasing stock of mature trees throughout this century. Contrary to the risk of wind damage, the occurrence of snow damage is likely to reduce: the number of risk days may reduce 11% by 2020, 23% by 2050 and 56% by 2100. At the same time, the risk of forest fires may increase 20% by 2100, especially in southern boreal forests, where in many places evaporation may exceed precipitation in summer time.

Increasing biotic damage

Biotic damage may be due to currently existing insects and fungi and/or alien insects and fungi expanding to cover new areas. In both cases, damaging insects are likely to benefit from the warming climate: population sizes are likely to increase due to successful breeding and an increasing number of generations per year. Increasing attacks on trees may also be due to an increase of abiotic damage in forests providing breeding platforms for bark beetles. The possible reduction of growth in Norway spruce in southern boreal forests increases further the risk of attacks of bark beetles. Risk of insect attacks is further increased by alien species, to which the current populations of different tree species have not yet been adapted. This is also true for alien fungal pests, and there is even a risk of increasing damage by domestic fungal pests. For example, the root rot common in Norway spruce and Scots pine in southern and middle boreal forests is likely to expand further north due to the elongation of the growing season and higher soil temperature.

Decline in biodiversity

According to IPCC (2007), global warming of 1.5–2.5°C implies that 20–30% of the known plant and animal species are likely to be at an increased risk of extinction. At high latitudes, climate warming is likely to affect most in northern and southern boreal areas. This implies that the special features of forests and terrestrial ecosystems in northernmost Europe may be diminished even at the current timber line. These forests may become suboptimal, e.g., for reindeer husbandry and recreation industry but they may provide many opportunities for forestry. In the southern parts of northern Europe, boreal features are likely to diminish and be partly replaced by boreonemoral features, e.g., with an increased dominance of herbs and grasses in ground cover, and an increasing share of several deciduous species that are currently of marginal importance, such as oak and lime on fertile sites. There is no evidence that any domestic tree species would become extinct under a warming climate.

In managed boreal forests, non-climatic stresses as cuttings are combined with climate warming affecting the resilience and adaptive capacity of ecosystems. The adaptation of trees and other species occurs through phenotypic plasticity, evolution or migration to suitable sites, the latter probably being the most common response to past climatic changes. Noss (2001) lists several options for sustaining forest functions and biodiversity, through land-use and management: (i) setting aside reserves of representative forest types across environmental gradients; (ii) protecting climatic refuges at multiple scales; (iii) protecting primary forests; (iv) avoiding fragmentation and maintaining connectivity parallel to climatic gradients; (v) providing buffer zones for the adjustment of reserve boundaries; (vi) using low-intensity forestry and avoiding the conversion of natural forests to plantations; (vii) maintaining natural fire regimes; (viii) maintaining diverse gene pools; and (ix) identifying and protecting functional groups and keystone species. This framework is applicable in management for mitigating long-term effects of climate warming on the functional and structural diversity of boreal forests.

Decline of terrestrial carbon stocks and enhancing warming

In boreal conditions, warming is likely to enhance the decay of soil organic matter in interaction with soil moisture defined by the balance between precipitation and evaporation. Kirschbaum (1995) showed that an increase of carbon loss was dependent on the prevailing climate conditions indicated by the temperature in organic soil. This implied that an increase of 1°C in the annual mean temperature is likely to increase the loss of soil carbon by 10% in the regions, where the mean annual temperature of organic soil is 5°C. However, the sensitivity of increase of carbon loss

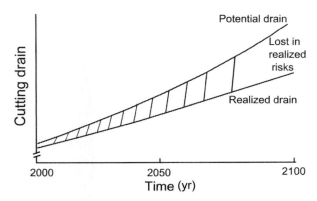

Fig. 20.2 Schematic presentations of the potential cutting drain and realized cutting drain considering the loss of cutting drain due to realized risks.

remained to only 3% per 1°C temperature increase, if the mean annual soil temperature is 30°C. Using the decay model of Meentemeyer (1978), Kellomäki et al. (2005) found carbon in the soil organic carbon would reduce 5% in the southern and mid-boreal conditions (currently the annual mean temperature 0–4°C) by 2050 and a further 10% by 2100. Such reduction was local and related to a reduction in the growth of Norway spruce and consequent reduction of litter caused by more frequent drought episodes. Regionally, the soil carbon continues to increase even in the south to the end of this century, but the rate of increase is likely substantially slower than in the northern boreal conditions.

20.4 Management of Boreal Forests under Climate Change

Opportunities vs. risks

In general, the productivity of managed boreal forests in northern Europe is likely to increase with climate warming but the risks of abiotic and biotic damage will also increase. Evidently, opportunities in terms of productivity clearly exceed risks in the short term (before 2050), but in the long term (beyond 2050) the risks increase more rapidly than the opportunities. Figure 20.2 shows that the potential cutting drain is likely to increase along with climatic warming. This implies that the opportunities probably provided by climate change can be realized only if the increased cutting potential is utilized. The regular and sustainable management and harvest of timber and biomass provides opportunities to redirect, step by step, the growth and development of forests to meet the gradual change in climate in a proper way to adapt to climate change.

Management under climate change—choices for the future

In the boreal zone, the time perspective of forestry spans decades, covering a gradual change in climate. This implies a gradual change in genotype/environment interaction, with a consequent need to gradual change in management and harvest to utilize the opportunities and to avoid the risks. In forest plantations, for example, the proper choice of tree species and their provenance are of primary importance (Table 20.2). Towards the end of this century, the growth of Norway spruce is likely reduced on sand-rich moraine soils in southern boreal sites making this species susceptible to increasing insect attacks. This is also true for Scots pine and birch, but natural regeneration provides an attractive choice in regenerating these species. Even in the northern boreal forests

Table 20.2 Possible ways to adapt to climate change in Finland in the short (2010–2039), medium (2040–2069) and long (2070–2099) term, as modified from Peltola et al. (2012).

Possible way to adapt to climate change	2010–2039	2040–2069	2070–2099
• Proper species and provenance choice	↑	↑↑	↑↑↑
• Management to increase forest carbon and control albedo of forest canopies	↑	↑↑	↑↑↑
• Proper management to avoid damage	↑	↑↑	↑↑↑
• Proper management of forest hygiene	↑	↑↑	↑↑↑
• Proper management to sustain biodiversity	↑	↑↑	↑↑↑
• Increasing use of forest biomass for energy and wood-based products	↑	↑↑	↑↑↑
• Proper harvest technology and logistic infrastructure to reduce damage to soil and trees	↑	↑↑	↑↑↑

The assessment is based mainly on model simulations using the FinAdapt SRES A2 climate scenarios, assuming the atmospheric-CO_2 to rise from 350 ppm to 840 ppm, with a mean annual temperature rise of 5°C by 2100. Legend for importance: small increase ↑, large increase ↑↑, very large increase ↑↑↑.

(above 63° N), climate change is likely to increase natural regeneration due to the increase of the crop of fully matured seeds, and the germination and establishment of seedlings.

The improved establishment of seedlings does not exclude the need for soil preparation in controlling the suppression due to grasses and herbs (Table 20.2). Similarly, early tending of coniferous seedling stands may even be more urgent than under the current climate due to the increased invasion and growth of deciduous trees under climate warming. In the short term, but especially in the long term, more frequent and/or more intensive thinnings are needed in order to utilize the increasing growth and to maintain the growth capacity and health of trees. Shorter rotation lengths, adapted to the increased growth and development of trees, are necessary in realizing climate change opportunities, but they are also likely to reduce the risk of abiotic and biotic damage.

20.5 Concluding Remarks

Managed boreal forests may adapt to the changing climate, but the rate of autonomous adaptation is probably slow to meet the future expectations of the use of forests for different purposes. Globally, the changes in CO_2, temperature, and precipitation may make boreal sites suboptimal for some tree species, whereas the conditions for some other species may become more optimal (Boisvenua and Running 2006). From the time perspective of 30–40 years (i.e., by 2050 and thereafter), the mean annual temperature elevation will probably exceed the year-to-year variability in the temperature under boreal conditions, with further changes in annual precipitation and its seasonal distribution. Under climate change, the current management provides many opportunities even beyond 2050, but the risks to forests and forestry are increasing in interaction. The future socio-economic context is also unknown, which makes it difficult to identity proper adaptive forest policy responsive to a wide variety of future economic, social, political and environmental circumstances (Jandl et al. 2015).

References

Bergh, J., U. Nilsson, H. Grip, P. -O. Hedwall and T. Lundmark. 2008. Effects of frequency of fertilization on production, foliar chemistry and nutrient leaching in young Norway spruce stands in Sweden. Silva Fennica 42(5): 721–733.

Boisvenua, C. and S. W. Running. 2006. Impacts of climate change on natural forests productivity–evidence since the middle of the 20th century. Global Change Biology 12: 862–882.

Bright, R. M., C. Antón-Fernández, R. Astrup, F. Cherubini, M. Kvalevåg and A. H. Strømman. 2014. Climatic change implications of shifting forest management strategy in a boreal forest ecosystem of Norway spruce. Global Change Biology 20: 607–621.

Bright, R. M., K. Zhao and R. B. Jackson. 2015. Quantifying surface albedo and other direct biogeophysical climate forcings in forestry activities. Global Change Biology 21: 3246–3266.

Cherubini, F., A. H. Strømman and E. Hertwich. 2011. Effects of boreal forest management practices on the climate impacts of CO_2 emissions from bioenergy. Ecological Modelling 223: 59–66.

Cherubini, F., R. M. Bright and A. H. Strømman. 2012. Site-specific global warming potential of biogenic CO_2 for bioenergy: contributions from fluxes and albedo dynamics. Environmental Research Letters 7: 1–11.

Eidman, H. H. 1992. Impact of bark beetles on forests and forestry in Sweden. Journal of Applied Entomology 114: 193–200.

IPCC. 2007. Climate Change 2007: Impacts, Adaptation and Vulnerability. Working Group II Contribution to the Fourth Assessment Report of the Intergovernmental Panel on Climate Change. Cambridge University Press, Cambridge, UK.

Jackson, R. B., J. T. Randerson, J. G. Canadell, R. G. Anderson, R. Avissar, D. D. Baldocchi et al. 2008. Protecting climate with forests. Environmental Research Letters 3: 1–5.

Jandl, R., J. Bauhus, A. Bolte, A. Schindlbacher and S. Schüler. 2015. Effect of climate-adapted forest management on carbon pools and greenhouse emissions. Current Forestry Reports 1: 1–7.

Kellomäki, S., H. Strandman, T. Nuutinen, H. Peltola, K. T. Korhonen and H. Väisänen. 2005. Adaptation of forest ecosystems, forests and forestry to climate change. Finnish Environment Institute, FinAdapt Working Paper 4: 1–50.

Kirschbaum, M. U. F. 1995. The temperature dependence of soil organic matter decomposition and the effect of global warming on soil organic C storage. Soil Biology and Biochemistry 27(6): 753–760.

Kirschbaum, M. U. F., D. Whitehead, S. M. Dean, P. M. Beets, J. D. Shephard and A. -G. Ausseil. 2011. Implications of albedo changes following afforestation on the benefits of forests as carbon sinks. Biogeosciences 8: 3687–3696.

Kuusinen, N., P. Kolari, J. Levula, A. Porcar-Castell, P. Stenberg and F. Berninger. 2012. Seasonal variation in pine forests albedo and effects of canopy snow on forest reflectance. Agricultural and Forest Meteorology 164: 53–60.

Kuusinen, N. 2014. Boreal forest albedo and its spatial and temporal variation. Ph. D. Thesis, University of Helsinki, Helsinki, Finland.

Lutz, D. A. and R. B. Howarth. 2014. Valuing albedo as an ecosystem service: implication for forest management. Climatic Change 124: 53–63.

Meentemeyer, V. 1978. Macroclimate and lignin control of litter decomposition rates. Ecology 59: 465–472.

Naudts, K., Y. Cheng, M. J. McGrath, J. Ryder, A. Valade, J. Otto et al. 2016. Europe's forest management did not mitigate climatic warming. Science 351: 597–599.

Noss, R. F. 2001. Beyond Kyota: forest management in a time of rapid climate change. Conservation Biology 15(3): 578–590.

Peltola, H., E. Vapaavuori, P. Niemelä, S. Kellomäki, H. Gregow, O. Huitu et al. 2012. Ilmastonmuutokseen sopeutuminen metsätaloudessa. Report to ISTO Project.

Sathre, R. and J. O'Connor. 2010. Meta-analysis of greenhouse gas displacement factors of wood product substitution. Environmental Science & Policy 13: 104–114.

Sathre, R., L. Gustavsson and J. Bergh. 2010. Primary energy and greenhouse gas implications of increasing biomass production through fertilization. Biomass and Bioenergy 34: 572–581.

Schwaiger, H. P. and D. N. Bird. 2010. Integration of albedo effects caused by land use change into the climate balance: should we still account in greenhouse units? Forest Ecology and Management 260: 278–286.

Seppälä, R., A. Buck and P. Katila. 2009. Execute summary and key message. In: R. Seppälä, A. Buck and P. Katila (eds.). Adaptation of Forests and People to Climate Change. A Global Assessment Report. IUFRO World Series 22: 6–14.

Sjølie, H. K., G. S. Latta and B. Solberg. 2013. Potential impact of albedo incorporation in boreal forest sector climate change policy effectiveness. Climatic Policy 13(6): 665–679.

Spracklen, D. V., K. S. Carslaw, M. Kulmala, V. -M. Kerminen, S. -L. Sihto, I. Riipinen et al. 2008. Contribution of particle formation to global cloud condensation nuclei concentrations. Geophysical Research Letters 35: 1–5.

Stendahl, J., M. -B. Johansson, E. Eriksson, Å. Nilsson and O. Landvall. 2010. Soil organic carbon in Swedish spruce and pine forests–differences in stock levels and regional patterns. Silva Fennica 44(1): 5–21.

Valsta, L. 2012. Boreaalisten metsien käytön kokonaisvaikutus ilmaston lämpenemiseen. Power Point presentation 1.11.2012. University of Helsinki, Finland.

Appendix

Units and Conversions

In studies of forest ecology and forest management, the units used vary even in the same publications. This is especially the case in papers dating back to the time before standards provided by the SI-system. The tables below therefore list a selection of frequently used dimensions in plant physiology and ecology with application in producing biomass and energy based on forest biomass. They are used in this book, and can also be used to convert earlier units of measurement into the SI-system, which was not used in older publications.

Prefixes and units

Prefix in SI-system	Explanation	Value
T	tera-	10^{12}
G	giga-	10^{9}
M	mega-	10^{6}
k	kilo-	10^{3}
h	hecto-	10^{2}
d	deci-	10^{-1}
c	centi-	10^{-2}
m	milli-	10^{-3}
µ	micro-	10^{-6}
n	nano-	10^{-9}
p	pico-	10^{-12}

Energy

Unit	Transformations
J	$1\ J = 1\ N \cdot m = 1\ kg \cdot m^2 \cdot s^{-2} = 1\ W \cdot s = 0.239\ cal = 10^7\ erg$
W·h	$1\ W \cdot h = 3.6\ kW \cdot s = 3.6\ kJ = 0.86\ kcal$
MJ	$1\ MJ = 0.278\ kWh$
cal	$1\ cal = 4.1868\ J$
kcal	$1\ kcal = 1.163\ W \cdot h$

Pressure

Unit	Transformations
MPa	$1\ MPa = 10^6\ Pa = 10\ bar$
bar	$1\ bar = 10^5\ N \cdot m^{-2} = 10^5\ Pa = 100\ J \cdot kg^{-1} = 10^6\ erg \cdot cm^{-3}$
bar	$1\ bar = 750\ Torr = 0.9869\ atm$
atm	$1\ atm = 1.0132\ bar = 760\ Torr$

Amount and concentration

Unit	Transformations
Molarity	mol · kg⁻¹ of liquid
ppm	1 ppm = 10^{-6} mol · mol⁻¹; 1 µg · g⁻¹; 1 µl · l⁻¹
ppb	1 ppb = 10^{-9} mol · mol⁻¹; 1 ng · g⁻¹; 1 nl · l⁻¹
ppm	1 ppm CO_2 = 1.82 mg · m⁻³ = 41.6 µmol · m⁻³ = 0.101 Pa (at the temperature of 20°C and pressure 101.3 kPa)

Radiation and energy

Transformations

1 W · m⁻² = 1 J · m⁻² s⁻¹ = 31.53 MJ · m⁻² · a⁻¹

1 mol photon = 1.8 · 10^5 J (when λ 650 nm) ... 2.7 · 10^5 J (when λ 450 nm)

1 cal · cm⁻² · min⁻¹ = 6.98 · 10^2 W · m⁻² = 6.98 · 10^5 erg · cm⁻² · s⁻¹

1 erg · cm⁻² · s⁻¹ = 1.43 · 10^{-6} cal · cm⁻² · min⁻¹ = 10^{-3} W · m⁻²

Gas exchange

Transformations

1 g CO_2 (exchange) ≈ 0.73 g O_2 (exchange)

1 g O_2 (exchange) ≈ 1.38 g CO_2 (exchange)

Diffusion D_{CO_2} = 0.64 D_{H_2O}

Diffusion D_{H_2O} = 1.56 D_{CO_2}

0.03 %$_{vol}$ CO_2 = 300 µ · l⁻¹ = 282 µbar = 28 Pa CO_2 (partial pressure)

1 µl · l⁻¹ = 1.963 µg CO_2 · l⁻¹ (at a pressure of 1013 mbar and temperature 0°C)

1 mg CO_2 · dm⁻² · h⁻¹ = 0.028 mg CO_2 · m⁻² · s⁻¹ = 0.63 µmol CO_2 · m⁻² s⁻¹

1 mg CO_2 · m⁻² · s⁻¹ = 36 mg CO_2 · dm⁻² · h⁻¹ = 22.7 µmol CO_2 · m⁻² · s⁻¹

1 µmol CO_2 · m⁻² · s⁻¹ = 0.044 mg CO_2 · m⁻² · s⁻¹ = 1.58 mg CO_2 · dm⁻² · h⁻¹

1 mg H_2O · dm⁻² · h⁻¹ = 1.54 µmol H_2O · m⁻² · s⁻¹

Conductance (at temperature of 20°C and pressure 101.3 kPa)

1 cm · s⁻¹ ≈ 0.416 mol · m⁻² · s⁻¹

1 mol · m⁻² · s⁻¹ ≈ 0.024 mm · s⁻¹

Biomass

Transformations

1 g DM · m⁻² = 10^{-2} Mg · ha⁻¹

1 g DM ≈ 0.42 – 0.51 g C ≈ 1.5 – 1.7 g CO_2

1 g C ≈ 2 – 2.22 g DM ≈ 3.1 – 3.4 g CO_2

1 g CO_2 ≈ 0.59 – 0.66 g DM ≈ 0.27 – 0.30 g C

1 g CO_2 = 3.67 g [=44/12] C

Biomass = Volume [m³] x Density of mass [kg m⁻³]

Carbon content in stem wood [kg C m⁻³]:
Scots pine 0.3091; Norway spruce 0.3715; birch 0.4152

DM = Dry Mass

Index

A

abiotic damage 245, 246, 348, 349
adaption to climate change 1
adaptive management 6, 273–277, 294, 295
adaptive measures in management 275
albedo 5, 8, 169, 298, 299, 331–336, 338, 347, 348, 351

B

basal respiration 90, 119
bedrock 23
biochemical model for photosynthesis 167
biodiversity 3, 4, 16, 18, 19, 28, 277, 294, 346, 349, 351
biomass growth 90, 126–129, 138, 241
biotic damage 245, 246, 262, 267, 273, 276, 277, 345, 348–351
boreal forest 1, 5, 6, 8, 11, 13–16, 18, 19, 22, 24–26, 28, 30–33, 38, 44, 52, 56–58, 90, 127, 149, 152, 155, 160, 165, 167, 183, 194, 197, 201, 203, 205, 208, 216, 218, 219, 221, 223, 226, 228, 231, 233, 235–237, 239, 245, 246, 252, 262, 263, 267, 271, 273, 276, 277, 279, 284, 286, 287, 290, 292, 294, 305, 306, 308–312, 314–319, 323, 324, 328, 331, 333, 334, 343, 345–351
boreal forest climate change 165, 273
boreal forest disturbance 245, 268
boreal forest resources 16
boreal tree species 17, 61, 63, 91, 278

C

canopy photosynthesis and respiration 115, 116, 284
carbon neutrality of bioenergy 331
carbon sequestration 8, 16, 18, 27, 28, 102, 166, 167, 207, 215, 239, 243, 294, 298, 299, 310, 311, 315, 316, 318–321, 331–333
carbon uptake 18, 63, 65, 66, 68, 69, 71, 74, 75, 79, 81, 83, 85, 102, 106, 110, 124, 141, 162, 166, 203, 284, 285, 299, 300, 302, 304, 305, 308, 312, 313, 320, 321, 326–328
changes in precipitation 5, 39, 42, 48, 49, 214, 262, 274
changes in temperature sum 47
climate change 1, 3–8, 13, 16, 19, 35, 37, 38, 43, 44, 47–49, 51, 54, 56, 57, 61, 63, 66–68, 70, 75, 80, 85, 89, 93, 102, 103, 106–108, 115, 124, 126, 131, 135, 138, 141, 142, 145, 148, 150, 152, 155, 157–161, 165–169, 192–197, 201, 203, 205, 207–209, 211, 214, 215, 218, 219, 223, 226, 228, 231, 233, 235, 236, 238–241, 243–245, 249, 250, 252, 257–260, 262, 263, 267, 271, 273–279, 281, 283–290, 292, 294, 298, 299, 304, 308, 318, 320, 326, 331–333, 338, 343, 345–348, 350, 351
climate change mitigation 8, 16, 298, 299, 326, 333, 338, 346, 348
climatic limits 13, 22, 278

D

diversity 2–4, 16, 18, 19, 28, 32, 267, 274, 275, 277, 294, 345, 346, 349, 351

E

ecosystem goods and services 3, 4, 274
effect of nitrogen on photosynthesis 71
emissions of volatile secondary compounds 141
evaporation in relation to precipitation 53
extreme weather episodes 44, 46

F

Finland 1, 8, 13–16, 22–29, 32, 43, 44, 47, 48, 50, 53, 55, 57, 67, 73, 84, 129–132, 135–137, 143, 148, 150, 152, 176, 180, 182, 184, 189–191, 205–208, 210–216, 221–223, 235, 236, 238–240, 242, 244, 249, 250, 252, 253, 255, 257–261, 263, 264, 266, 276, 278, 279, 282, 284, 287, 290, 291, 293, 306, 346, 351
forest ecosystem 1, 3–6, 8, 18, 19, 30–32, 44, 66, 90, 148–150, 155, 157–163, 165–167, 181, 192, 197, 201, 203, 204, 206, 218, 233, 234, 239, 240, 245, 252, 259, 266, 273–276, 288, 298–301, 304, 305, 309, 310, 313–315, 318, 320, 321, 323, 326–328, 330–332, 347, 348
forest management 4, 66, 85, 148, 234, 255, 257, 294
forest resources 16, 17, 24, 26, 284, 289, 331, 347
forking of branches 126
functioning 2, 3, 5, 6, 30, 32, 63, 80, 157, 159, 167, 234, 245, 273, 275, 345

G

Global boreal forests 13, 14, 16
Global warming 8, 41, 42, 63, 298, 332, 349
greenhouse gas 5, 8, 37, 57, 298, 347, 348
greenhouse gas emission 57
gross primary production 92, 102, 167, 168, 188, 189, 203, 204, 208, 210, 218, 286–289, 300, 323

growth 2, 5, 15, 17–19, 23, 26–28, 30–32, 40, 47, 49, 53, 56–58, 63, 66, 67, 72, 86, 89–97, 101–103, 106, 115, 120, 122–124, 126–133, 136, 138, 141–146, 150, 152, 157–163, 165–168, 170, 175, 179–186, 189–192, 194–197, 203–205, 208, 209, 211, 212, 216, 218–229, 231, 234–241, 244–248, 252, 253, 255, 257, 265, 267, 273–281, 283–290, 292, 294, 299, 302, 304, 306, 308–316, 318–321, 323, 325–328, 338, 345–351
growth and development of forest stands 19, 30, 150, 197, 345, 350
growth of branches and shoots 126
growth respiration 89–92, 96, 97, 101–103, 126, 167

I

impact of elevated CO_2 on respiration losses 122
impact of thinning on growth and timber yield 226, 231
impacts of climate change 4, 8, 103, 126, 165, 211, 274
IPCC 5–7, 13, 37–42, 168, 298, 332, 349

L

light interception in shoots 126, 136, 138

M

maintenance respiration 89–97, 101–103, 167
managed boreal forest 16, 22, 30–33, 165, 235, 236, 343, 345, 346, 349–351
management 1, 3, 4, 6, 8, 13, 16, 18, 19, 28–33, 53, 56, 66, 85, 148–150, 157, 158, 160, 162, 163, 165–168, 181, 185, 187, 190, 194, 197, 201, 205, 207, 211, 218–223, 226, 228–231, 233–236, 238–240, 242–245, 253, 255, 257, 258, 267, 271, 273–279, 282, 284, 288–295, 298, 299, 312–316, 318–324, 326–334, 336–338, 346–351
mitigation of climate change 8, 271, 298, 332
model validation 161, 187
modeling 1, 5, 40, 66, 83, 90, 150, 155, 160–163, 165, 167, 181, 186, 188, 197, 252, 302
modelling 162
models 1, 5, 6, 8, 40–43, 54, 57, 63, 70, 72, 74, 75, 77, 83–86, 96, 97, 108, 116, 118, 119, 122, 138, 150, 157, 160, 161, 163, 165–168, 170, 174, 175, 177, 181–183, 187–194, 197, 205–207, 211, 214, 216, 219, 222, 224, 226, 228, 235, 236, 239–241, 249, 250, 253, 257, 264, 265, 277, 278, 281, 284, 286, 288, 291, 292, 305, 306, 309–314, 317, 318, 322, 324, 328, 329, 346, 350, 351
mortality 5, 27, 30, 133, 142, 158, 159, 166, 167, 181, 182, 189, 218, 219, 226–229, 231, 236, 240, 246, 273, 277, 279, 281, 283, 284, 289, 306, 310, 311

N

natural boreal forests 18, 30, 33, 231, 310, 316, 338, 349
net ecosystem exchange 28, 188, 189, 204, 205, 209, 210, 214, 299–304, 311, 312, 316, 317
net primary production 5, 14, 28, 92, 102, 203, 204, 208–210, 214, 300, 323
nitrogen 5, 18, 56, 57, 58, 63, 67, 70–75, 77, 89–91, 93, 97, 98, 100–103, 116, 119, 120, 127, 128, 138, 146, 165–168, 170, 171, 177, 179, 186, 187, 191, 192, 194, 196, 203, 204, 223, 236, 265, 266, 276, 306, 309, 310, 315, 329–331, 333, 336, 337, 347

O

opportunities 3, 6, 8, 19, 157, 160, 273, 274, 276, 298, 345, 346, 349–351
optimized management 233, 234, 239, 244, 293

P

photosynthesis 5, 8, 63–66, 68–72, 74–76, 79–81, 83–86, 90, 102, 110, 111, 115–118, 126–128, 145, 150, 158, 160, 165–167, 177, 179, 180, 187, 194, 203, 205, 206, 208, 221, 250, 284, 285, 300–302, 310
physical and chemical properties of wood 141, 142, 152, 348
precipitation 2, 5, 15, 16, 22, 23, 37, 39, 42, 44, 46–50, 52–54, 57, 67, 85, 127, 150, 152, 157–159, 165–170, 172, 173, 194–196, 205, 207, 208, 211, 213, 214, 221, 223–226, 229, 231, 244–246, 250, 252, 258–262, 267, 273, 274, 276, 279, 282, 285–288, 293, 295, 303, 311, 318, 348, 349, 351
properties of needles 113, 117

Q

Q_{10} 90–93, 99, 100, 102, 120

R

radiative forcing 8, 37, 38, 39, 298, 299, 326, 331–338
regeneration 2, 4, 15, 30, 31, 33, 63, 69–71, 158–160, 163, 166, 167, 181, 183–185, 218–222, 231, 234, 236, 239, 240, 275–278, 291, 310, 311, 314–316, 327, 347, 350, 351
respiration 5, 28, 63, 66, 70, 73–75, 89–103, 116–122, 126, 128, 131, 145, 158, 160, 166–168, 179, 189, 203–205, 209, 214, 300–304, 308, 310, 323
risks 252, 255, 257, 262, 276, 277, 294, 298, 345, 346, 348, 350, 351
rule-based management 235, 244, 288, 292, 318

S

sap flow and crown transpiration 122
Scots pine 1, 8, 16, 17, 22, 24, 26, 49, 50, 52, 67, 69, 72–83, 85, 86, 92–103, 106, 108–113, 115–124, 126–135, 137, 138, 141, 142, 144–152, 158, 166, 167, 174, 177, 179–183, 188–196, 205–207, 211–213, 219, 221–224, 226–228, 230, 231, 235–240, 247–251, 257, 258, 267, 278–282, 284–286, 288–290, 292–294, 301, 303–306, 308, 309, 312–320, 323, 328, 335, 338, 346–350
seasonality of photosynthesis 70
secondary compounds 141, 145–148, 150, 262, 265, 266
sensitivity of respiration 93
soil moisture 31, 49, 50, 53, 58, 63, 85, 108, 158, 163, 165, 167, 174, 183, 188, 194, 219, 236, 257, 286, 302, 304, 349
soils 14, 15, 18, 22, 24, 27, 39, 57, 238–240, 276, 314, 350
stomatal conductance 63–65, 68, 69, 71, 74, 75, 78, 79, 81, 106–111, 113, 116, 166, 167, 175
structure 2–6, 8, 25, 30–33, 40, 66, 74, 89, 115–117, 119, 126, 127, 129, 135, 138, 141, 144, 145, 157–163, 166–168, 174, 181, 187–189, 211, 213, 233, 234, 243–245, 252, 253, 255, 257, 258, 273–277, 284, 290, 291, 294, 298, 310, 318–320, 326, 333, 334, 338, 345, 347, 348, 351

substitution of fossil fuels and materials 330
succession 4, 6, 13, 14, 30, 32, 33, 158–160, 162, 233, 234, 267, 274, 314, 338
sustainable management 4, 338, 345, 347, 350

T

temperature 2, 5, 6, 8, 13, 15, 18, 19, 22–25, 27, 28, 31, 37, 39–42, 44–58, 63, 65–86, 89–91, 93–103, 106–113, 115–124, 126–138, 141–145, 147, 148, 150–152, 157–160, 165–167, 169, 170, 173, 175, 176, 179, 181–183, 188, 191–196, 203–211, 213, 214, 216, 219–226, 229, 231, 236, 237, 240–242, 244–253, 258–267, 273, 274, 276, 278–282, 284, 286, 287, 293, 295, 301–306, 308–311, 318, 346, 348–351
temperature elevation 50, 51, 56, 67, 69, 80, 84, 85, 93, 97, 98, 117, 122, 128, 130, 131, 138, 141, 145, 152, 222, 231, 267, 308, 348, 351
temperature sum 23, 24, 27, 28, 44, 45, 47, 48, 53, 56, 57, 179, 181, 191–193, 207–210, 216, 236, 237, 241, 248, 249, 278–282, 287

thinning 6, 16, 30–33, 149, 162, 166, 167, 185, 187, 194–196, 219, 224–231, 233–235, 243, 252, 253, 255, 275, 276, 278, 286–290, 292, 293, 299, 308, 312, 315–324, 327–329, 333, 334, 336–338, 351
timber biomass 30, 162, 267, 345
timber production 18, 239, 242, 243, 252, 276, 288, 291, 292, 294, 316, 318–321
timber yield 32, 194–196, 223–231, 286, 288, 289, 292, 294, 295, 319, 320, 323, 329
total growth 27, 130, 131, 133, 166, 186, 190–192, 194, 211, 219, 226–229, 231, 238, 283–285, 289, 308
transpiration 5, 49, 50, 52, 53, 63, 64, 66, 68, 80, 81, 106–108, 111–113, 116, 122, 124, 158–160, 166, 167, 168, 172, 175, 214, 244, 285–289, 300

W

water use efficiency 64, 106, 109, 111–113, 122, 286
weather generator 50, 168
whole tree physiology 115, 116